ANNALS *of* THE NEW YORK ACADEMY OF SCIENCES

EDITOR-IN-CHIEF
Douglas Braaten

ASSOCIATE EDITOR
Rebecca E. Cooney

PROJECT MANAGER
Steven E. Bohall

EDITORIAL ADMINISTRATOR
Daniel J. Becker

Artwork and design by Ash Ayman Shairzay

T0351942

The New York Academy of Sciences
7 World Trade Center
250 Greenwich Street, 40th Floor
New York, NY 10007-2157

annals@nyas.org
www.nyas.org/annals

Published by Blackwell Publishing
On behalf of the New York Academy of Sciences

Boston, Massachusetts
2012

ANNALS *of* THE NEW YORK ACADEMY OF SCIENCES

VOLUME 1264

ISSUE

The Brain and Obesity

ISSUE EDITORS

Giovanni Cizza and Kristina I. Rother

National Institute of Diabetes and Digestive and Kidney Diseases, National Institutes of Health, Bethesda, Maryland

TABLE OF CONTENTS

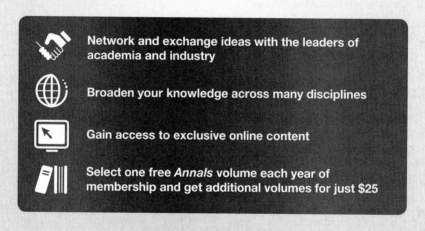

Ann. N.Y. Acad. Sci. ISSN 0077-8923

The Brain and Obesity Lectures Series – the beginning of a new field?

It is a great pleasure to introduce the Brain and Obesity Lecture Series in this issue of *Annals of the New York Academy of Sciences*. These lectures were initiated in 2006 when the Intramural Center for Obesity was about to open at the National Institutes of Health (NIH) Clinical Center, a state-of-the art facility that includes a 21-bed inpatient facility and three metabolic chambers and other equipment for studies on phenotyping subjects with obesity. An interdisciplinary group of clinical investigators was recruited that included endocrinologists, nutritionists, geneticists, exercise physiologists, and neuropsychologists. The fact that the Intramural Center for Obesity was part of the NIH Clinical Center, a 240-bed facility completely devoted to clinical research, offered a valuable opportunity for collaboration and cross-fertilization within clinical research and with translational and basic research.

We shared with other colleagues, mostly in the field of neurosciences, the conviction that the brain plays a pivotal role in obesity. This notion, largely accepted today, was quite controversial at the time. Rather, it was thought that obesity was a problem of eating too much and exercising too little. Without challenging this obvious tenet—derived from the law of thermodynamics—we thought that there was much more to it. Another commonly held notion was that "a calorie, is a calorie, is a calorie," a notion that has now been revisited. The Brain and Obesity Lecture Series started us on a journey of exploration, a search for ongoing work based on the idea that the brain had something to do with the development of obesity. In the following five years, there were a total of 16 lectures (Table 1). On March 22, 2012, ten of the original speakers, or members of their respective teams, returned to the NIH Clinical Center for a final round of talks and discussions. What is summarized below is the distillate of those lectures.

Beyond thermodynamic laws

The first lecture by David Allison provided an overview, mostly based on his group's original work, on multiple putative factors that may contribute to the obesity epidemic. He set forth the interesting idea that the perceived, rather than the actual, amount of calories available in the foreseeable future influences feeding behavior and the willingness to spend instead of saving energy. This novel idea is rooted in robust work conducted in animals and humans and has important clinical and epidemiological implications. Specifically, perceived energetic uncertainty, whether of mere food in animals or of economic resources in humans, is likely to set into motion behavioral responses that eventually lead to increased fat storage. This is compatible with the well-known observation in industrialized societies that obesity is more prevalent in the less-privileged strata of the society. Allison also pointed out that social status, both in animals and humans, has a role in determining the amount of adiposity and its distribution (i.e., visceral

doi: 10.1111/j.1749-6632.2012.06721.x
Ann. N.Y. Acad. Sci. 1264 (2012) vii–xiv © 2012 New York Academy of Sciences.

Table 1. Brain and Obesity Lecture series hosted by the NIH Clinical Center between 2007 and 2012

Speaker	Date	Affiliation	Title
Saverio Cinti	October 11, 2007	University of Ancona, Italy	Transdifferentiation properties of the adipose organ
Sabrina Diano	January 30, 2008	Yale School of Medicine, New Haven, Connecticut	Ghrelin's role in appetite and memory
Mark Mattson	February 20, 2008	National Institute on Aging, Baltimore, Maryland	BDNF as a regulator of systemic and brain energy metabolism
Marco Boscaro	April 2, 2008	University of Ancona, Italy	Visceral adipose tissue: emerging role of gluco- and mineralocorticoid hormones
Catherine M. Kotz	December 3, 2008	University of Minnesota, Minneapolis, Minnesota	Brain mechanisms underlying nonexercise activity thermogenesis and obesity
Luigi Fontana	December 10, 2008	Washington University School of Medicine, St Louis, Missouri	Adiposity, calorie restriction, and aging
David Gozal	February 25, 2009	University of Chicago, Illinois	Sleep deprivation and the obesity epidemic in children
Silvana Obici	March 25, 2009	University of Cincinnati, Ohio	Hypothalamic control of glucose secretion
Eve Van Cauter	April 9, 2009	University of Chicago, Illinois	Metabolic and endocrine consequences of sleep deprivation
William A. Banks	May 20, 2009	University of Washington, Seattle, Washington	Role of the blood–brain barrier in the control of food intake and body weight regulation
Stephen C. Benoit	November 4, 2009	University of Cincinnati, Ohio	Learned and cognitive controls of ingestive behavior
Renato Pasquali	November 18, 2009	University Alma Mater Studiorum, Bologna, Italy	Obesity, stress, and sex differences: the role of steroids and central neuroendocrine networks
David Allison	February 3, 2010	University of Alabama at Birmingham, Birmingham, Alabama	Conjectures on some curious connections among social status, hunger, fatness, and longevity

Continued

Table 1. *Continued*

Speaker	Date	Affiliation	Title
Paolo Sassone-Corsi	January 6, 2011	University of California, Irvine, California	Mammalian circadian clock and metabolism the epigenetic link
Weihong Pan	May 19, 2011	Pennington Biomedical Research Center, Baton Rouge, Louisiana	Leptin action on non-neuronal cells in the central nervous system; potential clinical implications.
Hans Rudi Berthoud	November 10, 2011	Pennington Biomedical Research Center, Baton Rouge, Louisiana	Metabolic and hedonic drives in the neural control of appetite: who is the boss?

versus subcutaneous). Furthermore, he reviewed the interdependence of energy intake and energy expenditure. When monkeys were calorie restricted by approximately 30%, a reduction in physical activity was observed, a clear attempt to defend the current body weight.

If stressful uncertainty inherent to a low socioeconomic status may lead to increased adiposity, then any intervention offering healthier diets and advice on proper portion size is destined to be less effective as long as the perceived or real uncertainty remains. In summary, Allison underlines the under-appreciated importance of psychological (i.e., perceived versus real food insecurity) and sociological (developing versus industrialized and post-industrialized societies) factors in determining adiposity and its distribution within an organism, as well as among members of a given society.

Is the blood–brain barrier truly a "barrier"?

According to the Merriam-Webster Dictionary a barrier is: "something material that blocks or is intended to block passage." Robert (Bill) Banks makes the cogent case that this term was never scientifically correct. Using leptin as an example, Banks elucidates the role of the blood–brain barrier (BBB)—this fastidious and sophisticated structure—in regulating body weight. Leptin is a large molecule, and its transport to the brain is highly regulated and facilitated by specific carriers. This regulation is not immutable over time but rather is influenced by short-term and long-term changes in physiological states. Thus, when peripheral levels of the fat-derived hormone leptin decline, as a result of decreased fat deposition and/or increased fat catabolism, the brain becomes "alerted" to the fact that times are getting more difficult. It may not be the right time to invest in reproduction or other expensive behaviors, but rather save energy.

The orchestrated series of neurobehavioral responses to feeding are the result of millions of years of evolution that have shaped the features of many different species. However, evolution is efficient but slow. Living beings are exquisitely adapted to survive in a world of limited and precarious food resources. They have not had time, however, to adapt equally efficiently to conditions in which the brain needs to know that too much fat is being accumulated, a relatively novel condition from an evolutionary standpoint. As a consequence, the BBB is not as effective in transporting leptin inside the brain when large concentrations of plasma leptin are pushing at "the gates" of the brain. In other words, the hormone leptin evolved in conditions of paucity and non-predictability of food availability.

Further interesting experimental observations summarized by Banks are the effects of triglyceride levels on the permeability of the BBB. During periods of fasting, plasma triglycerides increase as a result of mobilization of fat reserves; this may be teleologically advantageous to alert the brain that reserves are being consumed. Triglycerides apparently impair the transport of leptin into the brain. Elucidation of the structural and molecular mechanisms that allow modifications of the BBB is yet to come, but this knowledge may promote effective drug development and novel use of known drugs (such as leptin) for the treatment of obesity. On a more general note, a better understanding of the mechanisms regulating access to the brain may result in a better understanding on how to deliver therapeutic agents to the central nervous system.

Embracing the complexity of the stress response

Renato Pasquali provided an overview of the fascinating and rapidly evolving research field of the stress response, elegantly integrating his own work with that of well-known scholars, including Biorntorp, Dallman, McEwen, and Sapolski, in addition to Chrousos and Gold. The latter two codified the now classic functional description of the stress system in their seminal review article published in *JAMA* in 1992, defining the effector branches of the stress system as the hypothalamic–pituitary–adrenal (HPA) axis and the sympatho–adrenal–system (SAS). The complexity of these systems is, however, daunting. Where does this complexity come from? First of all, characterizing the level of activity of the effectors of the stress system, the HPA axis and SNS, is elusive, as Pasquali knows well, based on his clinical experience on patients with Cushing's syndrome, Addison's syndrome, pheochromocytoma, and similar conditions. In healthy subjects, this task is even more difficult, as abnormalities are more subtle and of less obvious clinical consequence, at least in the short-term. Borrowing from the principle of uncertainty originally formulated by Werner Heisenberg, the methods we use to measure a phenomenon (in Heisenberg's case particle momentum) perturb the system; *mutatis mutandis*, even the minor stress of venipuncture increases ACTH and catecholamine levels, for example. Thus, the stress system is rapid in its reaction, making any instantaneous picture already outdated. In addition, stress, or more properly, *stressors* are wide ranging; they may consist of physical factors, such as pain and fasting, or more complex challenges, such as sleep deprivation; they may be purely psychological, such as immobilization in rodents and acute anxiety in humans, or they can—and will—resonate differently in different individuals, a concept known as *coping*. In addition, there are sex differences, genetic differences, and species-dependent differences. *Stressology*—provided such a word exists—is an eminently interdisciplinary science that includes, and benefits from, contributions of various fields, from human genetics to biostatistics, history, sociology, and many others.

In reference to the application of these general principles to the field of obesity, Pasquali summarizes the now classic observations that body fat distribution is influenced by glucocorticoids—alas stress—and that there are differences between males and females that, albeit present in humans, seem to become less accentuated as we move down the phylogenetic scale. Obesity, similar to stress, is a phenotype difficult to characterize. This is exemplified by the measure of body mass index (BMI), which lumps together individuals with very heterogenous body compositions, all being equally labeled as obese when surpassing a certain number. In a similar fashion, the definition of metabolic syndrome (MS) encompasses many heterogeneous conditions. To have MS, one needs to have at least 3 of 5 components. We will call these components A–E. Therefore, one could have MS based on three components: A + B + C, A + B + D, A + B + C, or A + B + E; four components; or five components. Subjects may be labeled as having the same condition over time but in reality they have two different conditions (A + B + C versus, as an example, C + D + E). Furthermore, the definition is categorical and does not differentiate between a person with a marginally elevated cholesterol concentration (e.g., 201 mg/dL) versus a very high concentration, which is associated with serious medical risks. We wonder about the significance of such a vague definition. There may definitely be value *tout court* for those who have commercial interests in patentable remedies, since the vast majority of living human beings, and soon our pets, will be labeled as suffering from MS.

Returning to Pasquali's review, the effects of stress on MS are also described. Among the most interesting take-home messages are the notion that the stress response varies not only interindividually but also according to gender, and the introduction of the concept of *resiliency*, namely, the identification of the factors predicting coping at an individual level.

Of free will

Hans Rudi Berthoud reminds us that much of our actions, at least in the field of food behavior, is influenced by strong biological components, and thus may not be a reflection of our free will. The food reward response is redundant, involves many neurocircuits, and is represented within the mammalian brain at many neuroanatomical levels, from the more rudimental hindbrain, to the more recent and sophisticated areas, including the prefrontal cortex. In addition, the food reward system seems to be characterized by a unilateral plasticity, implying that more food calls for more food. Unfortunately, the reverse is not true, as indicated by the inability to comply with dietary restrictions for prolonged periods of time.

The experimental work of Berthoud's group aims at determining the effects of adiposity in rodents on the main components of the reward function, *liking* and *wanting*. Liking appears to be potentiated in obese animals, as clearly indicated by a switch in the dose–response curve of corn oil administration, with less response to the lower dose than to the higher dose of corn oil. Not surprisingly, wanting is heavily influenced by the dopaminergic system, as indicated by classic neuroimaging studies conducted in obese human subjects. The neurobiology of food seeking, especially the activation of the dopaminergic system, has prompted some to introduce the controversial concept of *food addiction*. Before officially codifying this characterization by adding it to the list of addictions in the upcoming DSM-V manual, the pros and cons should be carefully weighed. For example, defining obesity as a food addiction syndrome may inadvertently prompt an epidemic of anorexia nervosa in predisposed subjects.

Exercising, not reading books, will make you smarter?

Work conducted in the laboratory of Mark Mattson has depicted the fascinating biology of brain-derived neurotropic factor (BDNF), which is a member of the neurotrophin family of growth factors, a discovery that led to the Nobel Prize for Rita Levi-Montalcini and Stanley Cohen in 1986. Mattson and his team are exploring the important roles of BDNF in controlling energy metabolism, cardiovascular functions, learning, and memory. BDNF and its cognate receptor TrkB are highly represented in brain areas crucial for energy metabolism, including several hypothalamic nuclei and the brain stem. BDNF is an appetite suppressant, as indicated by targeted disruption of this peptide in rodents. Conversely, fasting, especially if intermittent, increases BDNF concentrations in various brain regions, including the hippocampus, an important area for memory and learning. BDNF not only modulates energy intake; it also affects energy expenditure, as suggested by the observation that subjects with obesity and diabetes have low circulating levels of BDNF. In addition, BDNF is involved in the stress reaction; ablation of BDNF at birth makes animals over-sensitive to stress, as shown by an exaggerated rise in plasma corticosterone levels after immobilization. Furthermore, BDNF controls the vagal component of the autonomic nervous system; exercise- or fasting-induced rises in BDNF stimulate the vagal system, leading to lower heart rate and increased heart rate variability. Conversely, BDNF ablation makes animals more prone to an exaggerated sympathetic response. True to its name, a rise in BDNF, usually stimulated by exercise, improves cognitive functions in rodents in parallel with a documented increase in dendritic spine and sprouting. Mattson *et al.* conclude that BDNF has the potential to keep us sharp, lean, and smart, as long as we continue to live in an environment that is challenging because food is scarce and its availability unpredictable. Unfortunately, our modern environment is anything but that. Until our biology "catches up" with our culture, we need to modify our environments and habits in non-obesogenic ways.

Unraveling a novel, and unexpected, neurobiology for the fat-derived hormone leptin

Wei Pan's review is complementary to Banks' review in that it builds upon the complex regulation of the BBB to describe a novel role for leptin. It is well known that the transport of leptin into the brain is highly regulated via the availability of five locally expressed leptin receptors ObRa–e, derived by alternative splicing and belonging to the cytokine receptor superfamily. These receptors are expressed not only in areas such as the arcuate nucleus of the hypothalamus and the median eminence, where their presence is expected because of the role of leptin in energy homeostasis, but also in the dentate gyrus and CA1 of the hippocampus, where they modulate learning and memory. Surprisingly, these receptors are also found in several cerebellar layers. Leptin effects are also observed in the nucleus of the tractus solitarius and in the dorsal vagal complex, with possible implications for autonomic nervous system (ANS) activities, including feeding and gastric motility. Some of the central functions of leptin, as they interrelate with circadian rhythms and sleep functions, are also described in Pan's article.

The distribution of the ObRs within different CNS cells is interesting; for example, they are present in astrocytes at a lower level ($\sim$ 20%) compared with the arcuate nucleus, and the concentration seems to be inducible, as it increases in mice with adult-onset obesity. Additional central functions of leptin include modulation of the threshold for seizures.

To fidget or not to fidget

Catherine Kotz and her colleagues provide an interesting summary of their preclinical work on *spontaneous physical activity* (SPA)—a term that refers to undirected movements in rodents and fidgeting in humans. It is known from work of Ravussin and other groups that SPA may account for a substantial portion of total energy expenditure, ranging in humans from 100 to up to 700 calories a day. SPA has a clear genetic component and is modulated at a central level by several neuropeptides, including orexin, cholecystokinin, CRH, neuromedin, and other peptides. Specifically, the neurons that produce the two closely related peptides, orexin A and orexin B, are mostly located in a small brain area within the lateral hypothalamus. These neurons, however, have widespread projections to the rest of the brain and thus modulate fundamental functions, such as energy homeostasis, reward reactions, stress responses, and, last but not least, sleep and arousal. Overall, the main role of orexin is to expend energy and to modulate food intake and macronutrient preferences. Consistent with the above, obesity-resistant rats have increased levels of SPA in the setting of increased activity of orexin A.

Another fundamental function of orexin is the modulation of sleep, a function underlined by the neuroanatomical connections between the endogenous clock, the suprachiasmatic nucleus, and orexin cell bodies. This function is intertwined with the role in energy homeostasis and supported by the observation that subjects with narcolepsy tend to be overweight and by an animal model of narcolepsy develops early-onset obesity. The neural networks underlying SPA represent promising targets for neuropharmacological agents of potential therapeutic value in the treatment of obesity.

The white fat–mineralocorticoid axis: a new endocrine axis?

Marco Boscaro presents preclinical and clinical evidence supporting the recent hypothesis that fat tissue and the mineralocorticoid axis cross talk. Until the discovery of leptin in 1994, fat tissue was merely regarded as a depot organ, but has now been reclassified as an endocrine organ, with "full dignity." In addition to leptin, this organ secretes a series of cytokines and other substances that function as hormones. Boscaro highlights different mechanisms and anatomical levels by which the stress reaction of the HPA axis may differ from one individual to another. Some of these mechanisms include the role of glucocorticoid receptor polymorphisms in the individual variability to glucorticorticoid action, as well as the differential regulation of the two isoenzymes (11-β-hydroxysteroid dehydrogenase type 1 and 2) that convert cortisol into cortisone, and vice versa. Of interest, 11-β-HSD1 is present in adipocytes, where its activity is influenced by energy metabolism.

Subjects with obesity, especially the visceral type, have an increased activity of the renin–angiotensin–aldosterone system (RAAS). In turn, increased RAAS prevents further differentiation of pre-adypocytes into more mature cells; thus, these cells grow in size, becoming larger than usual and secrete a greater amount of inflammatory cytokines. The pharmacological blockade of RAAS activity, induced by angiotensin converting enzyme (ACE) inhibitors and angiotensin receptor blockers (ARB), has been associated with an improvement in insulin resistance, possibly mediated by a facilitation of adipocyte differentiation. The idea that RAAS is involved in adipocyte differentiation paves the way for pharmacological manipulation of this system with the aim of exercising anti-adipogenetic activity. In addition, Boscaro hypothesizes the presence of a factor of adipocyte origin that would, in turn, stimulate the production of aldosterone, as indicated by *in vitro* studies of adrenal cells in culture exposed to fat cell–conditioned medium.

The clinical correlate to these preclinical observations consists in subjects with primary hyperaldosteronism who have decreased insulin sensitivity compared to matched controls, and increased expression of interleukin 6 (IL-6; a diabetogenic factor) in their white adipocytes. In summary, Boscaro eloquently describes the existence of a novel, bidirectional white fat–mineralocorticoid axis.

From complexity back to simplicity: links between chronobiology and molecular biology

It is well known that the internal pacemaker in mammals is represented by a small number of neurons, approximately 15,000, localized in the anterior hypothalamus, an area known as the suprachiasmatic nucleus. The work of Paolo Sassone-Corsi embraces the complexity of the regulation of the metabolic process at the level of peripheral tissue, and how the central clock, entrained by environmental signals including light and temperature, coordinates these processes.

The surprising observation that one out of 10 genes displays a circadian oscillation underlines the importance of chronobiology. The CLOCK (circadian locomotor output cycles kaput) gene (*Clock*) was discovered by Turek's group in 1994, the same year in which leptin was identified. *Clock* encodes for a basic helix-loop-helix-PAS transcription factor that affects both the persistence and period of circadian rhythms. CLOCK functions as an essential activator of downstream elements in the pathway critical to the generation of circadian rhythms. In their review, Sassone-Corsi and colleagues focus on how CLOCK and another core transcription factor BMAL1 (brain and muscle aryl hydrocarbon receptor nuclear translocator-like) regulate other genes, and simultaneously induce the synthesis of their own repressors, period (PER) and cryptochrome (CRY). Furthermore, they describe how CLOCK is able to induce chromatin remodeling, acting as an enzyme that opens the structure of chromatin and exposes it to transcriptional regulation. These activities require energy; and thus, the levels of nicotinamide adenine dinucleotide, (NAD^+), a coenzyme found in all living cells, exhibit circadian oscillations. In addition, sirtuin 1 (silent mating type information regulation 2 homolog (SIRT1 in human)), is an NAD-dependent deacetylase that targets proteins that contribute to important cell processes such as reaction to stress and cell survival (e.g., reaction to stressors, longevity). The expression of nicotinamide phosphoribosyltransferase (NAMPT), the rate-limiting step for the biosynthesis of NAD^+, is indirectly controlled by SIRT1. In summary, circadian clocks, energy, metabolism, and cell survival are intimately connected via direct molecular coupling.

Sleep deprivation, obesity, and insulin resistance – it all adds up

Eliane Lucassen and colleagues report on the mechanistic and epidemiological work conducted by many investigators in the last two decades, including the group of Eve Van Cauter. Epidemiological evidence indicates an association between short sleep/poor sleep quality and increased weight. Lucassen also reviews in detail the neuroendocrinology of sleep, that is, the circadian changes displayed by several hormones involved in metabolism, appetite, and energy expenditure. Increased levels of proinflammatory cytokines, as indicated by the work of Alexandros N. Vgontas and colleagues, are associated with sleep deprivation; in turn, elevated levels of IL-6 and other cytokines induce sleepiness and the "sick behavior" syndrome. Finally, the article highlights the need for future studies, prospective and interventional in nature, while underlying challenges of the classic approach implementing randomized, controlled trials that may prove inept in this field.

Sleep deprivation and obesity in children – our grandmother was right

David Gozal and colleagues describe the self-reverberating mechanisms that potentiate obesity and sleep apnea in children. The issues of obesity and sleep deprivation are qualitatively similar but quantitatively different in children and adults. Similar to adults, childhood obesity and sleep deprivation have reached epidemic proportions, but the clinical consequences and social costs are more serious. Children are more susceptible than adults (impact on growth and intellectual development), and due to their longer life expectancy, there is a longer duration of the negative consequences. Societal changes—a forced modernization—imposed on us by the mass media, including greater use of TV, especially in the bedroom, use of mobile phones, and other electronic media play an important role. Though children have a greater need for more sleep than adults, the exact amount is variable and cannot be directly determined. Lack of proper family routines regarding meals, bedtime, and regular exercise make things even worse, especially because they often result in a lifetime lack of healthy habits, which may then affect the next generation. In conclusion, Gozal argues, and we strongly agree, that sufficient knowledge has accumulated, both from the point of view of biological plausibility and epidemiological evidence, to inform health policies.

The Brain and Obesity Lecture Series was fully supported by the National Institutes of Health (NIH), Intramural Research Program: National Institute of Diabetes and Digestive and Kidney Diseases (NIDDK). We would like to thank the speakers for their contributions. We would also like to thank the staff of *Annals of the New York Academy of Sciences* and the New York Academy of Sciences for their support in preparing this publication.

GIOVANNI CIZZA[1] and KRISTINA I. ROTHER[2]

[1] *Section on Neuroendocrinology of Obesity and* [2] *Section on Pediatric Diabetes and Metabolism, NIDDK, National Institutes of Health Bethesda, Maryland*

Ann. N.Y. Acad. Sci. ISSN 0077-8923

ANNALS OF THE NEW YORK ACADEMY OF SCIENCES
Issue: *The Brain and Obesity*

Conjectures on some curious connections among social status, calorie restriction, hunger, fatness, and longevity

Kathryn A. Kaiser,[1,2] Daniel L. Smith, Jr.,[2,3] and David B. Allison[1,2,3]

[1]Office of Energetics, School of Public Health, University of Alabama at Birmingham. [2]Nutrition Obesity Research Center, University of Alabama at Birmingham. [3]Department of Nutrition Sciences, School of Health Professions, University of Alabama at Birmingham

Address for correspondence: David B. Allison, Ph.D., Ryals Public Health Building, Room 140J, University of Alabama at Birmingham, 1665 University Boulevard, Birmingham, Alabama 35294. Dallison@uab.edu

Many animal and human studies show counterintuitive effects of environmental influences on energy balance and life span. Relatively low social and/or economic status seems to be associated with and produce greater adiposity, and reduced provision (e.g., caloric restriction) of food produces greater longevity. We suggest that a unifying factor may be perceptions of the environment as "energetically insecure" and inhospitable to reproduction, which may in turn provoke adiposity-increasing and longevity-extending mechanisms. We elaborate on two main aspects of resources (or the perceptions thereof) on body weight and longevity. We first discuss the effects of social dominance on body weight regulation in human and animal models. Second, we examine models of the interactions between caloric restriction, body composition, and longevity. Finally, we put forth a relational model of the influences of differing environmental cues on body composition and longevity.

Keywords: hunger; fatness; caloric restriction; social status; longevity

Introduction

Socioeconomic status is also related to the incidence and prevalence of obesity, such that the poor are disproportionately affected by obesity, regardless of race/ethnicity. Research is needed to further understand the impact of socioeconomic status on the development of obesity.[1]

A large body of research in animal models indicates that substantially reducing caloric intake while maintaining optimal nutrition results in significant increase in life span. . .[2]

Is perceived energetic uncertainty (in the face of a metabolizable energy surfeit) a key mediating variable in the causal chain leading to increased fat stores? Although the link between objective economic aspects of socioeconomic status (SES) and health outcomes is a familiar topic in the literature, body composition changes may also be significantly influenced by aspects of SES such as subjective social status (SSS). SSS is the self-evaluation of one's place within a community or group that is not necessarily proportional to immediate income or material

resources. Despite the manifest implications in the contexts of obesity science and public health policy, the mechanisms of and the extent to which SSS is linked to obesity are unknown. We propose the novel hypothesis that it is the "socio" as much as or more than the "economic" in SES that increases obesity risk. That is, the self-perception of being low in a social hierarchy, independent of any specific, objective economic factors, may lead to physiological, cognitive, and behavioral changes that ultimately result in increased adiposity.

In this paper, we highlight and connect the current views on two aspects of energetics: (1) SES and obesity, and (2) caloric restriction and longevity. There is clear experimental and observational evidence for causative associations between SES and obesity as well as between caloric restriction and longevity. The implications of these two topics taken together are challenging to current paradigms. Both relationships have intriguing and complex research findings, and the exact mechanisms of these observed relationships remain elusive.

doi: 10.1111/j.1749-6632.2012.06672.x

In brief, we explore the idea that animals (including humans) respond to perceived threats to their energetic security by switching life strategies to build and preserve energy stores to the extent that they can so as to buffer against true food scarcity that may occur later, and to extend life span, so as to breed more slowly or in better times. We begin with some long-standing observations and then offer conjectures about how these relationships operate in different populations of humans and animals. These observational and experimental studies include a variety of settings relating to general macroeconomics, food-specific economics, living environments, social perceptions, social hierarchies, economic uncertainty, hunger state, chronic hunger, food restriction, and energy uncertainty. We conclude with recommendations on future research directions.

SES and obesity

Is there a relationship?

In 1962, the classic Midtown Manhattan Study (which was originally conceived to examine social class and mental illness) revealed an unexpected inverse relationship between socioeconomic status (measured as a composite of the father's occupation and education) and body weight.[3,4] This commentary from the 1962 paper describes the new observation and speculates excitably about some "seemingly obvious" implications:

> The fact that obesity is 7 times more frequent in lower-class than in upper-class women has profound implications for theory and for therapy. For it means that whatever its genetic and biochemical determinants, obesity in man is susceptible to an extraordinary degree of control by social factors. It suggests that a broad-scale assault on the problem need not await further understanding of the physiological determinants of obesity. Such an assault might be carried out by a program of education and social control designed to reproduce certain critical influences to which society has already exposed its upper-class members.[4]

In their 1989 review of studies published after 1941, Sobal and Stunkard[5] reported a strong, inverse relationship for women in developed societies, but inconsistent findings for men and children. For people in developing societies, Sobal and Stunkard found a strong positive relationship between SES and increasing body mass. More recent examinations of this relationship have focused on more specific aspects of the question. For example, in a 2005 study, the utility of SES measures in predicting significant weight increase was examined in 34 longitudinal studies performed in developed countries.[6] The 1989 findings of Sobal and Stunkard were generally supported when education and occupation were used as predictors, but income was less consistently associated with obesity in this later review. Also, no relationship using any SES indicators was observed in studies with predominantly African (not African American) samples.[6] In a 2008 review of cross-sectional studies of children reported between 1990 and 2005, a stronger inverse association was found between adiposity and parental education than between adiposity and parental occupation or income.[7] Furthermore, in examining these more recent studies, Shrewsbury and Wardle found that SES was inversely associated with adiposity in 10 of the 18 studies in younger children (5–11 years), which supports the proposal that SES-related gradients in adiposity develop early in the life course.[7] These results differed from what Sobal and Stunkard found for child studies in their 1989 review[5] in that it appears that during the latter part of the 20th century, positive associations between SES indicators and obesity were rarely observed. This difference may be the result of increasing weight in all social classes or perhaps more severe obesity being increasingly prevalent in lower SES groups.

In a more recent examination of this question in children, the National Longitudinal Survey of Youth (NLSY; a 15-year cohort reported in 2006), excess body weight was inversely related to childhood SES (using mother's education as a proxy) and this disparity increased with age. Using the NHANES (National Health and Nutrition Examination Survey) to compare White, Black, and Mexican American children, Wang and Zhang found that for white adolescent boys and girls, there was a significant inverse association between SES and obesity in the NHANES III (1988–1994) sample.[8] In contrast, Black girls with a high SES had a higher prevalence than did their low- and medium-SES counterparts in the 1988–1994 and 1999–2002 samples. For Mexican American children, the study sample was small and no consistent patterns were observed.[8] Explanations for the observed racial disparities and causal mechanisms remain obscured by nonexperimental approaches.

SES and obesity—is the relationship causal?

Thus far, the evidence for the SES-obesity link we have highlighted is based on strictly observational data, demonstrating correlation, but not necessarily causation. Other approaches are needed to assess causality and both quasi-experimental and experimental studies have recently been reported. One large randomized controlled trial examining living conditions and obesity was the Move to Opportunity (MTO) study.[9] Between 1994 and 1998, 4498 women with children who were living in urban, high-poverty census tracts ($\geq$40% of residents below the federal poverty level) were randomly assigned to one of three groups: one group received no vouchers, one received housing vouchers with no specific directions on where they might use them, and the third group were directed to use the housing vouchers they received to move to an area of low poverty (<10% of residents below the poverty level). At the 10- to 15-year follow-up, the group that moved to low-poverty areas had significantly lower percentages of women with body mass index (BMI) values above 35 and 40 and significantly lower glycated hemoglobin levels than did the control and nonspecific voucher groups. A follow-up analysis of the MTO study (Zhao, 2008; unpublished dissertation) examined several specific neighborhood factors of the locations the participants moved to, including food prices, availability of restaurants and food stores, availability of facilities for physical activity, crime, and population density. These factors had little impact on the intention-to-treat effects. This study supports the suggestion of Sobal and Stunkard in 1962 that perhaps the relative economic conditions of neighbors in the living environment are a "critical influence" on obesity.

In a recent analysis of two twin adoption cohort studies of over 2000 families, researchers sought to determine whether the association between rearing parent SES and adoptee BMI was statistically significant, even when controlling for the BMI of the rearing parents. In a form of "natural randomization," both datasets suggested that shared genetic diathesis and direct environmental transmission contribute approximately equally to the association between rearing parent SES and offspring BMI.[10] If the underlying mechanisms of these effects were understood, an intervention might be expected to reduce obesity at a level equivalent to the associated causal influence. In some cases, this might be a large effect.

One intriguing way in which causal inference about some aspects of the SES-obesity link could be achieved is through prospective studies using lottery and casino winners versus losers. These events are naturally occurring randomizations. Although such studies would address the single SES indicator (income) that is often least associated with weight, careful data collection and study design could yield new insights as to the effect of increased income on weight and other aspects of health. Randomized trials of income supplementation have been done, but without the examination of obesity or weight change as outcomes.[11] Connor *et al.* suggested that if trials could be performed on contest winners or by randomizing recipients by use of unclaimed public funds, the results could inform policy deliberations on taxes, public benefits or entitlements, and minimum wage levels.[11] Sudden changes in financial resources may or may not influence self-perceived social status.

Effects of social perceptions on food intake and body composition

In humans, lower SSS is associated with higher BMI and waist-to-hip ratio to an even greater extent than is objective income.[12] Differences in social status have also been shown to influence food intake patterns in animal studies. Some bases for this view include that subordinate status birds across many species (willow tit, great tit, greenfinch, chickadees, titmouse, nuthatch) carry greater fat reserves than do dominant status birds.[13,14] In contrast to naturalistic observations in birds, controlled laboratory studies in primates provide further insight to socially subordinate effects. In nonhuman primates, socially subordinate females have demonstrated chronic psychological stress, have reduced glucocorticoid negative feedback, and have higher frequencies of anxiety-like behavior than do the socially dominant females.[15] Twenty-four-hour intakes of both low- and high-fat diets were significantly greater in subordinates than in dominants, an effect that persisted whether standard monkey chow (13% of calories from fat) was present or absent. Additionally, feeding patterns were altered in subordinates: dominants restricted their food intake to daylight but subordinates continued to feed at night.[15]

Social status and body composition in rodents

Perceptions of food availability may have effects in settings where social status is involved. Whereas the previous example in primates demonstrates a situation in which more energy consumed results in greater adiposity in subordinate females, the social stress associated with group housing of male rats has complex effects on body weight and body composition that depend on the hierarchical status of the rat within the group. Upon initial exposure to the social group dynamic in a visual burrow system (VBS), all animals lose weight comprising both lean and fat mass, with the subordinate males losing more weight (and more lean mass) than their dominant counterparts.[16] This greater reduction in body weight is evident despite an equal or greater intake of food per gram of body weight;[16] however, on a per animal basis, the food intake is often reduced in stress situations or subordinate animals. When removed from the VBS system and allowed to recover, both dominant and subordinate males gain weight rapidly, recovering the weight lost and reaching a significantly higher body weight, with no significant difference between the groups by approximately three weeks.[16] Subsequent exposure to the social stress in the VBS produces similar weight loss patterns (subordinates lose more than do dominant animals), with subordinate males having a significantly greater body fat percentage after removal and recovery for three weeks, particularly in the visceral fat depots.[16,17] Importantly, all animals in this experimental design are fed *ad libitum* (ad lib); thus, this is not an "economic" effect, but rather a social effect. It appears that the perception of social stress alters the physiologic response, thus improving metabolic efficiency and leading to greater body fat gains when the animal is removed from the socially stressful environment. Would such a response be expected in humans as we interact in multiple social groups at varying levels throughout our daily life and over the course of life? Additionally, how would an individual's perception of being subordinate or dominant alter these physiologic responses, particularly in an environment where exposures are intermittent and energetic availability is not limited?

What if we all ate just a little bit less?

When mice are provided ad lib access to a palatable diet and limited exercise capacity, much like humans, they adopt a fairly sedentary lifestyle and overeat, which results in a positive energy balance and weight gain over time. To reduce this behavior, researchers sometimes limit the amount of food provided to control fed animals by ~5% compared to what their ad lib counterparts would voluntarily eat.[18] This is suggested to reduce excess weight gain, decrease variability among animals, and improve health. Somewhat unexpectedly, female mice in which intake was restricted by 5% of ad lib intake for one month of feeding exhibited a nonsignificant lowering of body weight but a significant reduction in lean mass and a significant increase in fat mass.[19] That is, the animals were fatter even though they ate less. This phenotype was present across fat pad depots, including subcutaneous and visceral sites, reflecting a shift in energy assimilation and storage in the animals exposed to mild calorie restriction (CR) animals.[19] Total energy expenditure and resting energy expenditure as measured by indirect calorimetry were both significantly lower in the 5% CR mice, although the diurnal pattern of energy expenditure remained intact with no significant difference in locomotor activity. Interestingly, feeding behavior was significantly different between groups despite this mild reduction, with 5% CR mice demonstrating gorging behavior immediately after food was provided, with no difference between groups for the rest of the dark cycle, but less intake during the day (light phase) in the 5% CR mice.[19]

One possible explanation for these results is that the small restriction resulted in daily depletion of food stores for the mice, which in turn induced daily hunger episodes that would not be present under ad lib conditions. Additionally, the gorging behavior could invoke a metabolic effect with the majority of energy intake coming in a single or limited number of meals, with little additional energy intake throughout the rest of the daily cycle. In line with these observations are the results of a study in which 41 recombinant inbred strains of mice were subjected to dietary restriction (fed 60% of ad lib intake) and then measured for body composition and longevity.[20] Whereas CR normally reduces adiposity and increases longevity, life span extension was associated with fat preservation in these strains, with one strain significantly increasing fat mass in both sexes despite the 40% calorie reduction.[20] How such a shift in the perception of food availability might translate to humans and body composition outcomes, in addition to more complex relationships of adiposity with health and longevity, merits further investigation.

A similar observation of social stress influencing energy intake (both caloric amount and timing of intake) with resulting body weight changes is also reported for nonhuman primates. Whereas rodents are found to eat less in stressed environments, excess consumption is proposed to be a major contributor to obesity development in humans.[15] In one study, socially housed female macaques with a long-term, established dominance hierarchy were tested for three weeks during exposure to low-fat or high-fat diets. Consumption of either diet type was higher in subordinate animals than in dominant animals, during both day and nighttime measures across the three-week period. The increase in body weight over the three-week period was associated with the amount of intake, as might be expected, and was independent of social status. Thus, monkeys of lower social status consumed more calories with increased day and nighttime feeding than did their dominant counterparts and gained more weight.

In studies of dietary restriction, shifting female monkeys fed a high-fat diet ad lib to a restricted (30% fewer calories) low-fat diet resulted in a significant reduction in physical activity, which offset the energetic deficit sufficiently to prevent significant weight loss over a one-month period.[21] A subsequent month of further restriction (60% fewer calories than ad lib) resulted in a further depression in physical activity, as well as a significant reduction in body weight, with no overall change in the percentage of body fat.[21] Because these animals were singly housed, interactions related to social dominant or subordinate status were not obtained. Whether social hierarchy would influence this energetic compensation by altering physical behavior patterns remains to be explored. As a model for humans, nonhuman primate studies such as these could combine the social and energetic uncertainty elements to more comprehensively assess determinants for obesity-related outcomes pertinent to human social and environmental conditions.

Perhaps it's the socio, not the economic

In contrast to food uncertainty or restriction in animals, in humans, Supplemental Nutrition Assistance Program (SNAP) participation (but not food insecurity) is associated with higher adult BMI in Massachusetts residents living in low-income neighborhoods.[22] In 2005, a survey was done of 435 adult residents of low-income census tracts in Massachusetts. After adjustment for age, sex, sociodemographic characteristics, and food insecurity, both participation in the SNAP and participation in any federal nutrition program 12 months before the survey were each associated with an approximate 3.0 kg/m^2 higher adult BMI. However, prolonged participation in the SNAP was associated with lower BMI. Persons who were eligible but did not participate had lower BMIs than did participants. This implies that perhaps short-term participation in the SNAP is associated with greater risks for obesity or perhaps that the psychosocial mechanisms associating food insecurity and obesity diminish over time in the SNAP.

Economic uncertainty

In their study of the 1979 cohort of the Longitudinal Study of Youth, Smith *et al.* found that economic variability and uncertainty (as measured by probability of unemployment, number of income drops, volatility of income, and probability of being in poverty) were related to an increase in body fat.[23] The authors asserted that body fat serves as an "insurance plan" against starvation, with greater risk requiring greater insurance. The economics of uncertainty and the relationship to food intake have been experimentally studied in animals.

Studies in mice indicate that two factors interact to determine the number of meals and meal size.[24] These cost factors have been termed *approach cost* (procurement) and *unit cost* (consummatory). Results from varying schedules of approach and unit costs indicate that meal patterns in mice are sensitive to approach cost, whereas the total amount consumed is more sensitive to the unit cost. It was shown previously that when mice are forced to pay a price for food (e.g., to work in the form of bar presses or nose pokes), their daily intake is maintained well at lower prices but declines at higher prices, which is a relatively inelastic but classic consumer demand function.[24] At relatively low costs (up to 25 responses), mice maintain body weight or slow growth that is comparable to no-cost conditions. The feeding patterns under these conditions show that mice "graze" during the first part of the night with undefined meals, consistent with a period of almost continuous home cage activity, but show more defined, small meals during the latter part of the night, eating very little by day.[25]

Economic disparity and obesity

The Gini coefficient, which was developed by Italian statistician and sociologist Corrado Gini, was proposed in 1912 as a measure of the inequality of income distribution.[26] A Gini value of 0 indicates a perfectly equal distribution of income, and a value of 1 indicates maximal inequality in which one person in the country has all the income. In analyses of 23 high-income and 10 middle-income countries with less economic dispersion, Due *et al.* (2009) found that economic inequality as measured by the Gini coefficient was more important in explaining both level of and inequality in overweight among adolescents than was absolute economic level.[27] These findings further suggest that inequality as disparity in wealth promotes weight gain.

Hungry for money and other valued resources

Some researchers assert that the desire for money is an evolutionary extension of our innate desire for food. Briers *et al.* conducted three studies showing the reciprocal association between the incentive value of food and money.[28] In study 1, hungry participants were less likely than were satiated participants to donate money to a charity. In study 2, participants in a room with an olfactory food cue, which is known to increase the desire to eat, offered less money in a "give-some" game than did participants in a room free of food smells. In study 3, the participants' desire for money was associated with an increase in the amount of candies they ate in a subsequent taste test, but only among participants who were not restricting their food intake in order to manage their weight. Perhaps in present-day societies, the attraction to money is so powerful that people who, relatively speaking, fail in their quest for increased financial resources become frustrated. Because financial and caloric resources are valued as exchangeable, people might tend to appease their desire for money by consuming more calories than is necessary to sustain body weight.[28]

Examining effects of hunger on other resources of high value, Pettijohn *et al.* performed a field test of the *environmental security hypothesis*.[29] This hypothesis is a context-dependent theory of attraction and preferences that draws on evolutionary theory and ecology. Researchers examined the effects of a hunger state on perceived attractiveness of idealized romantic partners.[30] The results indicated that hungry male participants preferred ideal romantic partners who were relatively heavier (>120 lbs) and satiated males preferred ideal partners who were relatively lighter (<120 lbs), to a statistically significant degree. In assessing the influence of perceived maturity of idealized partners, hungry male participants preferred those who were slightly older than themselves and full males preferred ideal partners who were slightly younger.[30] These findings together suggest a reliable influence of the evaluation of one's personal resources in a short-term context on perceptions of both food intake drive and mate preferences.

Hence, the economics of perceived food availability and hunger states seem related to food-seeking behavior as well as individual fat-storage strategies that are beyond conscious control. Next, we examine the effects of limited food (caloric restriction) on another important aspect of evolutionary fitness: longevity.

Caloric restriction and longevity

Calorie restriction remains the most highly researched, nongenetic intervention to improve health and increase life span in research organisms ranging from single-celled yeasts to nonhuman primate models.[31,32] These health and longevity benefits are proportional to the amount of restriction up to the point of malnutrition[33] and are generally independent of the macronutrient content being restricted (fat versus carbohydrates versus protein). The data, considered as a whole, suggest a clear, causal relationship between energy provision and mortality rate, as well as senescence and metabolic-related disease.

The mechanisms of action for these observations remain unclear. One consistent phenotype of CR across animal models is the reduction of body weight and particularly body fat.[32] In longevity studies using rodents, many ad lib-fed laboratory rodents develop age-associated obesity, even when fed a presumably healthy, low-fat diet. When adding a high-fat diet, most rodents exhibit a diet-induced obesity response similar to what is expected of humans who over-consume a calorie-rich, high-fat diet. With the switch from an ad lib diet to a restricted feeding paradigm, CR induces a rapid and sustained weight loss associated with the caloric deficit. This "negative energy balance" phase is followed by the establishment of a new equilibrium in which the reduced body weights are matched to the energy provision, i.e., relative energy balance. Therefore, the body weight and body composition changes associated with CR are more long term in nature than is the temporary energy deficit that is experienced during the initiation of CR. Thus, one could posit that the lower body weight or fat induced by CR partially mediates the effect of CR on life span. We tested this in a large sample of Wistar rats ($N = 1200$).[34] The relative contribution of body weight to the CR effect was approximately 11%, thus supporting this hypothesis of partial mediation by low body weight. In a related study, we tested the influence of intentional weight loss by CR on mortality rates in outbred rats after the establishment of obesity. CR resulted in increased longevity compared with that in animals that remained obese, which suggests that weight loss can increase longevity even after the onset of obesity.[33]

In addition to the rodent model, two long-term, randomized CR studies in nonhuman primates are providing the first evidence that age-related mortality may be significantly lower in CR animals, concomitant with significantly reduced metabolic-related disease.[35] Whether similar disease prevention and maximal longevity extension will be present in the CR monkeys is not yet established, but additional interim results for mortality are expected in the near term (1–3 years).

CR and longevity: is perceived energetic uncertainty (independent of immediate energy intake) a mechanism of causation?

One aspect of the CR paradigm is feeding less than would be voluntarily consumed under similar but ad lib conditions. This results in animals that are both acutely and chronically hungry,[36,37] which induces a gorging behavior by animals that are normally restricted once food is presented. Whereas ad lib fed animals have constant access to adequate food supplies, CR animals will often consume the majority of their daily food allotment shortly after it is given, resulting in extended periods of fasting during which food is neither available nor perceived by the senses. This means that not only are CR mice energetically restricted, but a host of neuroendocrine signals that

coordinate energy intake and expenditure are presumably altered for most of the animal's life span. Because this may be perceived as a stressful situation, it raises the possibility that a small amount of something that is harmful in large doses may in fact be beneficial, i.e., the idea of hormesis.[38–40] The idea of hormesis is often discussed in toxicology, where complex J- or inverted U-shaped curves—as opposed to a monotonic response—are observed with a dose-response pattern (i.e., just the right amount is beneficial, whereas either too little or too much is harmful). Whereas homeostasis is the process by which bodily functions are maintained, allostasis is the process by which bodily functions change in response to environmental challenges.[41] It is plausible that the perception of energetic uncertainty is a hormetic signal leading to allostatic adaptive responding that both increases fat deposition when sufficient metabolizable energy is available and leads to increased life span if conditions are otherwise permissive.

Regarding CR and hunger, it seems reasonable that organisms must judge the nutrient and energetic state of their environment on the basis of specific cues. If, as in the case of CR, those cues are altered owing to impaired access to food supplies the majority of the time, might this elicit a protective response to promote the maintenance and preservation of the organism until a more favorable nutritional environment is encountered? Although one might expect these cues to be external from the environment, it is possible that internal cues like fat stores might also be coordinated in the response to balance the need versus availability equation of energy perception. Related to this idea, visceral adipose tissue is associated with increased morbidity and mortality and is proposed to play a role in multiple metabolic-related diseases. Importantly, visceral adipose depots are sensitive to energetic needs and are reduced in negative energy balance, that is, the CR state. By surgically removing part of the visceral fat in rats, significant health and longevity benefits have been achieved independent of CR, which suggests a role for fat and particularly visceral adipose tissue in mediating health and longevity.[42] Another possible interpretation would suggest that visceral adipose tissue reduction by mechanical means removes a signal of surplus energy stores, which results in a perceived energetically "lean" time and contributes to the health and longevity benefit. Whether such a signal exists and is directly related to the specific fat depots themselves is unknown.

Perception of nutrient availability has also been studied in worms and flies in relationship to survival and mortality kinetics. *Drosophila melanogaster* has been used extensively in nutrition and aging research partly because of its relatively short life span and the ease with which researchers can produce and modify diets of known composition. Although it was previously known that CR increases life span in flies, recent studies have shown that mortality is acutely sensitive to nutrient availability.[43] This was modeled by switching flies from dietary restriction (DR) to ad lib feeding (or vice versa—ad lib to DR), which meant switching them from being housed with standard, rich media access (ad lib) to a dilute media composition. Within two days of switching the flies to the opposite nutrient state, mortality rates paralleled those of animals continuously exposed to the nutrient condition.[43] Although it is possible that dietary factors and intake amounts could produce a biological caloric effect resulting in survival modulations, the acute response suggests that additional factors such as nutrient or environmental perception were also involved beyond the proposed mechanisms of more long-term aging phenotypes such as accumulated cellular damage.[43] This idea is further supported by results showing that the extended life span of DR flies could be shortened by sensing (e.g., smelling) live yeast that were present but not available for consumption.[44] Importantly, the perception of live yeast did not shorten life span of ad lib fed flies, which suggests that the effect was not generally negative (like a toxic effect), but rather unique to the DR response. Further studies showed that disruption of olfactory receptor modulator (Or83b) alone was sufficient to increase life span and induce a number of stress-resistant phenotypes, thus supporting the role of perception in mediating multiple aspects of the DR related response.[45] In agreement with other work in *C. elegans*,[46,47] these studies demonstrate a critical role of sensory perception in the longevity response to DR and point to interactions and modulation of perception on complex phenotypes such as life span.

The hunger aspect of hormesis is further supported by work using intermittent feeding (IF) or every other day (EOD) feeding in rodents. Rather than daily restriction, the EOD or IF models permit ad lib feeding, but only for specific periods of time followed by complete fasting (e.g., one-day fed ad lib, one-day fasted, one-day fed ad lib, etc.). Because food intake during the ad lib period may not fully compensate for the fasting period, EOD feeding may result in mild energy restriction. Even in rodent strains that do not lose weight with EOD feeding, life span is increased when the protocol is started at young age.[48,49] Comparing strains of rodents for the body weight and longevity response, it appears that life span is not fully predicted from changes in body weight. This suggests that other factors, possibly daily hunger, may be contributing to the benefit.

If the perception of restriction were beneficial, would repeated bouts of weight loss be beneficial?

As mentioned previously, CR is known to have positive effects on health and longevity, whereas obesity and excess adiposity are detrimental. What would happen if all individuals who were overweight or obese lost weight to a normal body size? Although this is not likely to occur, most of those who successfully lost the weight would regain the amount within a few years.[50] If, as is commonly believed, excess weight and adiposity are harmful, would it be beneficial (or harmful or have no effect) to go from overweight to normal and back again? Yo-yo dieting and weight loss is certainly observed in humans, and although not normally encountered in laboratory nutrition studies, this question of the potential health consequences with weight cycling from repeated bouts of weight loss and regain is beginning to be addressed. A study is currently underway at the University of Alabama at Birmingham in which high-fat diet feeding (similar to the Western diet macronutrient composition) is utilized to establish an overweight/obese cohort of mice. The mice are subsequently randomized to interventions in which food intake is continued ad lib, restricted sufficiently to achieve a normal body weight, or food intake is restricted and then refed similar to a yo-yo dieting experience. With the CR phase, animals clearly lose weight, and when released to ad lib feeding, usually return to approximately their pre–weight loss size or greater. Thus, not only is this study a model of CR for intentional weight loss (and regain with yo-yo dieting), these animals experience varied amounts and time periods of imposed hunger by CR. It may

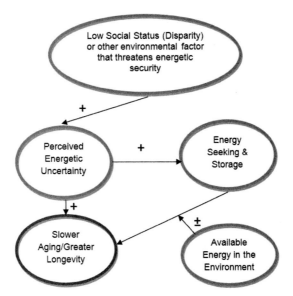

Figure 1. Hypothesized model of social and environmental influences on energy balance that affect aging.

be that treating overweight may be beneficial to do at regular intervals, even with modest results, similar to the way that oral health benefits from regular plaque removal (i.e., a "dental model of obesity treatment"). Although a person may not be able or motivated to prevent the buildup of plaque at the gum line (which is not viewed as a treatment failure), frequent plaque removal efforts by dental hygienists reduce the long-term, health-damaging effects.

Other evidence for the stress/hormesis hypothesis

Stress comes in a variety of forms and, as proposed in this work, does not necessarily require a specific physical or biological insult but rather may arise from the organism's perception of its relative state or environment. Multiple stressors have been discussed in association with life span modulation, including environmental temperature and osmotic stress.[39,51] The extent to which these survival-altering interventions operate through overlapping pathways and mechanisms that are critical for life span extension and that are mediated by a stress response remains to be demonstrated.

What might this imply about calorie restriction mimetics?

Considering that the perception of nutrients through hunger or other pathways may mediate the

effect of CR, might it be possible to illicit a similar physiologic response without requiring actual CR? This concept of mimicking the benefits of CR without actual caloric reduction was first proposed by Ingram *et al.*, in part, on the basis of their work with multiple compounds that would interact with metabolic pathways implicated in CR.[52–54] Although multiple compounds have been identified that induce a similar physiologic and transcriptional response to CR, few compounds have produced increased longevity in animal studies. Two of the earliest proposed CR mimetics were metformin and 2-deoxyglucose (2DG). Multiple rodent longevity studies are reported for metformin supplementation,[55–57] and there is one recent longevity report for 2DG.[58] One important aspect of these CR mimetics is how comprehensively the mimetic recapitulates the total CR response. Regarding the hunger associated with CR, there is little reason to suspect that metformin would elicit a hunger response. Thus, any hunger-associated hormetic response would be lacking. Similarly, with 2DG, no increase in food intake was observed,[58] which suggests a lack of overt hunger. The clear toxic effects of high 2DG make dosing a challenge; therefore, what dose of either compound, if any, could elicit a hunger response similar to CR is unclear. Whether additional CR mimetics should require demonstrated or reported hunger as part of the response is an open question. Current collaborations with the Interventions Testing Program of the National Institute on Aging testing an α-glucosidase inhibitor (acarbose) may provide useful information on a hunger-hormetic mechanism of action. Acarbose is expected to induce a hunger response in addition to metabolic and physiologic improvements. A related approach is being considered using ghrelin, a gut-derived peptide associated with feeding and hunger that could induce a chronic, artificial hunger state despite access to energetic requirements sufficient to meet the metabolic demands of the organism.

Implications if these hypotheses are true

We offer the diagram on the previous page (Fig. 1) as an integrated model of the observations and effects we have discussed. Using this model, future efforts to identify leverage points that affect perceptions of energetic uncertainty may be as important as or more important than identifying specific economic leverage points. Furthermore, environmental manipulations that lead to the perception of restricted food availability may have paradoxical effects. On a physiological level beyond conscious perceptions, some CR mimetics may need to work downstream of hunger signals to be effective. Because of the apparent cascade of processes in CR, even discounting side-effects, losing weight with an anorexigen may be less beneficial than losing an equivalent amount of weight with CR. Some CR-like interventions that do not actually reduce total energy intake may also lead to prolonged life. Our proposed model provides a testable framework in which future investigations might illuminate mechanistic pathways.

Conclusions

There is much to be understood if we are to reduce the undesirable human and economic costs that accompany unhealthy aging, reduced life span and obesity-related chronic illnesses. A large body of literature provides clues to the importance of how social and other environmental cues may operate in these contexts, based on evolutionary strategies that govern the protection of reproductive ability. We have provided a novel framework connecting social status, adiposity, and longevity that offers specific testable predictions and look forward to future research testing those predictions.

Acknowledgments

The authors were supported by Grant P30DK-056336.

Conflicts of interest

David B. Allison has, anticipates, or has had financial interests (e.g., paid consulting arrangements) with the following entities: the Frontiers Foundation; the Federal Trade Commission, Vivus, Inc.; Kraft Foods; University of Wisconsin; University of Arizona; Paul, Weiss, Wharton & Garrison, LLP; and Sage Publications.

References

1. NIH Publication No.11-5493. Strategic Plan for NIH Obesity Research. 3-31-2011. National Institutes of Health, Department of Health and Human Services. Available at: http://www.obesityresearch.nih.gov/about/strategic-plan.aspx
2. Hodes, R.J. Director's Statement: Fiscal Year 2008 Budget Request. 3-6-2007. FY2008 NIH Budget Hearing, House Appropriations Subcommittee on Labor, Health and Human Services, Education and Related Agencies. Available at: http://www.nia.nih.gov/about/budget/2007/fiscal-year-2008-budget/directors-statement-fiscal-year-2008-budget-request
3. Goldblatt, P.B., M.E. Moore & A.J. Stunkard. 1965. Social factors in obesity. *JAMA* 21: 1039–1044.
4. Moore, M.E., A.J. Stunkard & L. Srole. 1962. Obesity, social class and mental illness. *JAMA* 181: 962–966.
5. Sobal, J. & A.J. Stunkard. 1989. Socioeconomic status and obesity: a review of the literature. *Psychol. Bull.* 105: 260–275.
6. Ball, K. & D. Crawford. 2005. Socioeconomic status and weight change in adults: a review. *Soc. Sci. Med.* 60: 1987–2010.
7. Shrewsbury, V. & J. Wardle. 2008. Socioeconomic status and adiposity in childhood: a systematic review of cross-sectional studies 1990–2005. *Obesity* 16: 275–284.
8. Wang, Y. & Q. Zhang. 2006. Are American children and adolescents of low socioeconomic status at increased risk of obesity? Changes in the association between overweight and family income between 1971 and 2002. *Am. J. Clin. Nutr.* 84: 707–716.
9. Ludwig, J., L. Sanbonmatsu, L. Gennetian, *et al.* 2011. Neighborhoods, obesity, and diabetes—a randomized social experiment. *N. Engl. J. Med.* 365: 1509–1519.
10. Fontaine, K.R., H.T. Robertson, C. Holst, *et al.* 2011. Is socioeconomic status of the rearing environment causally related to obesity in the offspring? *PLoS One* 6: e27692.
11. Connor, J., A. Rodgers & P. Priest. 1999. Randomised studies of income supplementation: a lost opportunity to assess health outcomes. *J. Epidemiol. Community Health* 53: 725–730.

12. Adler, N.E., E.S. Epel, G. Castellazzo, J.R. Ickovics. 2000. Relationship of subjective and objective social status with psychological and physiological functioning: preliminary data in healthy white women. *Health Psychol.* **19:** 586–592.

13. Gosler, A.G. 1996. Environmental and social determinants of winter fat storage in the great tit Parus major. *J. Anim. Ecol.* **65:** 1–17.

14. Pravosudov, V.V. & T.C. Grubb. 1999. Effects of dominance on vigilance in avian social groups. *The Auk* **166:** 241–246.

15. Wilson, M.E., J. Fisher, A. Fischer, *et al.* 2008. Quantifying food intake in socially housed monkeys: social status effects on caloric consumption. *Physiol. Behav.* **94:** 586–594.

16. Tamashiro, K.L., M.M. Nguyen, M.M. Ostrander, *et al.* 2007. Social stress and recovery: implications for body weight and body composition. *Am. J. Physiol. Regul. Integr. Comp. Physiol.* **293:** R1864–R1874.

17. Tamashiro, K.L., M.M. Nguyen, T. Fujikawa, *et al.* 2004. Metabolic and endocrine consequences of social stress in a visible burrow system. *Physiol. Behav.* **80:** 683–693.

18. Pugh, T.D., R.G. Klopp & R. Weindruch. 1999. Controlling caloric consumption: protocols for rodents and rhesus monkeys. *Neurobiol. Aging* **20:** 157–165.

19. Li, X., M.B. Cope, M.S. Johnson, *et al.* 2010. Mild calorie restriction induces fat accumulation in female C57BL/6J mice. *Obesity* **18:** 456–462.

20. Liao, C.Y., B.A. Rikke, T.E. Johnson, *et al.* 2011. Fat maintenance is a predictor of the murine lifespan response to dietary restriction. *Aging Cell* **10:** 629–639.

21. Sullivan, E.L. & J.L. Cameron. 2010. A rapidly occurring compensatory decrease in physical activity counteracts diet-induced weight loss in female monkeys. *Am. J. Physiol. Regul. Integr. Comp. Physiol.* **298:** R1068–R1074.

22. Webb, A.L., A. Schiff, D. Currivan & E. Villamor. 2008. Food Stamp Program participation but not food insecurity is associated with higher adult BMI in Massachusetts residents living in low-income neighbourhoods. *Public Health Nutr.* **11:** 1248–1255.

23. Smith, T.G., C. Stoddard & M.G. Barnes. 2009. Why the poor get fat: weight gain and economic insecurity. *J. Health Econ. Pol.* **12:** 1–29.

24. Atalayer, D. & N.E. Rowland. 2009. Meal patterns of mice under systematically varying approach and unit costs for food in a closed economy. *Physiol. Behav.* **98:** 85–93.

25. Goulding, E.H., A.K. Schenk, P. Juneja, *et al.* 2008. A robust automated system elucidates mouse home cage behavioral structure. *Proc. Natl. Acad. Sci. USA* **105:** 20575–20582.

26. Pizetti, E., T. Salvemini & C. Gini. 1912. Variabilità e Mutabilità (Variability and Mutability), reprinted in Memorie di Metodologica Statistica. 1955. Rome, Libreria Eredi Virgilio Veschi.

27. Due, P., M.T. Damsgaard, M. Rasmussen, *et al.* 2009. Socioeconomic position, macroeconomic environment and overweight among adolescents in 35 countries. *Int. J. Obes.* **33:** 1084–1093.

28. Briers, B., M. Pandelaere, S. Dewitte & L. Warlop. 2006. Hungry for money: the desire for caloric resources increases the desire for financial resources and vice versa. *Psychol. Sci.* **17:** 939–943.

29. Pettijohn, T.F. & A. Tesser. 2005. Threat and social-choice: when eye size matters. *J. Soc. Psychol.* **145:** 547–570.

30. Pettijohn, T.F., D.F. Sacco & M.J. Yerkes. 2011. Hungry people prefer more mature mates: a field test of the environmental security hypothesis. *J. Soci. Evol. Cul. Psychol.* **3:** 216–232.

31. Weindruch, R., R.L. Walford, S. Fligiel & D. Guthrie. 1986. The retardation of aging in mice by dietary restriction: longevity, cancer, immunity and lifetime energy intake. *J. Nutr.* **116:** 641–654.

32. Weindruch, R. & R.L. Walford. 1988. *The Retardation of Aging and Disease by Dietary Restriction.* C. C. Thomas Publisher. Springfield, IL.

33. Merry, B.J. 2002. Molecular mechanisms linking calorie restriction and longevity. *Int. J. Biochem. Cell Biol.* **34:** 1340–1354.

34. Wang, C., R. Weindruch, J.R. Fernandez, *et al.* 2004. Caloric restriction and body weight independently affect longevity in Wistar rats. *Int. J. Obes. Relat. Metab. Disord.* **28:** 357–362.

35. Colman, R.J., R.M. Anderson, S.C. Johnson, *et al.* 2009. Caloric restriction delays disease onset and mortality in rhesus monkeys. *Science* **325:** 201–204.

36. Hambly, C., J.G. Mercer & J.R. Speakman. 2007. Hunger does not diminish over time in mice under protracted caloric restriction. *Rejuvenation Res.* **10:** 533–542.

37. Speakman, J.R. & C. Hambly. 2007. Starving for life: what animal studies can and cannot tell us about the use of caloric restriction to prolong human lifespan. *J. Nutr.* **137:** 1078–1086.

38. Masoro, E.J. 1998. Hormesis and the antiaging action of dietary restriction. *Exp Gerontol* **33:** 61–66.

39. Masoro, E.J. 2000. Caloric restriction and aging: an update. *Exp. Gerontol.* **35:** 299–305.

40. Neafsey, P.J. 1990. Longevity hormesis. A review. *Mech. Ageing Dev.* **51:** 1–31.

41. Pickering, T.G. 2003. The new science of stress—book review. *Nat. Med.* **9:** 640.

42. Muzumdar, R., D.B. Allison, D.M. Huffman, *et al.* 2008. Visceral adipose tissue modulates mammalian longevity. *Aging Cell* **7:** 438–440.

43. Mair, W., P. Goymer, S.D. Pletcher & L. Partridge. 2003. Demography of dietary restriction and death in Drosophila. *Science* **301:** 1731–1733.

44. Libert, S., J. Zwiener, X. Chu, *et al.* 2007. Regulation of Drosophila life span by olfaction and food-derived odors. *Science* **315:** 1133–1137.

45. Pletcher, S.D. 2009. The modulation of lifespan by perceptual systems. *Ann. N.Y. Acad. Sci.* **1170:** 693–697.

46. Alcedo, J. & C. Kenyon. 2004. Regulation of C. elegans longevity by specific gustatory and olfactory neurons. *Neuron* **41:** 45–55.

47. Antebi, A. 2004. Long life: a matter of taste (and smell). *Neuron* **41:** 1–3.

48. Goodrick, C.L., D.K. Ingram, M.A. Reynolds, *et al.* 1983. Differential effects of intermittent feeding and voluntary exercise on body weight and lifespan in adult rats. *J. Gerontol.* **38:** 36–45.

49. Goodrick, C.L., D.K. Ingram, M.A. Reynolds, *et al.* 1990. Effects of intermittent feeding upon body weight and lifespan in inbred mice: interaction of genotype and age. *Mech. Ageing Dev.* **55:** 69–87.

50. McGuire, M.T., R.R. Wing, M.L. Klem & J.O. Hill. 1999. Behavioral strategies of individuals who have maintained long-term weight losses. *Obes. Res.* **7:** 334–341.

51. Mattison, J.A., G.S. Roth, M.A. Lane & D.K. Ingram. 2007. Dietary restriction in aging nonhuman primates. *Interdiscip. Top. Gerontol.* **35:** 137–158.

52. Ingram, D.K., M. Zhu, J. Mamczarz, *et al.* 2006. Calorie restriction mimetics: an emerging research field. *Aging Cell* **5:** 97–108.

53. Lane, M.A., D.K. Ingram & G.S. Roth. 1998. 2-Deoxy-D-glucose feeding in rats mimics physiological effects of calorie restriction. *J. Anti Aging Med.* **1:** 327–337.

54. Lane, M.A., G.S. Roth & D.K. Ingram. 2007. Caloric restriction mimetics: a novel approach for biogerontology. *Methods Mol. Biol.* **371:** 143–149.

55. Smith, D.L., Jr., T.R. Nagy & D.B. Allison. 2010. Calorie restriction: what recent results suggest for the future of ageing research. *Eur. J. Clin. Invest.* **40:** 440–450.

56. Smith, D.L., Jr., C.F. Elam, Jr., J.A. Mattison, *et al.* 2010. Metformin supplementation and life span in Fischer-344 rats. *J. Gerontol. A Biol. Sci. Med. Sci.* **65:** 468–474.

57. Anisimov, V.N., T.S. Piskunova, I.G. Popovich, *et al.* 2010. Gender differences in metformin effect on aging, life span and spontaneous tumorigenesis in 129/Sv mice. *Aging* **2:** 945–958.

58. Minor, R.K., D.L. Smith, Jr., A.M. Sossong, *et al.* 2009. Chronic ingestion of 2-deoxy-d-glucose induces cardiac vacuolization and increases mortality in rats. *Toxicol. Appl. Pharmacol.* **243:** 332–339.

Ann. N.Y. Acad. Sci. ISSN 0077-8923

ANNALS OF THE NEW YORK ACADEMY OF SCIENCES
Issue: *The Brain and Obesity*

Role of the blood–brain barrier in the evolution of feeding and cognition

William A. Banks

GRECC, Veterans Affairs Puget Sound Health Care System and Division of Gerontology and Geriatric Medicine, Department of Medicine, University of Washington School of Medicine, Seattle, Washington

Address for correspondence: William A. Banks, M.D., Rm 810A, Bldg 1, GRECC-VAPSHCS, 1660 S. Columbian Way, Seattle, WA 98108. wabanks1@uw.edu

The blood–brain barrier (BBB) regulates the blood-to-brain passage of gastrointestinal hormones, thus informing the brain about feeding and nutritional status. Disruption of this communication results in dysregulation of feeding and body weight control. Leptin, which crosses the BBB to inform the CNS about adiposity, provides an example. Impaired leptin transport, especially coupled with central resistance, results in obesity. Various substances/conditions regulate leptin BBB transport. For example, triglycerides inhibit leptin transport. This may represent an evolutionary adaptation in that hypertriglyceridemia occurs during starvation. Inhibition of leptin, an anorectic, during starvation could have survival advantages. The large number of other substances that influence feeding is explained by the complexity of feeding. This complexity includes cognitive aspects; animals in the wild are faced with cost/benefit analyses to feed in the safest, most economical way. This cognitive aspect partially explains why so many feeding substances affect neurogenesis, neuroprotection, and cognition. The relation between triglycerides and cognition may be partially mediated through triglyceride's ability to regulate the BBB transport of cognitively active gastrointestinal hormones such as leptin, insulin, and ghrelin.

Keywords: Blood–brain barrier; leptin; feeding; cognition, obesity; central nervous system; evolution

Early experiments in obesity

In 1950, Ingalls *et al.* reported in *Journal of Heredity* a mouse that had a recessive trait for obesity and dubbed it the *ob/ob* mouse.[1] Differences between affected and unaffected littermates were obvious by 3 weeks of age, but by 10 months of age, unaffected littermates weighed about 29 g, whereas affected mice weighed about 90 g, a threefold difference. Parabiosis experiments 20 years later showed that obesity in these mice was caused by the absence of a circulating factor.[2] Some 20 years after the parabiosis experiments, Freidman *et al.* identified that missing factor as leptin.[3,4]

Unlike the *ob/ob* mouse, almost all obese humans have high, not low, levels of leptin.[5] Such conditions in which the regulatory hormone is high in the face of a deficient response are termed in endocrinology *resistance syndromes*.[6] Thus, obesity in humans is dominated by a resistance to, not a de-

ficiency of, leptin. In classic resistance syndromes such as pseudohypoparathyroidism, the first resistance syndrome to be described, resistance arose because of defects either in the receptor or in the receptor's intracellular machinery.

The *db/db* mouse lacks leptin receptors and is obese and so represents a classic case of receptor/postreceptor resistance.[7] However, the negative feedback loop formed between leptin and body fat has two steps not found in peripheral tissues. The first of these is at the blood–brain barrier (BBB). As leptin is a large molecule, a protein of 16 kDa, its transfer from blood into the brain is aided by a transport system. The second are the downstream neural circuitries: in other words, leptin acts on other cells and modulators such as melanocortin[8] to mediate its effects on body weight. Defects in either the ability of the BBB to transport leptin or in the response of the downstream neural circuitries to respond to leptin leads to the resistance

doi: 10.1111/j.1749-6632.2012.06568.x
Ann. N.Y. Acad. Sci. 1264 (2012) 13–19 © 2012 New York Academy of Sciences.

picture of high serum leptin levels in the face of obesity.

Transport of leptin across the BBB in obesity

Early work showed that acquired defects in the BBB transport of leptin played a role in obesity. In 1997, two publications showed that animals with diet-induced obesity went though a phase when they no longer responded to leptin given peripherally but still responded to leptin given directly into the brain.[9,10] This was evidence not only that leptin was crossing the BBB in ineffective amounts, but also there existed a phase in which resistance at the BBB (termed *peripheral resistance*) was functionally dominant to resistance at the receptor/postreceptor level (termed *central resistance*). Subsequent studies from several laboratories have shown pharmacokinetic impairment in the blood-to-brain transport in several models of obesity including obesity of maturity, diet-induced obesity, the obesity prone rat, and the Koletsky rat; the latter two being genetic models in which leptin receptors in brain are deficient or absent.[11–16] The obesity prone rat of Levin are born with defects in CNS leptin receptors but normal transport of leptin across the BBB.[11] As they develop obesity, they acquire impairments in leptin transport. As these studies would predict, pharmacologic induction of peripheral resistance with an agent that prevents endogenous leptin from crossing the BBB stimulates feeding and body weight gain.[17] These effects begin immediately with the onset of treatment and result in a body weight gain that is essentially all attributable to an increase in adiposity.

Calculations in the CD-1 mouse and experiments in sheep have suggested that peripheral resistance in diet-induced obesity accounts for all or the majority of leptin resistance.[18,19] In other models and in humans, results show that leptin resistance clearly has both peripheral and central components. The finding in humans that leptin levels in cerebrospinal fluid are higher in obesity than in thin persons clearly demonstrates resistance at the receptor level.[20,21] Mathematical modeling based on classic assumptions of receptor resistance suggests that at low to moderate levels of obesity, peripheral resistance dominates.[18] The modeling also suggests that as obesity increases, resistance at both sites progressively worsens.

The mechanisms of leptin resistance at the BBB are multiple. First, the saturable nature of leptin transport means that there is a serum level at which leptin transport into brain is at its maximum. CSF versus serum data from human studies[20–22] and brain versus serum data from perfusion studies in mice[23] independently show that the effects of saturation are seen at serum levels of 5–10 ng/mL (Fig. 1). In other words, the transport of leptin across the BBB is most efficient when serum levels are low.

Other mechanisms of resistance relate to the regulated nature of leptin. Several factors modulate the transport rate of leptin across the BBB (Table 1). In the case of lipopolysaccharide treatment, the effect is indirect in that it is mediated by an elevation in leptin levels.[24–26] In the case of triglycerides, the effect appears to be direct in that triglycerides inhibit leptin transport across *in vitro* models of the BBB.[27] Other factors give clues to the physiological role that leptin plays in energy homeostasis. For example, short-term fasting restores the leptin transport deficit in obese mice, whereas long-term fasting (starvation) inhibits transport.[14,28] These findings are thought to be driven by triglyceride levels that decrease in fasting but increase in starvation.

Therefore, the relation between body weight status and leptin transport rates across the BBB have both chronic and acute influences. At a stable steady state, obesity and serum leptin levels relate to BBB transport rates and brain levels of leptin. But in starvation, even before significant weight loss has occurred, serum leptin levels the become uncoupled to body weight.[29] With the elevation in triglycerides that occurs with extended periods of fasting, serum levels and transport rates become uncoupled.[30] Although a high level of serum leptin is a poor anorectic, a low brain level of leptin is a powerful orexigen. Hence, the leptin/BBB/brain axis seems geared to promote feeding, especially during times of low food intake, but not to inhibit feeding during times of plenty.

Leptin and energy balance: an evolutionary perspective

Early work that attempted to frame leptin actions as an adipostat were frustrating.[31,32] The perspective that leptin is most relevant when body fat reserves are low is a more versatile concept. In starving animals or animals with very low serum leptin levels, calorically expensive activities such as reproduction

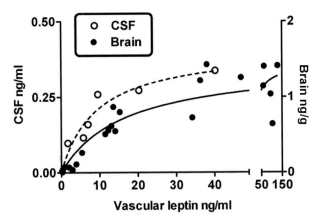

Figure 1. Relation between vascular and CNS levels of leptin. CSF versus blood levels of leptin (open circles, dotted line) are derived from several studies of humans in the literature.[20–22] Brain versus vascular levels of leptin (solid circle, solid line) are derived from a brain perfusion study.[23] Either approach shows a hyperparabolic relation between CNS and vascular levels of leptin that is explained by the saturable nature of the blood-to-brain transport of leptin across the BBB. With either approach, the curve shows significant saturation at relatively low vascular levels of leptin.

and support of the immune system are curtailed, and low doses of leptin will allow these activities to resume.[33–37] This suggests a view that leptin acts as an adipometer; that is, it informs the brain of the magnitude of caloric reserves. The brain then determines whether behavior and calories are best spent on seeking and ingesting food or whether caloric reserves are to be spent in other activities.

Various lines of evidence support the low-level permissive model of leptin action. As discussed earlier, it is at low serum levels of leptin that the BBB transporter for leptin works most efficiently. Most animals living in the wild have only 3–7% of their body weight as fat mass and serum leptin levels of 1–2 ng/mL.[38–42] Likewise, modern hunter gatherers as illustrated by the !Kung San have body mass indices of 19.1 for women and 19.4 for men.[43] These values are much lower than those considered optimal, normal, or ideal in Western societies, but are in the region where the BBB leptin transporter most efficiently conveys information to the brain.

There has been tremendous evolutionary pressure to safeguard against starvation, but probably little pressure to deal with problems of obesity.[44] As the Paleolithic diet was low in fat, simple carbohydrates, and salt, mechanisms from behavior to metabolic pathways adapted to obtain and maximize use of these important nutritional resources. An adipometer would have been needed to constantly inform the brain of whether calories were sufficient to devote to activities other than seeking food or whether caloric reserves were approaching the starvation range. Even though most modern hunter-gatherers live in tropical areas, they still face seasonal variations in food supply and, as a result, variations in body weight and fat mass.[45]

With this view, inhibition of transport by starvation and by triglycerides is explicable in metabolic and evolutionary terms. With starvation, triglycerides stored in adipose tissue are hydrolyzed to their free fatty acids.[46] Some of these free fatty acids are used to synthesize new triglycerides so that blood levels of both free fatty acids and triglycerides are elevated in starvation. A theoretical scenario can then be constructed from the premise that triglycerides evolved as a signal of stravation to the brain: a drop in caloric intake is sensed; adipose tissue responds by decreasing its production of leptin; thus serum leptin and brain levels of the leptin decrease. As starvation continues, serum triglycerides increase that impede leptin transport across the BBB; hence, brain levels of leptin decrease further. Decreasing leptin levels in the brain remove the permissive effect that allows calories to be spent on nonfeeding activities and redirects caloric use towards food-seeking activities.

Integration with other feeding signals

Many peptides and regulatory hormones have influences on feeding. Indeed, it may seem odd at first that so many factors control, regulate, or influence an act that at first seems as simple as eating. But

Table 1. Regulators of leptin transport across the BBB

State or substance	Effect on transport	Reference
Obesity	Decreased	12, 15, 16
Starvation	Increased	28
Fasting	Decreased	27
Epinephrine	Increased	67
Insulin	Increased	68
Glucose	Increased	68
Triglycerides	Decreased	27
Ovarectomy	Decreased	69
Alcohol administration	Increased	70
Lack of leptin receptor	Decreases	11, 15
Defects in leptin receptor	Decreased	11
Diurnal rhythm	Variable	71

ingestion, digestion, excretion, and related activities are, in reality, extremely complex acts. Digestion requires complex coordination of enzymatic, secretory, hormonal, neuronal, and muscular events that are under voluntary and involuntary controls. Many of the monoamines, peptides, and regulatory proteins known to affect feeding and digestion have specific functions that must be closely coordinated with other events.

Complexity of feeding also involves higher order cognitive functions. Berthoud has summarized neuronal circuitries that weld together ingestive, cognitive, and emotional aspects of feeding.[47] In Western societies, the decisions of when, where, and what to eat are not usually considered major intellectual challenges. But for animals living in the wild, one of the most complex cognitive-related behaviors in which they engage is the decision related to feeding. During periods when caloric use is high or when food resources are low, reproductive success will improve from even a primitive cost/benefit analysis of whether more calories are likely to be gained than expended in the search and acquisition of food. Primates, reptiles, and birds all illustrate situations in which cost/benefit analysis would be helpful in maximizing caloric acquisition/expenditure. Primates have been found to take recent trends in temperature and solar radiation into account in deciding whether to revisit trees for newly ripened fruit.[48] Iguanas balance acquisition of food located in cold areas against the disadvantage of leaving a warmer area with less palatable food.[49] In storing food for future use, mountain chickadees alter their caching strategies when potential pilferers are in the area.[50] This requires the chickadees to not only assess sites for security, but also evaluate the motives of other chickadees in the area. Studies with New Zealand robins illustrate how a degree of numerical competency can covey significant advantages in food storage, pilfering, and retrieval.[51]

Sickness behavior represents an especially good example of an adaption that balances feeding and caloric demand.[52,53] When animals are ill, short-term conservation of energy by reducing energy expending activities, including feeding behaviors in favor of rest can be beneficial. Calories can be used instead to increase body temperature and otherwise support immune activities.

Cost/benefit analysis is also productive regarding the probabilities of how safe it is to search for food. Whether a top predator or an occupier of a position low on the food chain, seeking and acquiring food can be a dangerous activity. Social rules about acquiring, sharing, or trading food exists in many animal societies and decisions on how to follow those rules or even when to violate them requires cognitive abilities.

Thus, feeding requires complex and contextual decisions. This may explain why so many of the feeding hormones have effects on learning, memory, attention, and other aspects of cognition. MSH, ghrelin, insulin, leptin, and glucagon-like peptide-1 are examples of gastrointestinal hormones with effects on cognition.[54–62] As with feeding, these hormones cross the BBB to exert their CNS effects. It is interesting that triglyceride levels not only affect transport of many of the feeding hormones across the BBB but also correlate with cognition.[63–66]

Conclusions

The BBB plays important roles in the regulation of the blood-to-brain transport of peptides and proteins involved in feeding. Leptin, insulin, and ghrelin are examples of hormones that are transported from blood-to-brain by saturable transport systems. Impaired transport of leptin across the BBB results in obesity and also occurs during starvation. Various mechanisms underlie transport defects of leptin. Triglycerides impair leptin transport and may be adaptive, having evolved to inhibit the CNS effects of leptin during starvation. Many gastrointestinal hormones have effects on cognition. Such effects may again have arisen from evolutionary pressures resulting from the complexity of feeding, especially those aspects that involve social and cost/benefit ratio decisions. These cognitive effects, like the feeding effects, of many of the gastrointestinal hormones are dependent on their abilities to cross the BBB.

Acknowledgments

Supported by VA Merit Review, R01NS051334 (WAB) and R01DK083485 (WAB).

Conflicts of Interest

The author declares no conflicts of interest.

References

1. Ingalls, A.M., M.M. Dickie & G.D. Snell. 1950. Obese, a new mutation in the house mouse. *J. Hered.* **41:** 317–318.

2. Coleman, D.L. 1973. Effects of parabiosis of obese with diabetes and normal mice. *Diabetologia* **9:** 297–298.

3. Zhang, Y., R. Proenca, M. Maffel, M. Barone, *et al.* 1994. Positional cloning of the mouse obese gene and its human homologue. *Nature* **372:** 425–432.

4. Halaas, J.L., K. S. Gajiwala, M. Maffei, S. L. Cohen, *et al.* 1995. Weight-reducing effects of the plasma protein encoded by the obese gene. *Science (Washington, DC)* **269:** 543–546.

5. Considine, R.V., M. K. Sinha, M. L. Heiman, A. Kriauciunas, *et al.* 1996. Serum immunoreactive-leptin concentrations in normal-weight and obese humans. *New England J. Med.* **334:** 292–295.

6. Verhoeven, G.F.M. & J.D. Wilson. 1979. The syndromes of primary hormone resistance. *New England J. Med.* **28:** 253–289.

7. Chen, H., O. Chatlat, L.A. Tartaglia, E.A. Woolf, *et al.* 1996. Evidence that the diabetes gene encodes the leptin receptor: identificaton of a mutation in the leptin receptor gene in db/db mice. *Cell* **84:** 491–495.

8. Zimanyi, I.A. & M.A. Pelleymounter. 2003. The role of melanocortin peptides and receptors in regulation of energy balance. *Curr. Pharmaceut. Des.* **9:** 627–42.

9. Halaas, J.L., c. Boozer, J. Blair-West, N. Fidahusein, *et al.* 1997. Physiological response to long-term peripheral and central leptin infusion in lean and obese mice. *Proc. Nat. Acad. Sci. USA* **94:** 8878–8883.

10. van Heek, M., D.S. Compton, C.F. France, R.P. Tedesco, *et al.* 1997. Diet-induced obese mice develop peripheral, but not central, resistance to leptin. *J. Clin. Invest.* **99:** 385–390.

11. Levin, B.E., A.A. Dunn-Meynell & W.A. Banks. 2004. Obesity-prone rats have normal blood–brain barrier transport but defective central leptin signaling before obesity onset. *Am. J. Physiol.* **286:** R143–R150.

12. Banks, W.A., C.R. DiPalma & C.L. Farrell. 1999. Impaired transport of leptin across the blood–brain barrier in obesity. *Peptides* **20:** 1341–1345.

13. Banks, W.A., B.M. King, K.N. Rossiter, R.D. Olson, *et al.* 2001. Obesity-inducing lesions of the central nervous system alter leptin uptake by the blood–brain barrier. *Life Sci.* **69:** 2765–2773.

14. Banks, W.A. & C.L. Farrell. 2003. Impaired transport of leptin across the blood–brain barrier in obesity is acquired and reversible. *Am. J. Physiol.* **285:** E10–E15.

15. Kastin, A.J., W. Pan, L.M. Maness, R.J. Koletsky, *et al.* 1999. Decreased transport of leptin across the blood–brain barrier in rats lacking the short form of the leptin receptor. *Peptides* **20:** 1449–1453.

16. Burguera, B., M.E. Couce, G.L. Curran, M.D. Jensen, *et al.* 2000. Obesity is associated with a decreased leptin transport across the blood–brain barrier in rats. *Diabetes* **49:** 1219–1223.

17. Elinav, E., L. Niv-Spector, M. Katz, T.O. Price, *et al.* 2009. Pegylated leptin antagonist is a potent orexigenic agent: preparation and mechanism of activity. *Endocrinology* **150:** 3083–3091.

18. Banks, W.A. & C. LeBel. 2002. Strategies for the delivery of leptin to the CNS. *J. Drug Target.* **10:** 297–308.

19. Adam, C.L. & P.A. Findlay. 2010. Decreased blood-brain leptin transfer in an ovine model of obesity and weight loss: resolving the cause of leptin resistance. *Int. J. Obes.* **34:** 980–988.

20. Caro, J.F., J.W. Kolaczynski, M.R. Nyce, J.P. Ohannesian, *et al.* 1996. Decreased cerebrospinal-fluid/serum leptin ratio in obesity: a possible mechanism for leptin resistance. *Lancet* **348:** 159–161.

21. Schwartz, M.W., E. Peskind, M. Raskind, E.J. Boyko, *et al.* 1996. Cerebrospinal fluid leptin levels: relationship to plasma levels and adiposity in humans. *Nat. Med.* **2:** 589–593.

22. Mantzoros, C., J. S. Flier, M. D. Lesem, T. D. Brewerton, *et al.* 1997. Cerebrospinal fluid leptin in anorexia nervosa: correlation with nutritional status and potential role in resistance to weight gain. *J. Clin. Endocrinol. Metabol.* **82:** 1845–1851.

23. Banks, W.A., C.M. Clever & C.L. Farrell. 2000. Partial saturation and regional variation in the blood to brain transport of leptin in normal weight mice. *Am. J. Physiol.* **278:** E1158–E1165.

24. Nonaka, N., S.M. Hileman, S. Shioda, P. Vo, *et al.* 2004. Effects of lipopolysaccharide on leptin transport across the blood–brain barrier. *Brain Res.* **1016:** 58–65.

25. Mastronardi, C.V., W.H. Yu, V. Rettori & S.M. McCann. 2000. Lipopolysaccharide-induced leptin release is not mediated by nitric oxide, but is blocked by dexamethasone. *Neuroimmunomodulation* **8:** 91–97.

26. Mastronardi, C.V., W.H. Yu, V.K. Srivastava, W.L. Dees, *et al.* 2004. Lipopolysaccharide-induced leptin release is neurally controlled. *Proc. Nat. Acad. Sci. USA* **98:** 14720–14725.

27. Banks, W.A., A.B. Coon, S.M. Robinson, A. Moinuddin, *et al.* 2004. Triglycerides induce leptin resistance at the blood–brain barrier. *Diabetes* **53:** 1253–1260.

28. Kastin, A.J. & V. Akerstrom. 2000. Fasting, but not adrenalectomy, reduces transport of leptin into the brain. *Peptides* **21:** 679–682.

29. Faggioni, R., A. Moser, K.R. Feingold & C. Grunfeld. 2000. Reduced leptin levels in starvation increase susceptibility to endotoxic shock. *Am. J. Pathol.* **156:** 1781–1787.

30. Banks, W.A., B.O. Burney & S.M. Robinson. 2008. Effects of triglycerides, obesity, and starvation on ghrelin

transport across the blood–brain barrier. *Peptides* **29**: 2061–2065.

31. Bray, G.A. 1996. Leptin and leptinomania. *Lancet* **348**: 140–141.

32. Wurtman, R.J. 1996. What is leptin for, and does it act on the brain? *Nat. Med.* **2**: 492–493.

33. Ahima, R.S., D. Prabakaran, C. Mantzoros, D. Qu, *et al.* 1996. Role of leptin in the neuroendocrine response to feeding. *Nature* **382**: 250–252.

34. Lord, G.M., G. Matarese, J.K. Howard, R.J. Baker, *et al.* 1999. Leptin modulates the T-cell immune respone and reverses starvation-induced immunosuppression. *Nature* **394**: 897–901.

35. Ziotopoulou, M., D.M. Erani, S.M. Hileman, C. Bjorbaek, *et al.* 2000. Unlike leptin, ciliary neurotrophic factor does not reverse the starvation-induced changes of serum corticosterone and hypothalamic neuropeptide levels but induces expression of hypothalamic inhibitors of leptin signaling. *Diabetes* **49**: 1890–1896.

36. Cheung, C.C., J. E. Thornton, J.L. Kuijper, D.S. Weigle, *et al.* 1997. Leptin is a metabolic gate for the onset of puberty in the female rat. *Endocrinology* **138**: 855–858.

37. Mounzih, K., R. Lu & F.F. Chehab. 1997. Leptin treatment rescues the sterility of genetically obese ob/ob males. *Endocrinology* **138**: 1190–1193.

38. Rutenberg, G.W., Coelho, Jr., D.S. Lewis, K.D. Carey, *et al.* 1987. Body composition in baboons: Evaluating a morphometric method. *Am. J. Primat.* **12**: 275–285.

39. Henson, M.C., V.D. Castracane, J.S. O'Neil, T. Gimpel, *et al.* 1999. Serum leptin concentrations and expression of leptin transcripts in placental trophoblast with advancing baboon pregnancy. *J. Clin. Endocrinol. Metabol.* **84**: 2543–2549.

40. Banks, W.A., J.E. Phillips-Conroy, C.J. Jolly & J.E. Morley. 2001. Serum leptin levels in wild and captive populations of baboons (Papio): implications for the ancestral role of leptin. *J. Clin. Endocrinol. Metabol.* **86**: 4315–4320.

41. Banks, W.A., J. Altmann, R. M. Sapolsky, J. E. Phillips-Conroy, *et al.* 2003. Serum leptin levels as a marker for a syndrome X-like condition in wild baboons. *J. Clin. Endocrinol. Metabol.* **88**: 1234–1240.

42. Ledger, H.P. 1968. Body composition as a basis for a comparative study of some East African mammals. *Symp. Zool. Soc. Lond.* **21**: 289–310.

43. Kirchengast, S. 1996. Weight status of adult !Kung San and Kavango people from northern Namibia. *Ann. Hum. Biol.* **25**: 541–551.

44. Eaton, S.B. & M. Konner. 1985. Paleolithic nutrition: a consideration of its nature and current implications. *New England J. Med.* **312**: 283–289.

45. Wilmsen, E.N. 1978. Seasonal effects of dietary intake on Kalahari San. *Fed. Proc.* **37**: 65–72.

46. MacDonald, R.S. & R.J. Smith. 2007. Starvation. In *Principles and Practice of Endocrinology and Metabolism*, Vol. 3rd. K.L. Becker, Ed.: 1247–1251. Lippincott Williams and Wilkins. Philadelphia.

47. Berthoud, H.R. 2007. Interactions between "cognitive" and "metabolic" brain in the control of food intake. *Physiol. Behav.* **91**: 486–498.

48. Janmaat, K.R. & K. Zuberbuhler. 2006. Primates take weather into account when searching for fruits. *Curr. Biol.* **16**: 1232–1237.

49. Balasko, M. & M. Cabanac. 1998. Behavior of juvenile lizards (*Iguana iguana*) in a conflict between temperature regulation and palatable food. *Brain Behav. Evol.* **52**: 257–262.

50. Pravosudov, V.V. 2008. Mountain chickadees discriminate between potential cache pilferers and non-pilferers. *Proc. Biol. Sci.* **275**: 55–61.

51. Hunt, S., J. Low & K.C. Burns. 2008. Adaptive numerical compentency in a food-hoarding songbird. *Proc. Biol. Sci.* **275**: 2373–2379.

52. Kelley, K.W., R.M. Bluthe, R. Dantzer, J.H. Zhou, *et al.* 2003. Cytokine-induced sickness behavior. *Brain Behav. Immunol.* **17**: S112–S118.

53. Larson, S.J. & A.J. Dunn. 2001. Behavioral effects of cytokines. *Brain Behav. Immunol.* **15**: 371–387.

54. Park, C.R. 2001. Cognitive effects of insulin in the central nervous system. *Neurosci. Biobehav. Rev.* **25**: 311–323.

55. Harvey, J., L.J. Shanley, D. O'Malley & A.J. Irving. *et al.* 2005. Leptin: a potential cognitive enhancer? *Biochem. Soc. Trans.* **33**: 1029–1032.

56. Paz-Filho, G.J., T. Babikian, R. Asarnow, T. Delibasi, *et al.* 2008. Leptin replacement improves cognitive development. *PLoS One* **3**: e3098.

57. Stranahan, A.M., E.D. Norman, K. Lee, R.G. Cutler, *et al.* 2008. Diet-induced insulin resistance impairs hippocampal synaptic plasticity and cognition in middle-aged rats. *Hippocampus* **18**: 1085–1088.

58. Kastin, A.J., N.P. Plotnikoff, C.A. Sandman, M.A. Spirtes, *et al.* 1975. The effects of MSH and MIF on the brain. In *Anatomical Neuroendocrinology*. W.E. Stumpf & L.D. Grant, Eds.: 290–297. Karger. Basel.

59. During, M.J., L. Cao, D.S. Zuzga, J.S. Francis, *et al.* 2003. Glucagon-like peptide-1 receptor is involved in learning and neuroprotection. *Nat. Med.* **9**: 1173–1179.

60. Farr, S.A., W.A. Banks & J.E. Morley. 2006. Effects of leptin on memory processing. *Peptides* **27**: 1420–1425.

61. Reger, M.A., G.S. Watson, P.S. Green, L.D. Baker, *et al.* 2008. Intranasal insulin administration dose-dependently modulates verbal memory and plasma amyloid-beta in memory-impaired adults. *J. Alzheimer's Dis.* **13**: 323–331.

62. Diano, S., S.A. Farr, S.E. Benoit, E.C. McNay, *et al.* 2006. Ghrelin controls hippocampal spine synapse density and memory performance. *Nat. Neurosci.* **9**: 381–388.

63. de Frias, C.M., D. Bunce, A. Wahlin, R. Adolfsson, *et al.* 2007. Cholesterol and triglycerides moderate the effect of apolipoprotein E on memory functioning in older adults. *J. Gerontol. B Psychol. Sci. Soc. Sci.* **62**: 112–118.

64. Perlmuter, L.C., D.M. Nathan, S.H. Goldfinger, P.A. Russo, *et al.* 1988. Triglyceride levels affect cognitive function in noninsulin-dependent diabetics. *J. Diabet. Complicat.* **2**: 210–213.

65. Rogers, R.L., J.S. Meyer, K. McClintic & K.F. Mortel. 1989. Reducing hypertriglyceridemia in elderly patients with cerebrovascular disease stabilizes or improves cognition and cerebral perfusion. *Angiology* **40**: 260–269.

66. Sims, R.C., S. Madhere, S. Gordon, E. Clark, Jr., *et al.* 2008. Relationships among blood pressure, triglycerides and verbal

learning in African Americans. *J. Nat. Med. Assoc.* **100:** 1193–1198.

67. Banks, W.A. 2001. Enhanced leptin transport across the blood–brain barrier by alpha1-adrenergic agents. *Brain Res.* **899:** 209–217.

68. Kastin, A.J. & V. Akerstrom. 2001. Glucose and insulin increase the transport of leptin through the blood–brain barrier in normal mice but not in streptozotocin-diabetic mice. *Neuroendocrinology* **73:** 237–242.

69. Kastin, A.J., V. Akerstrom & L.M. Maness. 2001. Chronic loss of ovarian function decreases transport of leptin into mouse brain. *Neurosci. Lett.* **310:** 69–71.

70. Pan, W., M. Barron, H. Hsuchou, H. Tu, *et al.* 2008. Increased leptin permeation across the blood–brain barrier after chronic alcohol ingestion. *Neuropsychopharmacology* **33:** 859–866.

71. Pan, W. & A.J. Kastin. 2001. Diurnal variation of leptin entry form blood to brain involving partial saturation of the transport system. *Life Sci.* **68:** 2705–2414.

Ann. N.Y. Acad. Sci. ISSN 0077-8923

ANNALS OF THE NEW YORK ACADEMY OF SCIENCES

Issue: *The Brain and Obesity*

The hypothalamic–pituitary–adrenal axis and sex hormones in chronic stress and obesity: pathophysiological and clinical aspects

Renato Pasquali

Division of Endocrinology, Department of Clinical Medicine, S. Orsola-Malpighi Hospital, University Alma Mater Studiorum of Bologna, Bologna, Italy

Address for correspondence: Renato Pasquali, M.D., Division of Endocrinology, Department of Clinical Medicine, S. Orsola-Malpighi Hospital, University Alma Mater Studiorum, Via Massarenti 9, 40138 Bologna, Italy. renato.pasquali@unibo.it

Obesity, particularly the abdominal phenotype, has been ascribed to an individual maladaptation to chronic environmental stress exposure mediated by a dysregulation of related neuroendocrine axes. Alterations in the control and action of the hypothalamic–pituitary–adrenal axis play a major role in this context, with the participation of the sympathetic nervous system. The ability to adapt to chronic stress may differ according to sex, with specific pathophysiological events leading to the development of stress-related chronic diseases. This seems to be influenced by the regulatory effects of sex hormones, particularly androgens. Stress may also disrupt the control of feeding, with some differences according to sex. Finally, the amount of experimental data in both animals and humans may help to shed more light on specific phenotypes of obesity, strictly related to the chronic exposure to stress. This challenge may potentially imply a different pathophysiological perspective and, possibly, a specific treatment.

Keywords: cortisol; androgens; stress; obesity

Introduction

Obesity, particularly its abdominal phenotype, is a key component of the metabolic syndrome, and represents an important risk factor for cardiovascular diseases (CVD) and type 2 diabetes (T2D).[1,2] Major pathophysiological events responsible for the association between obesity and metabolic or cardiovascular dysfunctions are likely to be the development of an insulin resistance state and systemic low-grade inflammation, which are characterized by mutual feeding.[3,4]

There is increasing evidence that chronic environmental stressors may play an important role in increasing individual susceptibility towards the development of chronic metabolic diseases, such as abdominal obesity and the metabolic syndrome,[5] as a consequence of an inadequate response to repeated or chronic stress stimuli and consequent whole body maladaptative behavior, according to the pioneeristic view of H Selye.[6] In fact, stress activates the hypothalamic–pituitary–adrenal (HPA) axis, the sympathetic nervous system (SNS), and the sympathoadrenal system.[7] Defense reactions involve the release of endogenous mediators, chiefly cortisol and catecholamines, and activate other endocrine systems, such as the renin–angiotension system and others.[8] In conditions of continuous stress exposure and poor coping, this hormonal response may be disrupted, making the response to stress inadequate and activating compensatory mechanisms. As a consequence, the allostatic load (the body price required for adaptation) may become overwhelming and the reactive processes may be maladaptive.[9]

These concepts were reinforced by the recent finding that glucocorticoids also play a dominant role in regulating stress-induced food intake and choice by interacting, in a feedback circuit, at the central brain neuroendocrine level.[10,11] Moreover, the

doi: 10.1111/j.1749-6632.2012.06569.x

Ann. N.Y. Acad. Sci. 1264 (2012) 20–35 © 2012 New York Academy of Sciences.

ability to adapt to both internal and particularly to environmental (external) stressors may differ between males and females, with specific pathophysiological events leading to the development of stress-related chronic diseases.[12] This seems to be mediated, at least in part, by the regulatory role of sex hormones on the activity of other hormonal systems, particularly the HPA axis, because a close crosstalk exists between sex hormones and glucocorticoids at both neuroendocrine and peripheral levels, with different specificities according to sex.[13]

Neuroendocrinology of stress response: roles of HPA axis and SNS

Appropriate regulatory control of the HPA stress axis is essential to health and survival. The parvocellular cells of the paraventricular nucleus (PVN) of the hypothalamus are the major information junction for the neuroendocrine response to stressors. There is a hierarchical organization of the stress-responsive neurocircuitries on the PVN, which is able to integrate information from multiple limbic sources with internally generated and peripherally sensed information, thereby tuning the relative activity of the adrenal cortex.[14] Corticotrophin-releasing hormone (CRH) neurons are regulated by sensory afferents that are relayed via brain-stem loci and transmit reactive stimuli that are generally excitatory and relatively direct. Conversely, limbic forebrain structures are hypothesized to convey anticipatory signals that involve processing within pathways proximal to the levels of the PVN and several hypothalamic regions. The outflow of the HPA axis is therefore a summation of integrated inputs from several forebrain regions, including the hippocampus, prefrontal cortex, amygdala, and septum. Evidence of this summation may be appreciated by the fact that many stressors produce parallel activation among numerous HPA-regulatory limbic regions,[15,16] and lesions to different regions can produce similar effects on stress response.[17] These networks are integrated in the PVN. In most organisms, the system efficiently modulates the HPA axis in accordance with needs, but there is considerable individual variation in the HPA response disposition in addition to some influence of genetic factors and early-life experience.[14]

In the hypothalamic areas, the cascade is, in turn, chiefly regulated by the CRH and arginine vasopressin (AVP),[18] whose release activates proopiomelanocortin in anterior pituitary corticotroph cells and the release of adrenocorticotrophic hormone (ACTH) into peripheral blood, from where it targets receptors in the adrenal cortex to release glucocorticoid hormones. CRH, the pivotal signaling hypothalamic molecule in the stress response, is under the control of several complex processes, For example, CRH gene regulation involves multiple activating and repressing transcription factors, specifically the glucocorticoid receptors and cyclic AMP.[19] Another important biological aspect is timing in the transcription of the signals, which is critical for effective glucocorticoid repression of the cyclic AMP-induced CRH gene. Recent studies performed in At-T20 cell lines have shown that a critical time window exists for effective repression of the CRH gene by glucocorticoids, whose disruption may result in a significant loss of glucocorticoid receptor-mediated repression, depending on specific situations.[20] Interestingly, similar interactions have been described for the secretion of ACTH from the pituitary.[21] Therefore, differences in timing of stimulatory and repression signals may be of consequence of adaptation of the organism to stress, thereby providing a molecular explanation for the variability in adaptation to stress. Additional data have shown that pulse frequency is increased under states of chronic stress in rats with genetically determined hyperresponsiveness of the HPA axis.[21]

Hyperactivation of the SNS also plays an important role in the body's response to both acute and chronic stress.[7] The complex physiological processes by which the SNS regulates the HPA axis at various levels in the suprahypothalamic and hypothalamic and pituitary have been extensively investigated in the past and there are excellent review articles summarizing the specific role of the SNS as a mediator of the stress response and its ability to increase the activity of the HPA axis.[7,9,22,23]

Finally, as mentioned above and briefly discussed below, the response of the HPA axis to chronic stress is under the control of sex hormones, including androgens and estrogens.[12,13,24] Overall, the complex interaction between different regulatory hormonal networks needs to be taken into consideration when investigating and interpreting how adaptive mechanisms to both acute and particularly chronic stress may be disrupted. On the other hand, we are facing an extraordinary challenge in the near future to better understand this complexity by using

Pathophysiological aspects of stress response and obesity in humans: from Bjorntorp's hypothesis onwards

Studies performed in animal models have clearly demonstrated the dominant role of a hyperactivated HPA axis on the development of obesity and associated dysmetabolic and cardiovascular comorbidities. Excess cortisol increases lipoprotein lipase levels (a lipid-storing enzyme) in adipose tissue and particularly in the visceral fat.[25] A series of studies performed in primates by Shively *et al.* provided an excellent model to investigate the responsibility of chronic stress in determining visceral obesity and associated metabolic and cardiovascular comorbidities.[26–28] Specifically, compared to nonstressed individuals, these authors found that primates (cynomolgus monkeys) exposed for two years to chronic physical and psychological stress developed pathological behavioral changes (aggressiveness), increased body weight and visceral fat deposition, insulin resistance and hyperinsulinemia, impaired glucose tolerance or diabetes, dyslipidemia, adrenal hypertrophy, increased cortisol response to ACTH stimulation test, and coronary atherosclerosis. Clinical and epidemiological studies in humans have shown that increased HPA axis activity together with activation of the SNS may be significantly related to long-term adverse stressful events during the life span,[7,9,29,30] and that the abdominal obesity phenotype may be strongly associated with stress-related or adverse life events and psychosocial conditions, low occupational and educational status, smoking habits, alcohol and/or drug abuse, fat overfeeding, psychiatric disorders, negative personality traits, and subjective abnormally perceived stress.[5,31,32] Other studies have shown that a mild increase in cortisol levels may occur in situations such as work stress and unemployment or poor demographic conditions.[33] A series of pathophysiological studies provided additional evidence documenting that subtle alterations of the HPA axis can be detected in abdominal obesity[34] (see further paragraph). Overall, these data suggest a central neuroendocrine dysregulation resulting, in turn, in slightly abnormal net cortisol production, either continuous or episodic.

Allostatic load conceptualizes the cumulative biological burden exacted on the body through attempts to adapt to life's demands, and refers to the price the body pays for being forced to adapt to adverse environmental stressors.[9,22,35] An individual's behavior can increase or decrease further risk for harm or disease, particularly in the long run. Behavior itself and cognition play an important role in determining what is stressful, which underlines the significant differences in how they respond to stressful situations. Key factors in this behavioral response include the interpretation of an event, differences in bodily conditions and, finally, the choices of compensation to stress. Therefore, the brain can also determine the behaviors and habits that make life more or less dangerous to the individual and may increase allostatic load in the long-term. Notably, several genetic factors have been shown to play a role in stress-related disorders and allostatic load.[35] Derangements in the hormonal mediators are key factors in the maladaptation process to chronic stress exposure and, in turn, may be deeply involved in the pathophysiology of several chronic diseases. Using a multisystem summary measurement of the allostatic load (including measures of blood pressure, metabolic functions, and hormonal parameters, chiefly cortisol and urinary norepinephrine and epinephrine) in a large sample of elderly individuals followed for seven years, Seeman *et al.*[36] found that higher baseline allostatic load scores were associated with significantly increased risk for seven-year mortality and with declines in cognitive and physical functioning, and were marginally associated with incident CVD events, independent of standard sociodemographic characteristics and baseline health status. Moreover, they found that the measure of allostatic load was a better predictor of mortality and decline in physical functioning than the metabolic syndrome. These findings have been extended to other inflammatory biomarkers,[37] suggesting that these should possibly be included in the measurement of pathological allostatic load. Overall, these findings suggest that allostatic load should be carefully considered as a potential risk factor for chronic metabolic and CVD and emphasize the need for further extensive research in this area.

Measurements of the activity of the HPA axis in obesity and the metabolic syndrome

The usual diagnostic procedures for investigating endogenous hypercortisolism in patients with Cushing's syndrome can be applied in obese dysmetabolic patients, although their sensitivity and specificity seem to be significantly reduced when dealing with subtle abnormalities of cortisol homeostasis.[38] With these limitations, a series of studies performed in either epidemiological or clinical settings have investigated ACTH and cortisol concentrations in basal conditions—with repeated measurements to investigate daily chronobiological changes, during dynamic studies following stimulation with different neuropeptides or psychological stress tests, or by suppression with dexamethasone—to challenge the responsiveness to the inhibitory feedback system.[34] In addition, there are studies evaluating glucocorticoid receptor density in different tissues, including adipose tissue. Finally, a growing amount of research in the last decade has focused on peripheral cortisol metabolism, particularly in the visceral fat tissues and liver, which specifically depends on the activity of enzymes controlling cortisol metabolism and reactivation from inactive compounds (cortisone), which include both 5α- and 5β-reductase and 11β-hydroxysteroid dehydrogenase type1 (11β-HSD1), respectively. These aspects have been extensively reviewed in recent years.[39–41] Only omental 11β-HSD1 has been independently associated with the amount of visceral, not subcutaneous, fat in women, which supports the concept that visceral fat is the major target tissue for expression of genes related to glucocorticoid action.[42]

Basal blood levels of ACTH and cortisol are usually normal in obese subjects (in fact some studies found slightly lower morning cortisol levels), as are ACTH and cortisol daily rhythms,[43] although some studies found either lower than normal single samples or lower 24-h integrated cortisol levels in adult obese men.[44] One study investigated the pulsatile secretion of cortisol and ACTH during daytime in women, and showed that those with visceral obesity had higher ACTH pulse frequency and lower ACTH pulse amplitude, particularly in the morning, but similar mean ACTH basal concentrations in comparison with their gluteofemoral type obese

counterpart and normal weight controls.[45] These data may support an increased sensitivity of cortisol secretion to non-ACTH-dependent pathways, such as the peripheral noradrenergic regulatory system, particularly during the zenith phase of the daily rhythm.

The assessment of free cortisol in saliva may have a promising role in the investigation of the HPA axis,[46] although cortisol concentrations in the saliva are approximately 30–50% lower than in the blood—its collection being noninvasive and stress free, and easy for frequent measurements, particularly for psychoneuroendocrinological and epidemiological studies.[47] The cortisol assay may also be laboratory-independent, particularly when the LM/MS–MS technology is applied.[48]

Several studies used urinary free cortisol (UFC) excretion as an integrated measurement of daily cortisol excretion rate[43] and reported higher than normal 24-h UFC excretion rates in women with abdominal obesity. Moreover, this measure has also been found to correlate positively with abdominal adiposity.[49–51] Interestingly, UFC excretion rates during nighttime seem to be particularly useful for detecting subtle alterations of cortisol secretion rates in obese individuals.[45] This may indirectly agree with the findings of the pulsatility study reported above, therefore suggesting that the night-time period and early morning hours are probably the best time to investigate subtle alterations of the HPA axis activity in obese individuals.

Dynamic studies in which the HPA axis was either stimulated or inhibited provide the most convincing evidence for a dysregulation of the HPA system in abdominal obesity. Specifically, there are studies demonstrating higher than normal cortisol responses after laboratory stress tests or after challenges of the HPA axis by administering CRH, or AVP alone or in combination,[34,52] in women with abdominal obesity, compared to women with the peripheral phenotype, regardless of phychiatric disorders such as anxiety and depression.[53] The strong reproducibility of the CRH test among individuals is undoubtedly a strong factor supporting the reliability of these data.[54] Similar differences between abdominal and peripheral obese women have been reported using the ACTH stimulatory test, with either maximal[55] or low[50] ACTH doses. Collectively, these findings support the conclusion of there being abnormal HPA axis activity in obesity, possibly because of hypersensitized and/or hyperresponsive hypothalamic and pituitary centers, leading to slightly but inappropriately elevated net cortisol production, either continuous or episodic. Increased central activity or responsiveness of the HPA axis may in turn be dependent on activated function of the SNS. Using a slow infusion of yohimbine, an α2-adrenoreceptor antagonist (whose activation inhibits ACTH release[56]) to achieve steady-state norepinephrine plasma levels near those observed during acute stress, we found that the ACTH and cortisol responses to combined CRH and AVP challenge were reduced in normal weight controls, whereas it was significantly amplified in obese women, particularly in those with the abdominal phenotype.[57] This suggests a specific synergy of increased noradrenergic pathways in favoring the increased responsiveness of the HPA axis to neuropeptide stimulation.

The suppressive challenge of the HPA axis using either standard or low dose dexamethasone has been investigated in the field of obesity. A standard dose (i.e., 1 mg overnight) unequivocally resulted in normal inhibition of cortisol secretion in obese individuals with different phenotypes,[58] whereas low doses (i.e., $\leq$0.5 mg or less overnight) have been suggested to provide potential insights into the different responsiveness of the HPA axis and obesity (in both sexes).[59] In this context, blunted inhibition of cortisol secretion has been reported, suggesting a reduced sensitivity to inhibition by low dose dexamethasone via downregulation of central glucocorticoid receptors.[60] Another study randomly administered different deaxamethasone doses (0.0035, 0.0070, and 0.015 mg/kg/body weight, and a standard 1 mg dose) to both obese and normal-weight individuals. The findings showed that obesity in men did not influence hormone response to each dose, whereas in women, with increasing amounts of abdominal fat (measured by waist circumference), the suppression of cortisol tended to be significantly lower than that expected on the basis of increasing doses,[61] further confirming the potential impairment of sensitive feedback signals in abdominally obese women.

Human stress response: interplay between sex hormones and the HPA axis

Sex difference may exist in the response to chronic stress, particularly in those individuals susceptible to developing an abnormal allostatic load, which intrinsically implies a maladaptation (pathological) syndrome to chronic stress exposure, either internal or related to environmental factors. Intriguingly, these mechanisms may also imply derangements in the regulation of neuroendocrine and peripheral actions of sex hormones. The two systems, the HPA axis and sex hormones, may in fact mutually interact in determining the abnormal response to chronic stress, thereby favoring not only the development of the abdominal obesity phenotype, but also its association with metabolic comorbidities.[12] Specific alterations in sex hormone secretion, transport, and metabolism are in fact present in obese males and females, and they are often associated with different patterns of body fat distribution. This topic has been extensively reviewed.[62] In brief, obesity in males is usually associated with a parallel increase in abdominal and visceral fat, whereas in women, a clear dichotomy of fat distribution occurs, with the abdominal phenotype being characterized by an enlargement of both subcutaneous and visceral fat depots and a modest increase in fat in the gluteofemoral areas. Moreover, the increase in body weight and the development of different phenotypes of obesity is associated with sex-specific differences in the production and metabolic clearance rates of major androgens. Both sexes show a significant decrease in the synthesis of sex-hormone binding globulin (SHBG), but this may lead to an increase in free androgen blood concentrations in women with the abdominal, but not the peripheral, phenotype.[63] Moreover, the production rates of SHBG-bound androgens (particularly testosterone) and androgens not bound to SHBG (such as dehydroepiandrosterone [DHEA] and androstenedione), have been found to be equally increased in female obesity.[64] Therefore, women with the abdominal obesity phenotype are characterized by a *mild relative hyperandrogenic state*.[62] Conversely, SHBG, as well as total and free testosterone blood levels, tends to progressively decrease with increasing body weight in the obese male, but 3-androstenediol glucuronide levels tend to be significantly higher, particularly in abdominally obese men, independently of circulating insulin concentrations.[62]

Sex-disease dimorphism can be observed in humans because men and women are at differential risk for a number of diseases. For example, autoimmune disorders are more common in women, whereas men are more prone to develop cardiovascular or infectious diseases.[65] The difference between the sexes is apparent in the prevalence of common psychiatric disorders, such as depression or anxiety, which are more prevalent in women, whereas antisocial behavior or substance abuse is more common in men.[65] Differences between the sexes in stress-related HPA axis responses have been investigated for many years, and it appears that they may not coincide with the subjective response to psychosocial stress.[66] Notably, it should be considered that experimental design issues may be relevant in the interpretation of the data and the conclusions. Brain limbic regions are presumed to be involved in the processing of psychological stress,[67] and there is evidence that hippocampal structures underlying higher-order cognitive processing may be sexually dimorphic,[68] as confirmed by recent brain imaging studies.[69,70]

Animal studies have repeatedly shown that glucocorticoid levels are higher in females than in males,[65] but studies in humans are still inconsistent. One of the main reasons is that the psychological approaches to investigate hormonal and metabolic dynamics associated with stress exposure have focused on acute stress tests in otherwise healthy subjects. On the other hand, data in these specific conditions support the concept that HPA axis activity, as defined in terms of cortisol response, tends to be higher in young or adult males than in their female counterparts, although contradictory results have also been reported.[65] Conceptually, sex differences in HPA axis responses to psychological stress might differ between clinical populations with chronic diseases and healthy volunteers. A larger cortisol response to negative daily events has been found in women with major depression than in their male counterparts.[71] By contrast, other studies have reported no sex difference or an increased HPA axis response to psychological challenges in males.[72] Notably, most studies performed on this topic included heterogeneous populations and did not analyze sex differences.[65] In addition, little attention has been paid to the role of genetics on the cortisol response to psychological stress, although heredity factors seemed to play a minor role in the cortisol response to psychological stress in one study.[73] Finally, there are few data on the impact of chronic stress exposure on major chronic diseases in males and females.

As reported above, the response to stress may be part of the biology of human sex difference, which suggests that sex hormones may be important regulators of sex-related individual responses to stress. A significant association between a list of chronic environmental stressors—such as anxiety and depressive traits (positive), alcohol abuse (positive), smoking (positive), demographic conditions (negative), occupational status (negative), educational levels (negative), personality, abuse of comfort foods (positive), and alteration of HPA axis dynamics associated with abdominal obesity—has been repeatedly described in both men and women.[30–32] Moreover, an association between abdominal fatness, excess cortisol, and subjectively perceived stress (positive) has also been described in many studies performed in males or females, although such differences have not been systematically investigated. In a previous review, we emphasized the potential role of sex hormones in the relationship between alterations of the HPA axis in abdominal obesity[12] (Fig. 1). The HPA axis is physiologically subjected to gonadal influence, as indicated by sex differences in basal and stress HPA function and by neuropathologies associated with HPA dysfunction.[14] Although there is still no clear definition of how sex hormones mutually interact with the HPA systems,[13] several studies performed in experimental animals have shown that a dual approach (for instance by manipulating sex hormones to evaluate effects on the HPA axis and *vice versa*) simultaneously manipulating the gonadal and adrenal axes can overcome this problem. ACTH release may be regulated by testosterone-dependent effects on AVP synthesis and by corticosterone-dependent effects on CRH synthesis in the PVN of the hypothalamus. In male rats, testosterone and corticosterone may act on stress-induced ACTH release that, ultimately, affects PVN motor neurons;[74] and peripherally, testosterone may interfere with cortisol metabolism in the liver and the adipose tissue.[75] In rodents, basal and stimulated ACTH and corticosterone levels tended to be greater in females than in males.[76,77] Human studies have additionally found that healthy women may be more responsive to CRH, with respect to ACTH secretion.[78–80] Similarly, cortisol responses to acute stress challenges have been found to be higher in abdominally obese women,[50] and the increase in cortisol levels was found to be positively correlated with abdominal fatness in middle-aged men.[47] We

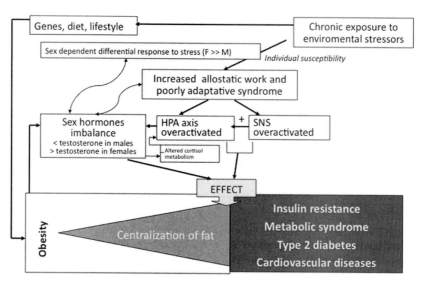

Figure 1. A general hypothesis on the coordinating role of the HPA axis, the SNS, and androgens in the development of central obesity and metabolic and cardiovascular comorbidities. Genes, excess diet, and unhealthy lifestyle are the main factors responsible for the development of obesity. In addition, chronic stress exposure may favor obesity by increasing allostatic work and developing a poor adaptive syndrome. Sex hormones regulate the response to stress differently in females and males. Obesity *per se* produces disparate effects on androgen production rates and metabolism according to sex. Increased *allostatic work* implies overactivation of the HPA axis and the SNS, which, in turn, further worsen androgen imbalance. The coordinating role of excess cortisol and noradrenergic tone, plus low testosterone in obese males and high testosterone in obese females, may therefore have a pathophysiological significance for the development of the abdominal phenotype of obesity and associated metabolic and cardiovascular comorbidities (from Ref. 12).

recently showed that whereas both normal-weight and obese men had significantly higher ACTH and cortisol concentrations than normal-weight and obese women, the hormone response to CRH plus AVP stimulation was conversely significantly higher in women than in men.[81] Therefore, even in the presence of obesity, a sex difference in the activity of the HPA axis may still persist and, possibly, be amplified. However, contrasting results were reported in another study that investigated the effect of long-term gonadal suppression by leuprolide on cortisol response to CRH stimulation in healthy males and females.[82] Additional studies performed in rats have shown that ovariectomy attenuated the response of the HPA axis, whereas estradiol replacement reversed it.[83,84] Interestingly, short-term estradiol treatment led to an enhanced ACTH and cortisol stress response in young men.[85,86] Another study performed in male rats found that ACTH and corticosterone responses to acute stress were increased by gonadectomy but completely reversed by testosterone replacement.[74]

Recently, one study focused on the androgen receptor mediated signaling and its interaction with corticosteroid action in an experimental model characterized by obesity, the male androgen receptor–null mutant (ARKO) mice.[86] The authors found that the obesity state in these animals was determined by an increased corticosteroid state due to impairment of a negative feedback regulatory system. In fact, both male and female ARKO mice exhibited hypertrophic adrenal glands and glucocorticoid overproduction. This was interpreted as secondary to increased ACTH stimulation because the same study found that glucocorticoid receptor expression in a pituitary gland cell line was under the positive control of the activated ARIs (androgen receptors). This suggested that locally activated androgen receptors may support the negative feedback regulation of glucocorticoid production via upregulation of their receptor expression in the pituitary gland.[86] These findings add significant support for the conclusion that, via their own receptors, androgens play an important role in the regulation of HPA axis activity, and that alterations of this regulatory pathway may lead to the development of obesity, at least in mice. Whether this may also occur in humans obviously needs to be defined.

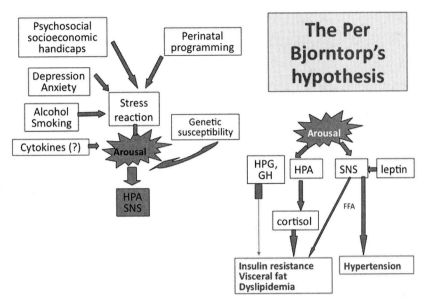

Figure 2. Arousal of the HPA axis and the SNS is induced by perceived stress from psychosocial and socioeconomic handicaps (related to fetal programming). Depression, anxiety, alcohol abuse, smoking (and cytokines) may cause direct activation, whereas genes may predispose for these reactions (see Refs. 29, 31).

The functional crosstalk between the HPA axis and sex steroids is, however, bidirectional, and may differ according to sex. Several years ago, Per Björntorp proposed the hypothesis that the combined alteration of the glucocorticoid pathway and androgens associated with abdominal obesity may have a role in the pathophysiology of metabolic syndrome and insulin resistance (Fig. 2).[29] His hypothesis considered available data supporting a strict local interaction between CRH and the gonadotropin releasing hormone (GnRH),[87] and evidence that an increased CRH secretion may inhibit GnRH secretion.[88] On this basis, combined HOA axis hyperactivity and low testosterone levels in obese men make sense, although many other factors are involved in determining hypotestosteronemia in male obesity.[12] Whether the same hypothesis may also apply to (abdominal) obesity in females remains unclear. As occurs in women with the polycystic ovary syndrome, there are theoretical bases to suggest that an insulin-mediated overstimulation of ovarian steroidogenesis may occur in females with abdominal obesity, as insulin acts as a true gonadotropic hormone, synergizing LH activity.[89] However, there are no consistent *in vitro* or *in vivo* data supporting a clear responsibility of insulin—or other factors mimicking insulin action, such as the

insulin growth factor-1 (IGF-1)—in women with abdominal obesity. Intriguingly, it should be considered that even in patients with Cushing's syndrome there are differences in the expression of altered gonadal function according to sex. Men with Cushing's syndrome are characterized by hypogonadotropic hypogonadism due to altered gonadotropin pulsatility, regardless of the extent of the hypercortisolism.[90] By contrast, women with Cushing's syndrome and mild hypercortisolism may present with androgen excess of both adrenal and ovarian origin, with reduced gonadotropins and polycystic ovaries; in the case of severe hypercortisolism, the HPG axis may be inhibited, similarly to what occurs in men with Cushing's syndrome.[91] Accordingly, in a recent study performed in obese males and females, we demonstrated a negative correlation between ACTH response to combined CRH plus AVP stimulation and testosterone, whereas cortisol response tended to be negatively correlated with the free androgen index in obese men, and positively correlated in obese women.[83]

To summarize, although there are clinical and experimental data supporting the hypothesis that the response to stress may differ to a some extent between men and women, this complex topic needs to be investigated more appropriately, possibly in selected well-defined individuals with well-defined phenotypes of obesity and chronic exposure to different stressors. In addition, the potential impact of chronic stress on eating behavior and subsequent weight changes should be investigated (see next paragraph). Finally, there is a need for prospective longitudinal studies in large cohorts of subjects of both sexes, to identify whether subtle alterations of the HPA axis (and, possibly, of the catecholaminergic system) may have a specific impact on the development of different phenotypes of obesity and associated metabolic, cardiovascular, and psychological comorbidities.

Alterations of HPA axis activity, insulin resistance, and metabolic syndrome

Although the pathophysiology of metabolic syndrome is only partially understood, it is clear that insulin resistance has a crucial pathophysiological role in the expression of all its features, particularly abdominal obesity.[92,93] Despite the lack of direct evidence in humans, several epidemiologic studies have provided evidence for a significant positive association between cortisol levels and surrogate measures of insulin resistance or metabolic alterations, other than indices of overweight or obesity.[94,95] Intriguingly, some of these studies were able to detect some difference according to sex.[94,96,97] A mild cortisol increase was identified as an early feature of essential hypertension.[98] Interestingly, cortisol blood levels may also represent an independent risk factor for CVDs, at least in South Asian individuals, but not in Caucasians,[99] suggesting that increased glucocorticoid action may contribute to ethnic differences in the prevalence of metabolic syndrome. Clinical studies performed in women with different obesity phenotypes have additionally shown that the cortisol response to hypothalamic neuropeptides or nightly UFC excretion rates were significantly correlated with fasting insulin or the HOMA index.[54] Finally, the reduced allostatic load seems to be associated with lower all-cause mortality.[100]

There is also evidence for a significant association between metabolic syndrome and hyperactivity of the SNS, an important mediator, together with the HPA axis, of the stress response. Brunner et al.[101] found that, compared to a nonaffected control group, middle-aged men with metabolic syndrome were characterized not only by significantly higher daily UFC, but also by increased urinary normetanephrine levels, together with more prevalent alterations of inflammatory markers and worsened indices of psychosocial distress, confirming the coordinated derangement of both major factors involved in stress adaptation and metabolic and cardiovascular abnormalities. Recently, data are emerging concerning association between the HPA axis and the occurrence of atheromatous disease.[102] The Caerphilly Heart Study,[103] investigating a large cohort of adult men, reported a positive association between incident CVDs and the plasma cortisol/testosterone ratio, a potential index of chronic stress exposure.[104,105] A more recent prospective study in approximately 6,500 healthy men and women found that the risk of CVDs increased in relation to psychological distress, and that behavioral factors rather than pathophysiological factors explained the largest proportion of variance in defining the amount of risk.[106] Other studies have reported that higher plasma cortisol was associated with the extent of coronary atherosclerosis, quantified by coronary angiography,[107] or predicted mortality in patients with heart failure.[105] Interestingly, recent epidemiological studies have also shown that polymorphisms in the glucocorticoid receptor gene may be associated with cardiovascular event rates.[108] Given the potential importance of the pathophysiology of metabolic syndrome, the International Diabetes Federation[109] recently recommended that a better understanding of the HPA axis function and activity in obese dysmetabolic individuals should be expanded.

Stress, cortisol, and eating

Appetite, satiety, and reward mechanisms regulate food intake by a complex network of internal mechanisms and environmental factors. Internal factors chiefly include hormones targeting several hypothalamic nuclei,[110,111] whose functions are to protect the adipose tissue stores by responding and interconnecting anorexigenic, and orexigenic signals from specific neuropeptides.[112] In peripheral tissues, leptin and insulin are long-term signals that relay the adequacy of adipose tissue stores to the hypothalamus via the anabolic or catabolic neurons.[112] External factors, including environmental and social conditions, and palatability of foods may in turn influence food intake.[113] In response to acute stress, sometimes perceived as something dangerous to personal safety, a rapid physiological response is often activated,[9] which reduces food intake by suppressing appetite.[114] By contrast, chronic stressful situations often cause one not to avoid food but to look for energy-dense foods,[10,115] which can favor rapid weight gain and lead to obesity. This depends on the magnitude, kind, and severity of events perceived as stressful, with some differences according to age and sex.[115]

Several studies have shown that women tend to increase their food consumption more frequently than males during chronic stressful conditions.[116] The HPA axis plays an important role in the regulation of energy homeostasis and, additionally, can directly influence caloric intake, by mechanisms not fully defined, in response to stress. After chronic stress or drug-treatment with a pharmacological stressor, increased food-seeking behavior and motivation for preferred foods, together with an increase in corticosterone blood levels, have been shown in experimental animals.[117] These changes may in turn decrease the negative impact of chronic stress on the reward pathways, depending on the action of other hormones, particularly insulin.[10,117] Interestingly, compared to males, female animals may show increased stress sensitivity and display a higher hormonal response and delayed stress recovery time,[118] suggesting that stress pathways are closely linked to brain reward centers and that some sex difference exists on the degree of reward engendered by preferred foods.

Many years ago, it was suggested for the first time that one of the contributing factors to obesity might be stress-induced eating, with a greater preference for hedonic, nutrient-dense foods, particularly those high in sugar and fat.[119] This hypothesis has been confirmed by more recent studies, further demonstrating that individuals with high values of perceived stress are characterized by higher cortisol blood levels and a significantly higher tendency to eat more than control nonstressed individuals.[120] This further supports the view that in vulnerable subjects, chronic stress may increase the need for increased reward functions, such as searching for energy-dense, high-calorie foods.[121] In agreement with this, we recently demonstrated a significant positive correlation between energy intake and food choice, particularly starchy foods and daily total lipid intake, with UFC per 24 h, regardless of body weight and waist-to-hip ratio (WHR).[122] Although

much more investigation is needed in humans, these findings may provide for human obesity indirect proof of the concept of what Dallman et al. described in chronically stressed animals with high blood levels of corticosterone.[10,117]

The SNS-neuropepdite Y network, diet, and obesity

The SNS has been implicated in the complex regulation of eating following long-term stress exposure. However, it is perplexing that not all individuals who are chronically stressed gain weight. Some people lose weight during stress, possibly because of an overactivation of the β-adrenergic lipolytic pathways.[123] Most individuals, however, gain weight when exposed to stressful chronic events, which often seems to be disproportionate to the number of calories consumed.[124] This may be a protective mechanism by which the body craves and stores fat because of perceived stress signals.[125]

As discussed above, an overactive HPA axis leading to elevated cortisol levels is to blame for an increased incidence of abdominal obesity, although previous studies have suggested that there are obese individuals who may have a desensitization to glucocorticoids, thereby not exhibiting increased blood cortisol.[126] It is known that SNS activity may increase in specific phenotypes of human obesity that, in turn, may dysregulate the activity of the HPA axis and glucocorticoid action in peripheral tissues. Recent data support evidence that this may be mediated by neuropeptide Y (NPY), a sympathetic neurotransmitter, directly in adipose tissue, the effects of which may be amplified by stress, coupled with high lipid and carbohydrate feeding. In fact, a recent study performed in 129SvJ mice found that chronic stress combined with a high fat/high sugar diet led to abdominal obesity by releasing NPY directly into the adipose tissue.[127] Moreover, the same study found that, in vitro, sympathetic neurons stressed with dexamethasone shifted toward expressing more NPY, which in turn stimulated endothelial cells, macrophage infiltration, and preadipocyte proliferation and lipid filling, by activation of Y2 receptors (NPY2R). Therefore, NPY is released from sympathetic nerves in animals exposed to stressors, which in turn upregulates receptors in abdominal fat in a glucocorticoid-dependent manner, thereby increasing growth and developing abdominal obesity. These findings suggest that glucocorticoids may act by priming adipose tissue for NPY effects by increasing peptide expression in sympathetic nerves. Interestingly, it was also found that the pharmacological inhibition of fat-targeted knockdown of NPY2R was antiangiogenic and antiadipogenic, with the consequence of reducing abdominal fat and associated metabolic alterations. These new findings may extend the results of Dallman's studies on the role of so-called comfort foods in the response to chronic stress, suggesting that by activating the sympathetic drive, leading in turn to peripheral NPY overexpression and action, the ingestion of high fat-high carbohydrate foods might be responsible for stress-related obesity and metabolic comorbidities. Whether this may apply to humans needs to be investigated, although some studies seem to corroborate this hypothesis. In fact, a Y2R silent mutation in a Swedish population has been associated with resistance to obesity,[128] whereas a gain-of-function polymorphism in the NPY gene seems to predispose to individual hyperlipidemia, atherosclerosis, and severe complications of T2D.[129] Taken together, these findings further support the metabolic impact of the SNS in the control of energy homeostasis. Among other effects, the SNS plays a major physiological role in the control of glucose metabolism, through activation of glycogen phosphorylase and inhibition of glycogen synthase, thereby increasing glycogen breakdown and glucose production, which combats the enhancement of glucose uptake by activation of the parasympathetic system.[130] All these effects occur directly in the liver, a terminal organ of both sympathetic and parasympathetic nerves. However, these nerves are also present in various nuclei of the hypothalamus[131–133] that contain many neurons expressing NPY receptors,[134] which are in turn involved in the control of autonomic activity. In the hypothalamus, NPY stimulates the activity of the HPA axis, which in turn promotes an increase in cortisol production by stimulating gluconeogenetic pathways in different organs, particularly the liver.[135] Recently, it was shown that intracerebroventricular administration of NPY acutely induces insulin resistance and endogenous glucose production via activation of sympathetic output to the liver.[136] These and many other studies on the effects of increased SNS activity, mediated by both central and peripheral NPY signals, coupled with hyperactivation of the HPA axis and with the occurrence of al-

tered feeding behaviors, may provide further pieces in the complex puzzle explaining the development of dysmetabolic obesity in subjects exposed to chronic stress.

Does a stress-dependent obesity phenotype exist?

Based on what has been discussed above, the next step should be to translate scientific knowledge into clinics that are working to define the individual phenotype of obesity, potentially dependent on chronic stress exposure. Because this phenotype of obesity includes specific pathophysiological events, it could be theoretically possible to evaluate future strategies for intervention and treatment. Unfortunately, affected patients do not usually perceive the association between weight gain and previous stress exposure, and thus tend to ascribe their state to specific hormonal and metabolic disorders as being responsible for rapid changes in body weight and shape. On the other hand, they can easily describe their weight history during a clinical interview or understanding chat with a physician. We have described a potential phenotype of obesity related to chronic stress exposure by investigating a cohort of women who developed weight gain after a well-defined stressful event. Compared with the nonstressed group, the women were characterized by significantly higher UFC per 24-h values and a much shorter time to achieve maximum weight gain.[137] Undoubtedly, well-defined longitudinal studies are needed to fully characterize the dynamics of weight gain and associated hormonal and metabolic alterations over time. Lessons from animal and human epidemiological and clinical studies could serve to reveal clinical, biological, and psychological aspects necessary to identify this phenotype. Other than careful personal history taking and investigating the dynamics and trajectories of weight gain, a physician should pay attention to well-defined stressful events in documenting relevant causes. Although this may be a difficult task, an appropriate investigation of eating disorders (by questionnaires, etc.), food craving patterns, or consumption of comfort foods (to improve reward pathways), particularly during the initial phases of weight gain, and any psychiatric disorders, particularly melancholic depression and the so-called stress-induced depression—a recently identified new subtype of depression[138–140]

—should be investigated. Finally, all these data should be coupled with a detailed measurement of fat distribution and an adequate metabolic investigation. Measurement of sex hormones and simple parameters of HPA axis functioning (as discussed above) may in turn support relevant information on the pathophysiology of this specific obesity phenotype.

Conclusions and perspectives

The existence of stress-related obesity is based on both clinical perspectives and available studies performed in experimental animals. Specific pathophysiological mechanisms may include hyperactivation of the HPA axis, which may persist if the maladaptation syndrome takes a long time to settle, and if psychiatric problems, such as depression, occur. Because of the complexity of neuroendocrine, behavioral, and metabolic adaptation to chronic stress exposure, much more research should be performed on this topic in order to plan specific therapeutic strategies.

Conflicts of interest

The author declares no conflicts of interest.

References

1. Haslan, D.W. & W.P.T. James. 2005. Obesity. *Lancet* **366:** 1197–1209.
2. Roth, J., X. Quiang, S.L. Marban, *et al.* 2004. The obesity pandemic: where have we been and where are we going? *Obes. Res.* **12**(Suppl): 88S–101S.
3. Dandona, P., A. Aljada, A. Chandhuri, *et al.* 2005. Metabolic syndrome. A comprehensive perspective based on interaction between obesity, diabetes, and inflammation. *Circulation* **111:** 1448–1454.
4. Dandona, P., A. Aljada & P. Mohanty. 2002. The anti-inflammatory and potential anti-atherogenic effect of insulin: a new paradigm. *Diabetologia* **45:** 924–930.
5. Rosmond, R. 2005. Role of stress in the pathogenesis of the metabolic syndrome. *Psychoneuroendocrinology* **30:** 1–10
6. Selye, H. 1936. Syndrome produced by diverse nocuous agents. *Nature* **1:** 38–42.
7. Plotsky, P.M., E.T. Cunningham & E.P. Widmaier. 1989. Catecholaminergic modulation of corticotropin-releasing factor and adrenocorticotropin secretion. *Endocr. Rev.* **10:** 437–445.
8. Hjemdahl, P. 2002. Stress and the metabolic syndrome. An interesting but enigmating association. *Circulation* **106:** 2634–2636
9. McEwen, B.S. 1998. Protective and damaging effects of stress mediators. *N. Engl. J. Med.* **338:** 171–179.
10. Dallman, M.F., S.E. La Fleur, N.C. Pecoraio, *et al.* 2004. Minireview: glucocorticoids-food intake, abdominal obesity, and wealthy nations in 2004. *Endocrinology* **145:** 2633–2638.
11. Steiner, M.A. & C.T. Wotjack. 2008. Role of the endocannabinoid system in regulation of the hypothalamic-pituitary-adrenocortical axis. *Prog. Brain Res.* **170:** 397–432.
12. Pasquali, R., V. Vicennati, A. Gambineri & U. Pagotto. 2008. Sex dependent role of glucocorticoids and androgen in the pathophysiology of human obesity. *Int. J. Obes. (Lond)* **38:** 1764–1779.
13. Viau, V. 2002. Functional cross-talk between the hypothalamic-pituitary-gonadal and -adrenal axes. *J. Neuroendocrinol.* **14:** 506–513.
14. Herman, J.P., H. Figueiredo, N.K. Mueller, *et al.* 2003. Central mechanisms of stress integration: hierarchical circuitry controlling hypothalamic-pituitary-adrenocortical responsiveness. *Front. Neuroendocrinol.* **24:** 151–180.
15. Cullinam, We, J.P. Herman, D.F. Battaglia, *et al.* 1995. Pattern of time course of immediate early gene expression in rat brain following acute stress. *Neuroscience* **64:** 477–505.
16. Sawchenko, P.E., H.Y. Li & A. Ericsson. 2000. Circuits and mechanisms governing hypothalamic responses to stress: a tale of two paradigms. *Prof. Brain Res.* **122:** 61–78.
17. Figueiredo, H.F., B.L. Bruestle, B.L. Bodie, *et al.* 2003. The medial prefrontal cortex differentially regulates stress-induced c-fos expression in the forebrain depending on the type of stressor. *Eur. J. Neurosci.* 1–8.
18. Lightman, S.L. 2008. The neuroendocrinology of stress: a never ending story. *J. Neuroendocrinology* **20:** 880–884.
19. Liu, Y., A. Kamitahara, A.J. Kim & J. Aguilera. 2008. Cyclic adenosine 3′, 5′-monophosphate responsive element binding phosphorylation is required but not sufficient for activation of corticothropin-releasing hormone transcription. *Endocrinology* **149:** 3512–3520.
20. van deer Laan, S., E.R. De Kloet & O.C. Meijer. 2009. Timing is critical for effective glucocorticoid receptor mediated repression of the camp-induced CRH gene. *PlosONE* **4:** e4327.
21. Shipston, M.J. & F.A. Antoni. 1992. Inactivation of early glucocorticoid feedback by corticothropin-releasing factor in vitro. *Endocrinology* **30:** 2213–2218.
22. Chrousos, G.P. 1998. Stressors, stress, and neuroendocrine integration of the adaptive response. *Ann. N.Y. Acad. Sci.* **851:** 311–335.
23. Sapolsky, R.M., L.M. Romero & A.U. Munck. 2000. How do glucocorticoids influence stress responses? Integrating permissive, suppressive, stimulatory, and preparative actions. *Endocr. Rev.* **21:** 55–89.
24. Williamson, M., B. Bingham & V. Viau. 2005. Central organization of androgen-senstive pathways to the hypothalamic-pituitary-adrean axis:implications for individual differences in responses to homeostatic threat and predisposition to disese. *Prog. Neuro-Psychopharmacol. & Biol. Psych.* **29:** 1239–1248.
25. Rebuffe-Scrive, M. 1991. Neuroendocrie egulation of adipose tissue: molecular oand hormona! mechanisms. *Int. J. Obes. Rel. Metab. Dis.* **15**(Suppl 2):83–86.
26. Shively, C. & T. Clarkson. 1989. Regional obesity and coronary atherosclerosis in females: a non-human primate model. *Acta. Med. Scand. Suppl.* **723:** 71–78.
27. Watson, S.L., C.A. Shively, J.R. Kaplan & S.W. Line. 1998. Effects of chronic social separation on cardiovascular disease

risk factors in female cynomolgus monkeys. *Atherosclerosis* **137**: 259–266.

28. Shively, Ca, T.C. Register & T.B. Clarkson. 2009. Social stress, visceral obesity, and coronary artery atherospdscerosis in female primates. *Obesity (Silver Spring)* **17**: 1516–1520.

29. Bjorntorp, P. & R. Rosmond. 2000. Neuroendocrine abnormalities in visceral obesity. *Int. J. Obes.* **24**: 580–585.

30. Chrousos, G.P. 1998. Editorial: a healthy body in a healthy mind and vice versa. The damaging power of "uncontrollable" stress. *J. Clin. Endocrinol. Metab.* **83**: 1842–1845.

31. Björntorp, P. 1993. Visceral obesity: a "civilization syndrome". *Obes. Res.* **1**: 206–222.

32. Rosmond, R., L. Lapidus, P. Marin & P. Björntorp. 1996. Mental distress, obesity and body fat distribution in middle-aged women. *Obes. Res.* **4**: 245–252.

33. Cota, D., V. Vicennati, L. Ceroni, *et al.* 2001. Relationship between socio-economic and cultural status, psychological factors and body fat distribution in middle-aged women living in Northen Italy. *Eating Weight Dis.* **6**: 205–213.

34. Pasquali, R., V. Vicennati, M. Cacciari & U. Pagotto. 2006. The hypothalamic-pituitary-adrenal axis activity in obesity and the metabolic syndrome. *Ann. N.Y. Acad. Sci.* **1083**: 111–128.

35. McEwen, B.S. 2000. Allostatis and allostatic load: implications for neuropsychopharmacology. *Neuropsychopharmacol.* **22**: 108–124.

36. Seeman, T.E., B.S. McEwen, J.W. Rowe & B.H. Singer. 2001. Allostatic load as a marker of cumulative biological risk: MacArthur studies of successful aging. *Proc. Natl. Acad. Sci. USA* **98**: 4770–4775.

37. Gruenewald, T.L., T.E. Seeman, C.D. Ryff, *et al.* 2006. Combinations of biomarkers predictive of later life mortality. *Proc. Natl. Acad. Sci. USA* **103**: 14158–14163.

38. Baid, S.K., D. Rubino, N. Sinaii, *et al.* 2009. Specificity of screening test for Cushing syndrome in an overweight and obese population. *J. Clin. Endrinol. Metab.* **94**: 3857–3864.

39. Sandeep, T.C. & B.R. Walker. 2001. Pathophysiology of modulation of local glucocorticoid levels by11 ß - hydroxysteroid dehydrogenase. *Trends Endocrinol. Metab.* **12**: 446–453.

40. Tomlinson, J.W., E.A. Walker, I.J. Bujalska, *et al.* 2004. 11ß-hydroxysteroid dehydrogenase type 1: a tissue specific regulation of glucorticoid response. *Endocr. Rev.* **25**: 831–866.

41. Seckl, J.R. & B.R. Walker. 2001. Minireview: 11beta-hydroxysteroid dehydrogenase type 1- a tissue- specific amplifier of glucocorticoid action. *Endocrinology* **142**: 1371–1376.

42. Veilleux, A., P.Y. Laberge, J. Morency, *et al.* 2010. Expressions of genes related to glucocorticoid action in human subcutaneous and omental adipose tissue. *J. Steroid Biochem. Mol. Biol.* **122**: 28–34.

43. Chalew, S., H. Nagel & S. Shore. 1995. The hypothalamic-pituitary-adrenal axis in obesity. *Obes. Res.* **3**: 371–382.

44. Smith, S.R. 1996. The endocrinology of obesity. *Endocr. Metab. Clin. North Am.* **25**: 921–942.

45. Pasquali, R., D. Biscotti, G. Spinucci, *et al.* 1998. Pulsatile secretion of ACTH and cortisol in premenopausal women: effect of obesity and body fat distribution. *Clin. Endocrinol. (Oxf)* **48**: 603–612.

46. Rosmond, R. & P. Bjorntorp. 2001. New targets for the clinical assessment of salivary cortisol secretion. *J. Endocrinol. Invest.* **24**: 639–641.

47. Rosmond, R., M.F. Dallman & P. Bjorntorp. 1998. Stress-related cortisol secretion in men: relationships with abdominal obesity and endocrine, metabolic and hemodynamic abnormalities. *J. Clin. Endocrinol. Metab.* **83**: 1853–1859.

48. Fanelli, F., I. Belluomo, V.D. Di Lallo, *et al.* 2011. Large serum steroid profiling by isotopic dilution-liquid chromatography-mass spectrometry: comparison with current immunoassas and reference intervals in healthy subjects. *Steroids* **76**: 244–253.

49. Pasquali, R., S. Cantobelli, F. Casimirri, *et al.* 1993. The hypothalamic-pituitary-adrenal axis in obese women with different patterns of body fat distribution. *J. Clin. Endocrinol. Metab.* **77**: 341–346.

50. Duclos, M., J-B. Corcuff, N. Etcheverry, *et al.* 1999. Abdominal obesity increases overnight cortisol excretion. *J. Endocrinol. Invest.* **22**: 465–471.

51. Epel, E.E., A.E. Moyer, C.D. Martin, *et al.* 1999. Stress-induced cortisol, mood, and fat distribution in men. *Obes. Res.* **7**: 9–15.

52. Vicennati, V., L. Ceroni, L. Gagliardi, *et al.* 2004. Response of the hypothalamic-pituitary-adrenal axis to small dose arginine-vasopressin and daily urinary free cortisol before and after alprazolam pre-treatment differs in obesity. *J. Endocrinol. Invest.* **27**: 541–547.

53. Vicennati, V. & R. Pasquali. 2000. Abnormalities of the hypothalamic-pituitary-adrenal axis in women with the abdominal obesity and relationship with insulin resistance: evidence for a central and peripheral alteration. *J. Clin. Endocrinol. Metab.* **24**: 416–422.

54. Bertagna, X., J. Coste, M.C. Raux-Demay, *et al.* 1994. The combined corticotropin releasing hormone/lysine vasopressin test discloses a corticothoph phenotype. *J. Clin. Endocrinol. Metab.* **79**: 390–394.

55. Marin, P., N. Darin, T. Amemiya, *et al.* 1992. Cortisol secretion in relation to body fat distribution in obese premenopausal women. *Metabolism* **41**: 882–886.

56. Al-Damluji, S. 1993. Adrenergic control of the secretion of anterior pituitary hormones. *Bailliere's Clin. Endocrinol. Metab.* **7**: 355–392.

57. Pasquali, R., V. Vicennati, F. Calzoni, *et al.* 2000. α2-adrenoreceptor regulation of the hypothalamic-pituitary-adrenocortical axis in obesity. *Clin. Endocrinol. (Oxf)* **52**: 413–421.

58. Anagnostis, P., V.G. Athyros, K. Tziomalos, *et al.* 2009. The pathhognetic role of cortisol in the metabolic syndrome: a hypothesis. *J. Clin. Endocrinol. Metab.* **94**: 2692–2701.

59. Huizenga, N.A.T.M., J.W. Koper, P. de Lange, *et al.* 1998. Interperson variability but intraperson stability of baseline plasma cortisol concentrations, and its relation to feedback sensitivity of the hypothalamo-pituitary-adrenal axis to a low dose dexamethasone in elderly individuals? *J. Clin. Endocrinol. Metab.* **83**: 47–54.

60. Ljiung, T., B. Andersson, B.A. Bengtsson, *et al.* 1996. Inhibition of cortisol secretion by dexamethasone in relation to body fat distribution: a dose-response study. *Obes. Res.* **4**: 277–282.

61. Pasquali, R., B. Ambrosi, D. Armanini, *et al.* 2002. Cortisol and ACTH response to oral dexamethasone in obesity and effects of sex, body fat distribution, and dexamethasone concentrations: a dose-response study. *J. Clin. Endocrinol. Metab.* **87:** 166–175.

62. Pasquali, R. 2006. Obesity and androgens: facts and perspectives. *Fertil Steril* **85:** 1310–1340.

63. Samoilik, E., M.A. Kirscner, D. Silber, *et al.* 1984. Elevated production and metabolic clearance rates of androgens in morbidly obese women. *J. Clin. Endocrinol. Metab.* **59:** 949–954.

64. Kirshner, M.A., E. Samojlik, M. Drejka, *et al.* 1990. Androgen-estrogen metabolism in women with upper vs lower body obesity. *J. Clin. Endocrinol. Metab.* **70:** 473–479.

65. Kudielka, B.M. & C. Kirschbaum. 2005. Sex differences in HPA axis response to stress: a review. *Biol. Psychol.* **69:** 113–132

66. Kirschbaum, C., B.M. Kudielka, J. Gaab, *et al.* 1999. Impact of gender, menstrual cycle phase, and oral contraceptives on the activity of the hypothalamic-pituitary-adrenal axis. *Psychosom. Med.* **61:** 154–162.

67. Herman, J.P. & W.E. Culligam. 1997. Neurocircuitry of stress: central control of the hypothalamic-pituitary-adrenocortical axis. *Trends Neurosci.* **20:** 78–84.

68. Wizemann, T.M. & M-L. Pardue. 2001. Exploring the biological contribution to human health. In *Does Sex Matter?* National Academy Press. Washington, DC.

69. Kilgore, W.D. & D.A. Yurgelun-Todd. 2001. sex differences in the amygdala activation during the perception of facial affect. *Neuro Report* **12:** 2543–2547.

70. Wingenfeld, K., C. Spitzer, N. Rullkötter & B. Löwe. 2010. Borderline personality disorder: hypothalamus pituitary adrenal axis and findings from neuroimaging studies. *Psychoneuroendocrinology* **35:** 154–170.

71. Peeters, F., N. Nicholson & I. Berkhof. 2003. Cortisol response to daily events in major depressive disorders. *Psychosom. Med.* **65:** 836–841.

72. Young, E.A. 1998. Sex differences in the HPA axis: implications for psychiatric diseases. *J. Gend. Specif. Med.* **1:** 21–27.

73. Kirschbaum, C., S. Wust, H-S. Faig & D.H. Hellahammer. 1992. Heritability of cortisol response to human corticotropin-releasing hormone, ergometry, and psychological stress in humans. *J. Clin. Endocrinol. Metab.* **15026–15030.**

74. Viau, V. & M.J. Meaney. 1996. The inhibitory effect of testosterone on hypothalamic-pituitary-adrenal responses to stress is mediated by medial preoptic area. *J. Neurosci.* **16:** 1866–1876.

75. Barat, P., D.E. Livingstone, C.M. Elferink, *et al.* 2007. Effects of gonadectomy on glucocorticoid metabolism in obese Zucker rats. *Endocrinology* **48:** 4836–4843.

76. Tilbrok, A.J., A.I. Turner & I.J. Clarke. 2000. Effects of stress on reproduction in non-rodent mammals: the role of glucocorticoids and sex differences. *Biol. Reprod.* **5:** 105–113.

77. Rivier, C. & S. Rivest. 1991. Effect of stress on the activity of the hypothalamic-pituitary-gonadal axis: peripheral and central mechanisms. *Biol. Reprod.* **45:** 523–532.

78. Gallucci, W.T., A. Baum, L. Laue, *et al.* 1993. Sex difference in the sensitivity of the hypothalamic-pituitary-adrenal axis. *Health Psychol.* **121:** 420–425.

79. Streeten, D.H., G.H. Anderson, Jr., T.G. Dalakos, *et al.* 1984. Normal and abnormal function of the hypothalamic-pituitary-adrenocortical axis in man. *Endocr. Rev.* **5:** 371–394.

80. Pasquali, R., L. Gagliardi, V. Vicennati, *et al.* 1999. ACTH and cortisol response to combined corticotropin-releasing hormone-arginine vasopressin stimulation in obese males and its relationship to body weight, fat distribution and parameters of the metabolic syndrome. *Int. J. Obes. Relat. Metab. Disord.* **23:** 419–424

81. Vicennati, V., L. Ceroni, S. Genghini, *et al.* 2006. Sex differences in the relationship between the hypothalamic-pituitary-adrenal axis and sex hormones in obesity. *Obesity* **14:** 235–243.

82. Roca, C.A., P.J. Schmidt, P.A. Deuster, *et al.* 2005. Sex-related differences in stimulated hypothalamic-pituitary-adrenal axis during induced gonadal suppression. *J. Clin. Endocrinol. Metab.* **90:** 4224–4231.

83. Lesniewska, B., B. Miskowiak, M. Nowak & L.K. Malendowicz. 1990. Sex differences in adrenocortical structure and function. XXVII. The effect of ether stress on ACTH and corticosterone in intact, gonadectomized, and testosterone- or estradiol-replaced rats. *Res. Exp. Med. (Berl)* **190:** 95–103.

84. Norman, R.L., C.J. Smith, J.D. Pappas & J. Hall. 1992. Exposure to ovarian steroids elicits a female pattern of cortisol levels in castrated male macaques. *Steroids* **57:** 37–43.

85. Kirschbaum, C., N. Schommer, I. Federenko, *et al.* 1996. Short-term estradiol treatment enhances pituitary-adrenal axis and sympathetic responses to psychosocial stress in healthy young men. *J. Clin. Endocrinol. Metab.* **81:** 3639–3643.

86. Miyamoto, J., T. Matsumoto, H. Shiina, *et al.* 2007. The pituitary function of androgen receptor constitutes a glucocorticoid production circuit. *Mol. Cell Biol.* **27:** 4807–4814.

87. Chrousos, G. & P.W. Gold. 1992. The concept of stress and stress system disorders. *JAMA* **267:** 1244–1252.

88. Olsen, D.H. & M. Ferin. 1987. Corticotropin-releasing hormone inhibits gonadotropin secretion in ovariectomized rhesus monkey. *J. Clin. Endocrinol. Metab.* **65:** 262–267.

89. Poretsky, L., N.A. Cataldo, Z. Rosenwaks & L.C. Giudice. 1999. The insulin-related ovarian regulatory system in health and disease. *Endocr. Rev.* **20:** 535–582.

90. Vierhapper, H., P. Nowotny & W. Waldhausl. 2000. Production rates of testosterone in patients with Cushing's syndrome. *Metabolism* **49:** 229–231.

91. Lado-Abeal, J., J. Rodriguez-Arnao, J.D. Newell-Price, *et al.* 1998. Menstrual abnormalities in women with Cushing's disease are correlated with hypercortisolemia rather than raised circulating androgen levels. *J. Clin. Endocrinol. Metab.* **83:** 3083–3088.

92. Kahn, R., E. Ferrranini, J. Buse & M. Stern. 2005. The metabolic syndrome: time for a critical reappraisal. Joint statement from the American Diabetes Association and the European Association for the study of Diabetes. *Diabetes Care* **28:** 2289–2304.

93. Reaven, G.M. 1988. Insulin resistance in human disease. *Diabetes* **37:** 1595–1607.

94. Phillips, D.I., D.J. Barker, C.H. Fall, *et al.* 1998. Elevated plasma cortisol concentrations: a link between low birth

weight and the insulin resistance syndrome? *J. Clin. Endocrinol. Metab.* **83:** 757–760.

95. Dinneen, S., A. Alzaid, J. Miles & R. Rizza. 1993. Metabolic effects of the nocturnal rise in cortisol on carbohydrate metabolism in normal humans. *J. Clin. Invest* **92:** 2283–2290.

96. Stolk, R.P., S.W.J. Lamberts, F.H. de Jong, *et al.* 1996. Gender difference in the association between cortisol and insulin sensitivity. *J. Endocrinol.* **149:** 313–318.

97. Filipowsky, J., P. Ducimetière, E. Eschwege, *et al.* 1996. The relationship between blood pressure, with glucose, insulin, heart rate, free fatty acids and plasma cortisol levels according to degree of obesity in middle-aged men. *J. Hypertens.* **14:** 229–235.

98. Walker, B.R., D.I. Phillips, J.P. Noon, *et al.* 1998. Increased glucocorticoid activity in men with cardiovascular risk factors. *Hypertension* **31:** 891–895.

99. Ward, A.M., C.H. Fall, C.E. Stein, *et al.* 2003. Cortisol and the metabolic syndrome in South Asians. *Clin. Endocrinol. (Oxf).* **58:** 500–505.

100. Karlamangla, A.S., Bh. Singer & T.E. Seeman. 2006. Reduction of allostatic load in older adults is associated with lower all-cause mortality: MacArthur studies of successful aging. *Psychosm Med.* **68:** 500–507.

101. Brunner, E.J., H. Hemingway, B.R. Walker, *et al.* 2002. Adrenocortical, autonomic and inflammatory cause of the metabolic syndrome. Nested case-control study. *Circulation* **106:** 2659–2665.

102. Kanaya, A.M., E. Vittinghoff, M.G. Shlipack, *et al.* 2003. Association of central obesity with mortality in post-menopausal women with coronary artery disease. *Am. J. Epidemiol.* **158:** 1161–1170.

103. Smith, G.D., Y. Ben-Shlomo, A. Beswick, *et al.* 2005. Cortisol, testosterone, and coronary heat disease. Prospective evidence from the Caerphilly Study. *Circulation* **112:** 332–340.

104. Lac, F. & P. Berthon. 2000. Changes in cortisol and testosterone levels and the T/C ratio during an endurance competition and recovery. *J. Sports Med. Phys. Fitness* **40:** 139–44.

105. Filare, E., X. Benain, M. Sagnol & G. Lac. 2001. Preliminary results on mood state, salivary testosterone:cortisol ratio and team performance in a professional soccer team. *Eur. J. Appl. Physiol.* **86:** 179–184.

106. Hamer, M., G.J. Molloy & E. Stamatakis. 2008. Psychological distress as a risk factor for cardiovascular events. *J. Am. Coll. Cardiol.* **52:** 2156–2162.

107. Alevizaki, M., A. Cimponeriu, J. Lekakis, *et al.* 2007. High anticipatory stress plasma cortisol levels and sensitivity to glucocorticoids predict severity of coronary artery disease in subjects undergoing coronary angiography. *Metabolism* **56:** 222–226.

108. Lin, R.C.Y., X.L. Wang & B.J. Morris. 2003. Association of coronary artery disease with glucocorticoid receptor N363S variant. *Hypertension* **41:** 104–407.

109. International Diabetes Federation. 2005. The IDF consensus worldwide definition of the metabolic syndrome. April 14. Available at http://www.idf.org/webdata/docs//metab syndrome def.pdf. Accessed 10/6/2005.

110. Levine, A.S. & C.J. Billington. 1997. Why do we eat? A neural systems approach. *Annu. Rev. Nutr.* **17:** 597–619.

111. Bellar, A., P.A. Jarosz & D. Bellar. 2008. Implications of the biology of weight regulation and obesity on the treatment of obesity. *J. Am. Acad. Nurse Pract.* **20:** 128–135.

112. Porte, D., Jr., D.G. Baskin & M.W. Schwartz. 2002. Leptin and insulin action in the central nervous system. *Nutr. Rev.* **60:** S20–S29.

113. Popkin, B.M., K. Duffey & P. Gordon-Larsen. 2005. Environmental influences on food choice, physical activity and energy balance. *Physiol. Behav.* **86:** 603–613.

114. Charmandari, E., C. Tsigos & G. Chrousos. 2005. Endocrinology of the stress response. *Annu. Rev. Physiol.* **67:** 259–284.

115. Oliver, G., J. Wardle & E.L. Gibson. 2000. Stress and food choice: a laboratory study. *Psychosom Med.* **62:** 853–865.

116. Adam, T.C. & E.S. Epel. 2007. Stress, eating and the reward system. *Physiol. Behav.* **91:** 449–458.

117. Dallman, M.F., N.C. Pecoraro & S.E. la Fleur. 2005. Chronic stress and comfort foods: self-medication and abdominal obesity. *Brain, Behav. Immun.* **19:** 275–280.

118. Bale, L. 2006. Stress sensitivity and the development of effective disorders. *Horm. Behav.* **50:** 529–533.

119. Kaplan, H.I. & H.S. Kaplan. 1957. The psychosomatic concept of obesity. *J. Nerv. Ment. Dis.* **125:** 181–201.

120. la Fleur, S.E., S.F. Akana, S. Manalo & M.F. Dallman. 2004. Interaction between corticosterone and insulin in obesity: regulation of lard intake and fat stores. *Endocrinology* **145:** 2174–2185.

121. Epel, E., R. Lapidus, B. McEwen & K. Brownell. 2001. Stress may add bite on appetite in women: laboratory study of stress-induced cortisol and eating behavior. *Psychoneuroendocrinology* **26:** 37–49.

122. Vicennati, V., F. Pasqui, C. Cavazza, *et al.* 2011. Cortisol, energy intake, and food frequency in overweight/obese women. *Nutrition* **27:** 677–680.

123. Dodt, C., P. Lönnroth, J.P. Wellhöner, *et al.* 2003. Sympathetic control of white adipose tissue in lean and obese humans. *Acta. Physiol. Scand.* **177:** 351–357.

124. Ludwig, D.S. 2003. Novel treatments for obesity. *Asia Pac. J. Clin. Nutr.* **12**(Suppl): S8.

125. Dallman, M.F. *et al.* 2003. Chronic stress and obesity: a new view of "comfort foods". *Proc. Natl. Acad. Sci. USA* **100:** 11696–11701.

126. Bose, M., B. Olivan & B. Laferrère. 2009. Stress and obesity: the role of the hypothalamic-pituitary-adrenal axis in metabolic disease. *Curr. Opin. Endocrinol. Diabetes Obes.* **16:** 340–346.

127. Kuo, L.E., J.B. Kitlinsla, J.U. Tilan, *et al.* 2007. Neuropeptide Y acts directly in the periphery on fat tissue and mediates stress-induced obesity and metabolic syndrome. *Nat. Med.* **13:** 803–811.

128. Lavebratt, C., A. Alpman, B. Persson, *et al.* 2006. Common neuropeptide Y2 gene variant is protective against obesity among Swedish males. *Int. J. Obes. Metab. Relat. Dis.* **30:** 453.459.

129. Karvonen, M.K., V.P. Valkonen, T.A. Lakka, *et al.* 2001. Leucine 7 to proline 7 polymorphism in the neuropeptide Y is associated with the progression of carotic atherosclerosis,

blood pressure and serum lipid in Finnish men. *Atherosclerosis* **159**: 145–151.

130. van den Hoek, A.M., C. van Heijningen, J.P. Schröder-van der Elst, *et al.* 1998. Intracerebroventricular administration of neuropeptide Y induces hepatic insulin resistance via sympathetic innervation. *Diabetes* **57**: 2304–2310.

131. Kalsbeek, A., S. La Fleur, C. Van Heijningen & R.M. Buijs. 2004. Suprachiasmatic GABAergic inputs to the paraventricular nucleus control plasma glucose concentrations in the rat via sympathetic innervation of the liver. *J. Neurosci.* **24**: 7604–7613.

132. Strack, A.M., W.B. Sawyer, J.H. Hughes, *et al.* 1989. A general pattern of CNS innervation of the sympathetic outflow demonstrated by transneuronal pseudorabies viral infections. *Brain Res.* **491**: 156–162.

133. Loewy, A.D. & M.A. Haxhiu. 1993. CNS cell groups projecting to pancreatic parasympathetic preganglionic neurons. *Brain Res.* **620**: 323–330.

134. Luiten, P.G., G.J. ter Horst, H. Karst & A.B. Steffens. 1985. The course of paraventricular hypothalamic efferents to autonomic structures in medulla and spinal cord. *Brain Res.* **329**: 374–378.

135. Wolak, M.L., M.R. DeJoseph, A.D. Cator, *et al.* 2003. Comparative distribution of neuropeptide Y Y1 and Y5 receptors in the rat brain by using immunohistochemistry. *J. Comp. Neurol.* **464**: 285–311.

136. Sainsbury, A., F. Rohner-Jeanrenaud, I. Cusin, *et al.* 1997. Chronic central neuropeptide Y infusion in normal rats: status of the hypothalamo-pituitary-adrenal axis, and vagal mediation of hyperinsulinaemia. *Diabetologia* **40**: 1269–1277.

137. Vicennati, V., F. Pasqui, C. Cavazza, *et al.* 2009. Stress-related development of obesity and cortisol in women. *Obesity (Silver Spring)* **17**: 1676–1683.

138. van Praag, H.M. 2004. Can stress cause depression? Prog Neuropsychopharmacol. *Biol. Psychiatry* **28**: 891–907.

139. Rydmark, I., K. Wahlberg, P.H. Ghatan, *et al.* 2006. Neuroendocrine, cognitive and structural imaging characteristics of women on longterm sickleave with job stress-induced depression. *Biol. Psychiatry* **60**: 867–873.

140. Bartolomucci, A. & R. Leopardi. 2009. Stress and depression: preclinical research and clinical implications. *PLoS ONE* **4**: e4265.

Ann. N.Y. Acad. Sci. ISSN 0077-8923

ANNALS OF THE NEW YORK ACADEMY OF SCIENCES
Issue: *The Brain and Obesity*

Food reward in the obese and after weight loss induced by calorie restriction and bariatric surgery

Hans-Rudolf Berthoud, Huiyuan Zheng, and Andrew C. Shin

Neurobiology of Nutrition Laboratory, Pennington Biomedical Research Center, Louisiana State University System, Baton Rouge, Louisiana

Address for correspondence: Hans-Rudolf Berthoud, Pennington Biomedical Research Center, Neurobiology of Nutrition Laboratory, 6400 Perkins Road, Baton Rouge, LA 70808. berthohr@pbrc.edu

Increased availability of tasty, energy-dense foods has been blamed as a major factor in the alarmingly high prevalence of obesity, diabetes, and metabolic disease, even in young age. A heated debate has started as to whether some of these foods should be considered addictive, similar to drugs and alcohol. One of the main arguments for food addiction is the similarity of the neural mechanisms underlying reward generation by foods and drugs. Here, we will discuss how food intake can generate reward and how behavioral and neural reward functions are different in obese subjects. Because most studies simply compare lean and obese subjects, it is not clear whether predisposing differences in reward functions cause overeating and weight gain, or whether repeated exposure or secondary effects of the obese state alter reward functions. While studies in both rodents and humans demonstrate preexisting differences in reward functions in the obese, studies in rodent models using calorie restriction and gastric bypass surgery show that some differences are reversible by weight loss and are therefore secondary to the obese state.

Keywords: obesity; diabetes; palatable food; hedonic eating; food addiction; liking; wanting; motivation; mesolimbic dopamine system

Introduction

Obesity and its comorbidities have been recognized as a major global health problem. Pharmacological treatment has come to an impasse, with only moderately effective drugs available now and few promising ones on the horizon. This has prompted reconsideration of other treatment avenues, including behavioral therapies, bariatric surgery, and various electrical stimulation devices. Dieting, although effective in producing weight loss and health improvements, has been proven very difficult to adhere to over time because calorie restriction–induced biological adaptations evoke strong feelings of hunger and craving for food, eventually overpowering restraint (e.g., Ref. 1). These strong adaptive responses do not seem to kick in after weight loss induced by bariatric surgeries, particularly Roux-en-Y gastric bypass (RYGB) surgery, so that obesity surgery has rapidly advanced to the most effective available long-term treatment for morbidly obese and

moderately obese patients with type 2 diabetes (e.g., Ref. 2).

Ideal treatments attack the cause, rather than the symptoms, of a disease. Clearly, the cause of obesity in a large majority of patients cannot easily be identified because it is multifactorial. Although it is widely accepted that an environment of readily accessible processed foods and a sedentary lifestyle are major causative factors, it is the interaction of these factors with genetic predisposition that is important for the expression of obesity.[3] These predisposing traits are likely due to a great number of gene variants, as well as epigenetic and other early life-programming processes.[4] Ultimately, only systematic analyses with controlled environmental conditions in genetically identified individuals using prospective, rather than cross-sectional, approaches will be able to dissect the true causes of obesity; further, such requirements are much easier met in animals than in human subjects.

One of the key ingredients for the development of obesity in the modern world appears to be

doi: 10.1111/j.1749-6632.2012.06573.x

Ann. N.Y. Acad. Sci. 1264 (2012) 36–48 © 2012 New York Academy of Sciences.

overindulgence in foods that are palatable, easily available, rich in calories from fat and sugar, and often poor in micronutrients. Because such foods are rewarding by generating pleasure and satisfaction, particularly in a world that for many is becoming increasingly stressful, the term "food addiction" has been used in analogy to drug addiction.[5,6] It has been suggested that obesity can result from an attraction or addiction to, and overconsuming of, such high calorie junk food. However, cause and effect in the relationship between the availability and overconsumption of such foods and the development of obesity are far from clear. Is the increased availability and/or the repeated exposure to highly rewarding foods more important than differences in the brain reward system? Does the obese state with its wide-ranging alterations in hormonal and inflammatory signaling secondarily affect the food reward system? These are questions that ultimately have to be answered in order to reach a clearer understanding of cause and effect and that will allow the design of better behavioral, surgical, and pharmacological treatments for obesity.

In the following review, we will first briefly discuss the basic concepts and the neural underpinnings of food reward, as well as the limited human data on the relationship between obesity and food reward processes. As an initial attempt to dissociate cause and effect, we will then summarize and discuss some of our own published data on this relationship in several rat models of obesity and weight loss.

Definition of food reward

It is thought that instinctual behaviors that are essential for survival have evolved over millions of years, and that their neural control mechanisms are particularly powerful.[7,8] Especially in warm-blooded animals, finding and eating food is a daily necessity that is very high in the hierarchy of instinctive behaviors even when scarce and dangerous environmental conditions must be overcome. Food reward has been suggested to provide the necessary motivation to overcome such conditions. Thus, food is a powerful natural reinforcer that outcompetes most other behaviors, particularly when metabolically hungry. Ingestive behavior consists of procurement, consummatory, and postconsummatory phases,[9] and each of these three phases contributes to reward, which then can guide future behaviors.

In the procurement phase, the decision-making process responsible for switching attention is central to the modern field of neuroeconomics, and reward expectancy is perhaps the main factor determining the outcome of this process–response selection. To make this choice, the brain uses representations of reward expectancy and effort/risk-requirement from prior experiences to optimize cost benefit.[10–14]

During the consummatory phase, immediate sensory attributes of the goal object such as seeing, smelling, and ultimately tasting the first bite of the food provide the first feedback to its predicted reward value and may acutely enhance its motivating power. Appetite is typically augmented by the generation of cephalic phase responses such as gastric acid and insulin secretion.[15] During eating, immediate, direct pleasure is derived from mainly gustatory and olfactory sensations, driving consumption throughout the meal until satiation signals dominate.[16] The length of the consummatory phase is highly variable, as it takes only a few minutes to devour a hamburger, but it may take hours to savor a five-course meal. During such longer meals, ingested food increasingly engages postoral reward processes that interact with oral reward.

The postingestive phase is probably the most complex and least understood phase of ingestive behavior in terms of reward processing. Nutrient sensors in the gastrointestinal tract and elsewhere in the body also contribute to the generation of food reward during and after a meal.[17,18] The same taste receptors found in the oral cavity are also expressed in gut epithelial cells [19] and in the hypothalamus.[20] But even when all taste processing is eliminated by genetic manipulation, mice still learn to prefer sugar over water, suggesting the generation of food reward by processes of glucose utilization.[18] Rather than the acute pleasure of tasty food in the mouth, there is a general feeling of satisfaction that lingers on long after termination and most likely contributes to the reinforcing power of a meal. Thus, a variety of sensory stimuli and emotional states or feelings with vastly different temporal profiles make up the rewarding experience of eating, and the underlying neural functions are only beginning to be understood.

Components of reward functions and their neural mechanisms

The neural mechanisms and behavioral manifestations of reward functions as they pertain to natural food reward and its similarities to drug reward have been subject to excellent reviews[5–7,21] and are discussed only briefly here. Berridge and Robinson have parsed reward into separable psychological and neural components, liking, wanting, and learning.[22] The characteristic orofacial expressions displayed by decerebrate rats[23] and anencephalic infants[24] in response to sweet taste strongly suggest that the forebrain is not the only brain area involved in experiencing the hedonic impact or liking of pleasant stimuli. Berridge and Robinson[22] refer to these expressions as objective affective reactions or implicit affect and to the psychological process as implicit *liking*. Besides neural circuits in the hindbrain, the nucleus accumbens and ventral pallidum in the limbic forebrain appear to be some of the other key components of the distributed neural network mediating liking of palatable foods. The mu-opioid receptor appears to play a crucial role. Local injection of the selective mu-opioid agonist DAMGO into the nucleus accumbens elicits voracious food intake, particularly of palatable sweet or high-fat foods.[25–27] This increased consumption of highly palatable foods appears to be due to increased liking, as morphine microinjections into this area increased the number of positive affective reactions,[28] and microinjection of a selective mu-opioid antagonist reduced sucrose drinking.[29] The most sensitive area for this effect was the caudal shell of the nucleus accumbens, near the border with the adjacent core.[28] We have recently demonstrated that nucleus accumbens injection of a mu-opioid receptor antagonist transiently suppressed such sucrose-evoked positive hedonic orofacial reactions.[30]

In humans, subjective liking can be assessed by questionnaires and visual analog scales. In the Power of Food Scale (PFS), appetite for palatable food items is estimated by asking subjects how much they would like to eat certain foods when they were available, when they are present in front of their eyes, and when they are actually tasted, but not ingested.[31] These three levels of proximity clearly generate different neural response patterns, involving more or less visual, taste, and olfactory processing. To consciously experience and give subjective ratings of pleasure from palatable foods (liking), humans very likely use areas in the prefrontal and cingulate cortex.[32] Thus, the neurological substrate responsible for liking palatable food items is distributed throughout the neuraxis and cannot be conveniently eliminated by lesions. One of the common denominators of the distributed network may be opioidergic transmission, particularly through the mu-opioid receptor.

Another component of reward is motivation or *wanting*. Typically, motivation comes to fruition by "going for" something that has generated pleasure in the past through a learning process—wanting what we like. However, wanting can also be dissociated from liking as demonstrated by sodium-depleted rats wanting hypertonic saline, a taste they had never "liked" before and also by drug addicts that no longer like to inject themselves.[33,34] Dopamine signaling within the mesolimbic dopamine projection system appears to be a crucial component of this process. Phasic activity of dopamine neuron projections from the ventral tegmental area to the nucleus accumbens in the ventral striatum are specifically involved in the decision-making process during the preparatory (appetitive) phase of ingestive behavior.[10,35] In addition, when palatable foods such as sucrose are actually consumed, a sustained and sweetness-dependent increase occurs in nucleus accumbens dopamine levels and turnover.[36–38] Dopamine signaling in the nucleus accumbens thus appears to play a role in both the preparatory and consummatory phases of an ingestive bout. The nucleus accumbens shell is thereby part of a neural loop including the lateral hypothalamus and the ventral tegmental area, with orexin neurons playing a key role.[8,39–46] This loop is likely important for the attribution of incentive salience to goal objects by metabolic state and other need signals available to the lateral hypothalamus, as discussed below.

In summary, although there have been excellent recent attempts to separate its components, the functional concept and neural circuitry underlying food reward is still poorly defined. Specifically, it is not well understood how reward, generated during anticipation, consummation, and satiation, are computed and integrated. Future research with modern neuroimaging techniques in humans and invasive neurochemical analyses in animals will be necessary for a more complete understanding.

Reward functions in the obese

Liking

A popular assumption is that obese individuals like food more than lean individuals and that this increased liking results in overeating and eventually obesity. Obese subjects report higher hedonic hunger, as measured with the PFS[47–49] and higher

liking for a given sweetness[50] compared with normal weight subjects. Interestingly, this is in spite of decreased perceived sweetness in obese subjects.[50] Thus, as concluded by Bartoshuk *et al.*,[50] liking increases as a function of sweetness more in obese subjects and more as BMI increases, and for the same perceived sweetness, liking increases as BMI increases. Importantly, in underweight subjects with

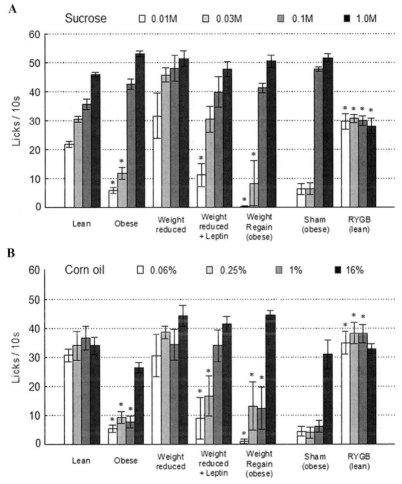

Figure 1. Brief access lick performance as a test of taste-guided liking of sucrose (A) and corn oil (B) in lean and obese rats. Lick performance was first compared between separate groups of chow-fed lean rats ($n = 7$) and high fat–fed obese rats ($n = 7$). Obese rats fed a high-fat diet throughout ($n = 6$) were then subjected to a cycle of weight loss ($\sim$20% in three weeks by means of calorie restriction) and regain (two weeks, as shown in Fig. 3). During the weight-reduced state they received either saline or leptin (1 mg/kg, ip, 1 h before test). In another experiment, lick performance was assessed three to five months after sham surgery ($n = 6$) or RYGB surgery ($n = 5$). Note that the significantly reduced response performance to low concentrations of both sucrose and corn oil in the obese versus lean, weight-reduced + leptin versus weight-reduced, and weight-regain versus weight-reduced groups (*$P < 0.05$ compared with the same concentration). Also note the significantly increased response to the two low concentrations of sucrose and the three low concentrations of corn oil, but the significantly reduced response to the highest concentration of sucrose in RYGB rats compared to sham-operated rats (*$P < 0.05$, compared with the same concentration). Statistics are based on two separate ANOVAs for each taste stimulant, one for the sham-operated and RYGB rats and one for all other conditions, and Bonferroni-corrected multiple comparisons.

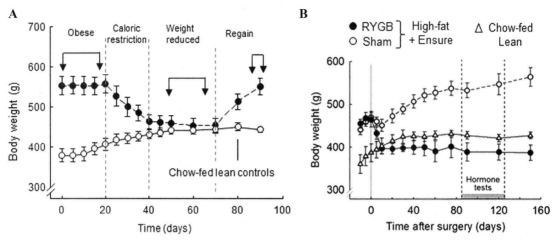

Figure 2. Body weight of rats used for testing reward behaviors. (A) Outbred Sprague–Dawley rats were either fed chow throughout (lean controls, $n = 6$) or were made obese by 12–16 weeks of high-fat feeding. One cohort of obese rats was then calorie restricted on a high-fat diet for three weeks and kept at a 20% lower body weight for four weeks, before the full amount of high-fat diet was restored and most of the lost body weight was regained. Arrows indicate the time of behavioral testing. (B) Outbred Sprague–Dawley rats made obese with a choice diet consisting of high-fat chow, Ensure, and chow were subjected to RYGB surgery ($n = 6$) or sham surgery ($n = 8$), and compared with age-matched, chow-fed lean controls. Behavioral testing was performed three to five months after surgery. Body composition was monitored throughout each experiment to verify the designations "lean" and "obese."

a BMI of <18.5, liking did not increase as a function of perceived sweetness.[50] Furthermore, subjects with a tendency for binge eating showed increased liking for all food categories.[51]

We have started to more systematically evaluate the specific contributions of the obese state on measures of liking by comparing high-fat diet–induced obese outbred Sprague–Dawley rats with (1) never obese, lean rats, (2) formerly obese rats after calorie restriction–induced weight loss, (3) formerly obese rats after RYGB surgery–induced weight loss, (4) weight-reduced rats treated with leptin, and (5) formerly obese, weight-reduced rats after renewed weight gain.[52,53] We also compared liking in the fed and fasted state of obese and lean rats.[52] Furthermore, we have measured liking in genetically selected lines of obesity-resistant and obesity-prone rats before and after a period of high-fat feeding.[52] As shown in Figure 1, while lean outbred Sprague–Dawley rats exhibit a near linear concentration-response curve in their brief-access sucrose- and corn oil–licking behavior, their obese counterparts show a right shift with less responding to the lowest concentrations but more responding to the highest concentrations. To determine whether the right shift was due to preexisting differences in the reward system or secondary effects of the obese state, the obese rats were then subjected to weight loss

induced by restricting their access to a high-fat diet to 50–70%. After weight loss of about 20% over a period of three weeks, body weight was maintained at this lower plateau, and liking of sucrose and corn oil was reassessed in the fed state (Fig. 2). Weight loss resulted in a prompt shift of the concentration-response curve back to the left, not much different from the never obese, lean rats, suggesting that most of the difference in brief access responding between lean and obese SD rats was due to secondary effects of the obese state, not to preexisting differences in reward processing. The reversibility of the phenotype was further underscored by a renewed right-shift when the restricted animals were again allowed unlimited access to a high-fat diet (Fig. 1). Although we cannot rule out a weight loss-independent effect of the calorie-restriction procedure, it is impossible to separate such an effect from weight loss.

Because human patients undergoing bariatric surgery were reported to decrease preference for fatty and sweet foods,[54–59] we tested rats with sucrose and corn oil after RYGB or sham surgery in the brief access paradigm. While sham-operated rats that remained obese exhibited a similar right shift in the concentration–response curves for both sucrose and corn oil as observed in nonoperated obese rats, RYGB rats that lost about 20% of body weight five months after surgery (Fig. 2) exhibited a flat

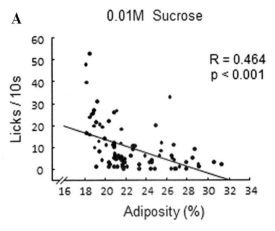

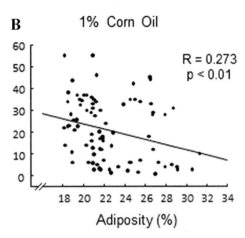

Figure 3. Regression analysis showing relationship between brief access lick performance and adiposity (as measured by NMR) across rats of all lean and obese conditions as shown in Figure 1. Note that lick performance for 0.01M sucrose (A) and 1% corn oil (B) was negatively correlated with adiposity.

concentration–response curve, with much more responding to the lowest concentrations, but less responding to the highest concentrations compared with sham-operated rats for both tastants (Fig. 1). This response pattern was different not just from the sham-operated rats, but also from the never obese, lean, as well as the weight-reduced, formerly obese rats, suggesting that, in addition to effects of weight loss, the surgery had weight loss–independent effects. During the early postsurgical phase, before much weight loss has occurred, aversive conditioning could play an important role in reduced food intake in both humans[60] and rats.[61] At later postsurgical times, changes in the pattern of circulating gut hormones acting on the brain are thought to be major candidates for reduced appetite and food intake.[62] However, as indicated by significant correlations between brief access responding to 0.01M sucrose as well as 1% corn oil and body fat content (as measured by NMR) across all lean and obese conditions of our studies, adiposity seems to be at least one factor determining the hedonic response to sweet and oily foods (Fig. 3).

In another model of obesity, the OLETF rat, which has a deficient CCK1-receptor, RYGB surgery also leads to selective reduction of brief access responding to high concentrations of sucrose,[63] and a similar effect was shown in chow-fed rats after RYGB surgery.[64]

To confirm the changes in taste-guided licking behavior as measured in the brief access test with a more specific measure of hedonic liking, we compared chow-fed lean, RYGB, and sham-operated rats in the taste reactivity test that quantitates the positive hedonic orofacial reactions to the taste of sucrose. The results were almost identical to the brief access test, with sham-operated obese rats showing a right shift of the concentration–response curve and RYGB rats an essentially flat curve, with more responding to the lowest, and less responding to the highest, sucrose concentration (Fig. 4). These findings suggest that the brief access test measures something very similar to the taste reactivity test, and that this liking is reversibly changed by the obese state and additionally by some unknown mechanism induced by RYGB surgery, independent of weight loss.

Commensurate with the reported shift in preference away from fatty and sweet foods in human patients after bariatric surgery,[56,57,61] we have also observed a shift in long-term preference for a low-fat over a high-fat diet after RYGB surgery in rats.[65] There is a significant decrease in fat preference after RYGB compared to sham-operated rats when given a choice between a low-fat (10%) and a high-fat (30%) liquid diet in 12-h tests. There was also a gradual shift in preference away from the 60% high-fat solid diet to regular (low-fat) chow. This relative avoidance of fat was accentuated in lean rats fed regular chow before they were subjected to RYGB in that, while sham-operated lean rats readily switched to a high-fat diet, RYGB rats consumed very little of the high-fat diet (Fig. 5). These results are consistent with the idea that changes in liking, as

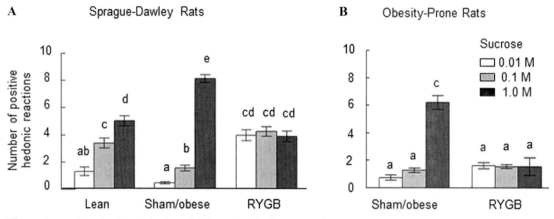

Figure 4. Number of positive hedonic orofacial reactions (Grill & Norgren's taste reactivity test) as a measure of liking of sucrose in Sprague–Dawley rats five months after RYGB or sham surgery and in age-matched, chow-fed lean rats (A), and in genetic lines of obesity-prone rats five months after RYGB or sham surgery (B). Bars that do not share the same letters are significantly ($P < 0.05$) different from each other (based on Bonferroni-corrected multiple comparisons following separate ANOVAs).

measured in our acute test paradigms, translate into long-term preference for more healthy foods low in fat and sugar after RYGB surgery.

Wanting

Berridge and Robinson's liking–wanting distinction[22] has recently been considered for studying humans.[66,67] In lean subjects, implicit wanting, as estimated by measuring reaction time in a forced choice paradigm, was not downregulated by food consumption in the same way as hunger, suggesting that it operates independent of homeostatic regulation.[68] Increased implicit wanting of specifically sweet high-fat foods is found in subjects with high binge eating scores,[69] suggesting that implicit wanting is a strong predictor of food intake. Using a different approach for estimating implicit wanting, Lemmens *et al.* found increased wanting for, and intake of, desserts and snacks in subjects with mild visceral obesity (BMI = 28 ± 1 kg/m^2) compared with lean subjects.[70]

While these limited observations appear to agree with the commonly held view that obese subjects want palatable foods more than lean subjects, these cross-sectional studies cannot distinguish cause and effect. They cannot distinguish whether preexisting differences cause the obese phenotype or whether the obese phenotype alters reward functions and behavior. Furthermore, if food reward processing is particularly important for the consumption of highly palatable foods, to the extent that these foods can become addictive to some individuals,[71,72] then

it is also not clear whether, in analogy to drug addiction, repeated exposure to, and abstinence from, these addictive foods independently lead to alterations in reward processing.[73,74] To determine the relative contribution of these three factors, they have to be selectively manipulated, which is not an easy task, particularly in human studies. The strongest evidence for predisposing, obesity-inducing differences in reward processing so far in humans has come from subjects carrying point mutations in genes known to be involved in brain reward processing. Comparing lean and obese subjects carrying different alleles of either the dopamine D2-receptor or mu-opioid receptor genes does reveal differences in behavioral and neural responses to palatable food.[75–78] In a prospective study, the dorsal striatal response to food and subsequent six-month weight gain was measured in subjects either carrying or not the Taq A1 allele for the dopamine D2 receptor gene that leads to reduced receptor expression and likely diminished dopamine D2 signaling. While in carriers, weight gain was positively correlated with the magnitude of the food-induced striatal response, a negative correlation was found in noncarriers, suggesting that blunted activation of food reward circuitry to food and food cues increases the risk for weight gain if coupled with genetic risk for attenuated dopamine signaling, and exaggerated activation increases risk for weight gain in individuals without such a risk.[79]

In selectively bred lines of obesity-prone (OP) and obesity-resistant (OR) rats,[80] as well as in

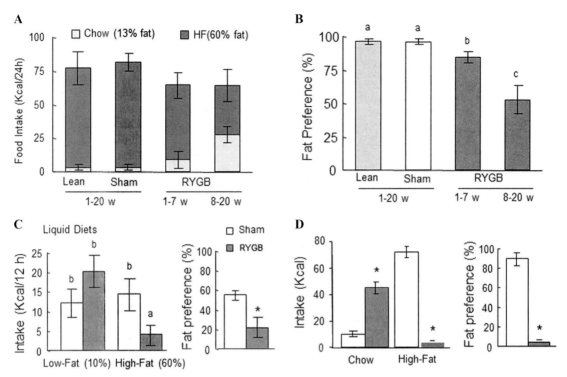

Figure 5. Gastric bypass surgery reduces fat preference in rats. (A, B) Gradual development of reduced fat preference after RYGB surgery in Sprague–Dawley rats as assessed with a choice of two complete solid diets low (13%) or high (60%) in fat. (C) Reduced fat preference measured three months after RYGB or sham surgery in Sprague–Dawley rats given 12-h access to two complete liquid diets, one low (10%) and one high (60%). Bars that do not share the same letter are significantly ($P < 0.05$) different from each other (based on Bonferroni-corrected multiple comparisons after appropriate ANOVA). (D) Almost complete avoidance of solid high-fat (60%) diet in chow-fed, nonobese Sprague–Dawley rats, three months after RYGB. *$P < 0.05$ compared to sham-operated rats.

Long-Evans rats,[81] several differences in mesolimbic dopamine signaling and in progressive ratio responding, a behavioral measure of wanting, have been reported. We have used both progressive ratio lever press responding and the incentive runway paradigms to assess wanting in various rat models of obesity and weight loss. Using the incentive runway, it was shown that mice made hyperdopaminergic by genetically attenuating the dopamine transporter (DAT) function found the goal box significantly faster than wild-type mice.[82] Thus, completion speed in the incentive runway is a measure of wanting and reinforcement learning to obtain a food or drug reward.[82,83]

Compared with lean chow-fed rats, high-fat diet–induced obese SD rats learned the task much slower, with initially significantly reduced completion speed, resulting in a lower *wanting index*[52] (Fig. 6A). This was not due to nonspecific motor impairment, as the net running speed was not differ-

ent, but obese rats spent significantly more time being distracted on the way, as indicated by significantly increased latency to leave the start box and duration of pauses and reversals. Similarly, young, chow-fed, genetically selected lines of OP rats exhibited a significantly reduced wanting index compared with their OR counterparts, which was further aggravated after eight weeks on a high-fat diet (Fig. 6A). Furthermore, progressive ratio lever press performance, as determined by the break point, was also lower in OP rats that had become overtly obese during 16 weeks on a high-fat diet compared to OR rats, both in the fed and fasted state (Fig. 6B).

Because it was reported that bariatric surgery patients have a diminished desire to eat[49] and exhibit changes in striatal dopamine D2 receptor availability,[84,85] we also measured wanting in our rat model of RYGB surgery. Sham-operated obese rats showed the familiar significant reduction of

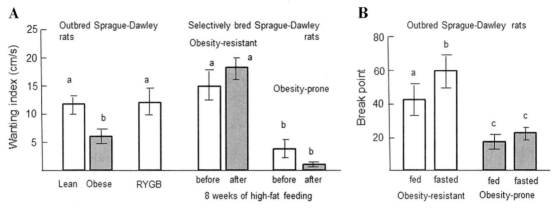

Figure 6. Motivation to obtain food reward (wanting) as measured in the incentive runway and progressive ratio lever press tests. (A) A wanting index was calculated as the mean completion speed averaged over trials 4–7. Note that in outbred Sprague–Dawley rats, high fat–fed obese rats showed reduced wanting compared to chow-fed lean rats and that wanting was restored after weight loss induced by RYGB surgery. Also, genetic lines of young obesity-prone rats exhibited reduced wanting compared to obesity-resistant rats. Eight weeks of high-fat feeding in these genetic lines did not produce further significant changes in wanting (B) Break point in the progressive ratio paradigm was significantly lower in obesity-prone versus obesity-resistant rats both in the fed and fasted condition. Bars that do not share the same letter are significantly ($P < 0.05$) different from each other (based on Bonferroni-corrected multiple comparisons after appropriate ANOVA).

wanting in the incentive runway compared with lean controls (Fig. 6), and the slower completion time was due to increased duration of distractions, not to decreased running speed. Most importantly, this reduced wanting was fully reversed by RYGB surgery. RYGB rats wanted the food reward in the goal box just as much as lean control rats. Our findings are in agreement with the work of Davis and Benoit,[81] demonstrating drastically reduced break points of progressive lever press responding for food reward in both high-fat diet–induced obese Long-Evans rats and young, genetically selected OP rats. This reduced wanting seen in obese rats by us and other authors seems at odds with the increased liking of the highest sucrose concentrations discussed above, as well as the increased implicit wanting of palatable foods observed in obese humans.[69,70] Because the reward obtained in the incentive runway and progressive ratio tests was in the form of solid foods (Fruit Loops® and sucrose pellets, respectively), it could be that liquid sucrose, as used in the brief access and taste reactivity tests, is a more salient stimulus that may not have reduced wanting in obese rats. This possibility should be further explored in future studies. The apparent discrepancy in wanting of obese rats and humans could lie in methodological differences. Specifically, different levels of effort required to obtain the reward could be

important. This is supported by the observation that genetically obese mice with MC4 receptor deficiency respond more when two lever presses were required (FR2) to obtain one small sucrose pellet, but they responded less when 50 lever presses were required (FR 50).[86]

Together, these findings suggest that the willingness of obese rats and mice to work for food depends on the effort required. Only if the effort is low, as for example in the brief access test or with low fixed ratio schedules of reinforcement, will obese rodents work for food reward. If the effort is higher, such as in the progressive ratio lever press and incentive runway paradigms, obese rodents stop working for food reward. Effort-dependent learning to work for food reward in operant schedules in normal weight rats has been comprehensively studied by Salamone and others, and it might have a neural basis in dopamine and adenosine signaling in the nucleus accumbens.[87–91] That is, interference with dopamine signaling in this pathway makes normal weight rats work less hard for food rewards, the same effect that is seen in obese rats. This strongly supports the reward-deficiency hypothesis, which suggests that individuals with low dopamine signaling compensate by engaging in more eating, thereby restoring a set point for reward generation (Fig. 7). There is considerable evidence for decreased dopamine

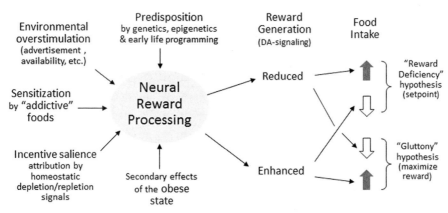

Figure 7. Schematic diagram showing the relationship between food reward and obesity.

signaling in both in obese rodents[80,81,92–96] and humans.[75–77,79,97,98] However, because many studies simply compared lean with obese subjects, it has not been clear whether reward (dopamine) deficiency is the cause or consequence of obesity. One possible explanation of our findings of reversible reduced wanting in obese rats is that weight loss by either calorie restriction or gastric bypass surgery is able to restore dopamine signaling to normal levels, and experiments to test this hypothesis are underway.

Finally, the behavioral phenomenon known as *delayed discounting*—a measure of immediate over-delayed gratification—could play a role in the reduced wanting observed in our study, in that obese rats overconsume readily available high concentrations of sucrose and corn oil from the spout but under-perform if the reward is delayed in the runway or lever press paradigms. This possibility is supported by observations in obese women showing significantly greater delayed discounting,[99] and children with difficulties in delaying gratification being more likely to become obese.[100]

Conclusions

In conclusion, obesity is associated with complex alterations in food reward functions at the neural and behavioral level (Fig. 7). In general, obese subjects like and want palatable foods more than lean subjects, but these effects appear to be strongly dependent on the salience of the food stimuli and on the effort necessary to obtain these foods. Obese rodents *like* sucrose and corn oil more than their lean or weight-reduced counterparts when the concentrations are above a certain level—at low concen-

trations, they actually like sugar and fat less. Similarly, obese rodents *want* palatable food more only when it is easy to obtain, but they do not want to work for it. There is considerable evidence in both rodents and humans that reward deficiency with defective mesolimbic dopamine signaling is an important mechanism underlying at least some of these alterations in the obese. If reward generation through this system is diminished, either at an early age due to genetic and nongenetic predisposition or later in life due to secondary effects of diets high in sugar and fat and/or the ensuing obesity, increased food intake and the pursuit of other pleasures is used in an attempt to restore a reward set point. Thus, obesity-associated alterations of reward behaviors and neural functions are the result of both predisposing traits and secondary effects of repeated exposure to palatable diets and/or the obese state. The mechanisms by which repeated exposure and the obese state separately contribute to changes in reward behaviors, such as liking and wanting, and their respective neural pathways remain to be demonstrated.

Conflicts of interest

The authors declare no conflicts of interest.

References

1. Rosenbaum, M. *et al.* 2008. Leptin reverses weight loss-induced changes in regional neural activity responses to visual food stimuli. *J. Clin. Invest.* **118:** 2583–2591.
2. Sjostrom, L. *et al.* 2007. Effects of bariatric surgery on mortality in Swedish obese subjects. *N. Engl. J. Med.* **357:** 741–752.

3. Ravussin, E. & C. Bogardus. 2000. Energy balance and weight regulation: genetics versus environment. *Br. J. Nutr.* **83**(Suppl 1): S17–S20.

4. Gluckman, P.D., M.A. Hanson & A.S. Beedle. 2007. Early life events and their consequences for later disease: a life history and evolutionary perspective. *Am. J. Hum. Biol.* **19**: 1–19.

5. Corwin, R.L. & A. Hajnal. 2005. Too much of a good thing: neurobiology of non-homeostatic eating and drug abuse. *Physiol. Behav.* **86**: 5–8.

6. Avena, N.M., P. Rada & B.G. Hoebel. 2009. Sugar and fat bingeing have notable differences in addictive-like behavior. *J. Nutr.* **139**: 623–628.

7. Kelley, A.E. & K.C. Berridge. 2002. The neuroscience of natural rewards: relevance to addictive drugs. *J. Neurosci.* **22**: 3306–3311.

8. Liedtke, W.B. *et al.* 2011. Relation of addiction genes to hypothalamic gene changes subserving genesis and gratification of a classic instinct, sodium appetite. *Proc. Natl. Acad. Sci. USA* **108**: 12509–12514.

9. Berthoud, H.-R. 2002. Multiple neural systems controlling food intake and body weight. *Neurosci. Biobehav. Rev.* **26**: 393–428.

10. Schultz, W., P. Dayan & P.R. Montague. 1997. A neural substrate of prediction and reward. *Science* **275**: 1593–1599.

11. Hare, T.A. *et al.* 2008. Dissociating the role of the orbitofrontal cortex and the striatum in the computation of goal values and prediction errors. *J. Neurosci.* **28**: 5623–5630.

12. McClure, S.M., G.S. Berns & P.R. Montague. 2003. Temporal prediction errors in a passive learning task activate human striatum. *Neuron* **38**: 339–346.

13. O'Doherty, J.P. *et al.* 2006. Predictive neural coding of reward preference involves dissociable responses in human ventral midbrain and ventral striatum. *Neuron* **49**: 157–166.

14. Rolls, E.T., C. McCabe & J. Redoute. 2008. Expected value, reward outcome, and temporal difference error representations in a probabilistic decision task. *Cereb. Cortex* **18**: 652–663.

15. Powley, T.L. 1977. The ventromedial hypothalamic syndrome, satiety, and a cephalic phase hypothesis. *Psychol. Rev.* **84**: 89–126.

16. Smith, G.P. 1996. The direct and indirect controls of meal size. *Neurosci. Biobehav. Rev.* **20**: 41–46.

17. Sclafani, A. & K. Ackroff. 2004. The relationship between food reward and satiation revisited. *Physiol. Behav.* **82**: 89–95.

18. de Araujo, I.E. *et al.* 2008. Food reward in the absence of taste receptor signaling. *Neuron* **57**: 930–941.

19. Rozengurt, E. & C. Sternini. 2007. Taste receptor signaling in the mammalian gut. *Curr. Opin. Pharmacol.* **7**: 557–562.

20. Ren, X. *et al.* 2009. Sweet taste signaling functions as a hypothalamic glucose sensor. *Front. Integr. Neurosci.* **3**: 12.

21. Levine, A.S., C.M. Kotz & B.A. Gosnell. 2003. Sugars: hedonic aspects, neuroregulation, and energy balance. *Am. J. Clin. Nutr.* **78**: 834S–842S.

22. Berridge, K.C. & T.E. Robinson. 2003. Parsing reward. *Trends Neurosci.* **26**: 507–513.

23. Grill, H.J. & R. Norgren. 1978. The taste reactivity test: II. Mimetic responses to gustatory stimuli in chronic thalamic and chronic decerebrate rats. *Brain Res.* **143**: 281–297.

24. Steiner, J.E. 1973. *The Gustofacial Response: Observations on Normal and Anencephalic Newborn Infants.* U. S. Department of Health, Education, and Welfare. Bethesda, MD.

25. Zhang, M. & A.E. Kelley. 2000. Enhanced intake of high-fat food following striatal mu-opioid stimulation: microinjection mapping and fos expression. *Neuroscience* **99**: 267–277.

26. Kelley, A.E. *et al.* 2002. Opioid modulation of taste hedonics within the ventral striatum. *Physiol. Behav.* **76**: 365–377.

27. Will, M.J., E.B. Franzblau & A.E. Kelley. 2003. Nucleus accumbens mu-opioids regulate intake of a high-fat diet via activation of a distributed brain network. *J. Neurosci.* **23**: 2882–2888.

28. Pecina, S. & K.C. Berridge. 2000. Opioid site in nucleus accumbens shell mediates eating and hedonic 'liking' for food: map based on microinjection Fos plumes. *Brain Res.* **863**: 71–86.

29. Kelley, A.E., E.P. Bless & C.J. Swanson. 1996. Investigation of the effects of opiate antagonists infused into the nucleus accumbens on feeding and sucrose drinking in rats. *J. Pharmacol. Exp. Ther.* **278**: 1499–1507.

30. Shin, A.C. *et al.* 2010. Reversible suppression of food reward behavior by chronic mu-opioid receptor antagonism in the nucleus accumbens. *Neuroscience* **170**: 580–588.

31. Lowe, M.R. *et al.* 2009. The Power of Food Scale. A new measure of the psychological influence of the food environment. *Appetite* **53**: 114–118.

32. Kringelbach, M.L. 2004. Food for thought: hedonic experience beyond homeostasis in the human brain. *Neuroscience* **126**: 807–819.

33. Wyvell, C.L. & K.C. Berridge. 2000. Intra-accumbens amphetamine increases the conditioned incentive salience of sucrose reward: enhancement of reward "wanting" without enhanced "liking" or response reinforcement. *J. Neurosci.* **20**: 8122–8130.

34. Tindell, A.J. *et al.* 2009. Dynamic computation of incentive salience: "wanting" what was never "liked." *J. Neurosci.* **29**: 12220–12228.

35. Carelli, R.M. 2002. The nucleus accumbens and reward: neurophysiological investigations in behaving animals. *Behav. Cogn. Neurosci. Rev.* **1**: 281–296.

36. Hernandez, L. & B.G. Hoebel. 1988. Feeding and hypothalamic stimulation increase dopamine turnover in the accumbens. *Physiol. Behav.* **44**: 599–606.

37. Hajnal, A., G.P. Smith & R. Norgren. 2004. Oral sucrose stimulation increases accumbens dopamine in the rat. *Am. J. Physiol. Regul. Integr. Comp. Physiol.* **286**: R31–R37.

38. Smith, G.P. 2004. Accumbens dopamine mediates the rewarding effect of orosensory stimulation by sucrose. *Appetite* **43**: 11–13.

39. Stratford, T.R. & A.E. Kelley. 1999. Evidence of a functional relationship between the nucleus accumbens shell and lateral hypothalamus subserving the control of feeding behavior. *J. Neurosci.* **19**: 11040–11048.

40. Zheng, H., L.M. Patterson & H.R. Berthoud. 2007. Orexin signaling in the ventral tegmental area is required for high-fat appetite induced by opioid stimulation of the nucleus accumbens. *J. Neurosci.* **27**: 11075–11082.

41. Harris, G.C., M. Wimmer & G. Aston-Jones. 2005. A role for lateral hypothalamic orexin neurons in reward seeking. *Nature* **437:** 556–559.

42. Peyron, C. *et al.* 1998. Neurons containing hypocretin (orexin) project to multiple neuronal systems. *J. Neurosci.* **18:** 9996–10015.

43. Nakamura, T. *et al.* 2000. Orexin-induced hyperlocomotion and stereotypy are mediated by the dopaminergic system. *Brain Res.* **873:** 181–187.

44. Balcita-Pedicino, J.J. & S.R. Sesack. 2007. Orexin axons in the rat ventral tegmental area synapse infrequently onto dopamine and gamma-aminobutyric acid neurons. *J. Comp. Neurol.* **503:** 668–684.

45. Korotkova, T.M. *et al.* 2003. Excitation of ventral tegmental area dopaminergic and nondopaminergic neurons by orexins/hypocretins. *J. Neurosci.* **23:** 7–11.

46. Borgland, S.L. *et al.* 2006. Orexin A in the VTA is critical for the induction of synaptic plasticity and behavioral sensitization to cocaine. *Neuron* **49:** 589–601.

47. Cappelleri, J.C. *et al.* 2009. Evaluating the Power of Food Scale in obese subjects and a general sample of individuals: development and measurement properties. *Int. J. Obes.* **33:** 913–922.

48. Thomas, J.G. *et al.* 2011. Ecological momentary assessment of obesogenic eating behavior: combining person-specific and environmental predictors. *Obesity (Silver Spring)* **19:** 1574–1579.

49. Schultes, B. *et al.* 2010. Hedonic hunger is increased in severely obese patients and is reduced after gastric bypass surgery. *Am. J. Clin. Nutr.* **92:** 277–283.

50. Bartoshuk, L.M. *et al.* 2006. Psychophysics of sweet and fat perception in obesity: problems, solutions and new perspectives. *Philos. Trans. R. Soc. Lond. B. Biol. Sci.* **361:** 1137–1148.

51. Finlayson, G. *et al.* 2011. Implicit wanting and explicit liking are markers for trait binge eating. A susceptible phenotype for overeating. *Appetite* **57:** 722–728.

52. Shin, A.C. *et al.* 2011. "Liking" and "wanting" of sweet and oily food stimuli as affected by high-fat diet-induced obesity, weight loss, leptin, and genetic predisposition. *Am. J. Physiol. Regul. Integr. Comp. Physiol.* **301:** R1267–R1280.

53. Shin, A.C. *et al.* 2011. Roux-en-Y gastric bypass surgery changes food reward in rats. *Int. J. Obes.* **35:** 642–651.

54. Burge, J.C. *et al.* 1995. Changes in patients' taste acuity after Roux-en-Y gastric bypass for clinically severe obesity. *J. Am. Diet. Assoc.* **95:** 666–670.

55. Kenler, H.A., R.E. Brolin & R.P. Cody. 1990. Changes in eating behavior after horizontal gastroplasty and Roux-en-Y gastric bypass. *Am. J. Clin. Nutr.* **52:** 87–92.

56. Thirlby, R.C. *et al.* 2006. Effect of Roux-en-Y gastric bypass on satiety and food likes: the role of genetics. *J. Gastrointest. Surg.* **10:** 270–277.

57. Thomas, J.R. & E. Marcus. 2008. High and low fat food selection with reported frequency intolerance following Roux-en-Y gastric bypass. *Obes. Surg.* **18:** 282–287.

58. Tichansky, D.S., J.D. Boughter, Jr. & A.K. Madan. 2006. Taste change after laparoscopic Roux-en-Y gastric bypass and laparoscopic adjustable gastric banding. *Surg. Obes. Relat. Dis.* **2:** 440–444.

59. Olbers, T. *et al.* 2006. Body composition, dietary intake, and energy expenditure after laparoscopic Roux-en-Y gastric bypass and laparoscopic vertical banded gastroplasty: a randomized clinical trial. *Ann. Surg.* **244:** 715–722.

60. Sugerman, H.J. *et al.* 1992. Gastric bypass for treating severe obesity. *Am. J. Clin. Nutr.* **55:** 560S–566S.

61. le Roux, C.W. *et al.* 2011. Gastric bypass reduces fat intake and preference. *Am. J. Physiol. Regul. Integr. Comp. Physiol.* **301:** R1057–R1066.

62. le Roux, C.W. *et al.* 2006. Gut hormone profiles following bariatric surgery favor an anorectic state, facilitate weight loss, and improve metabolic parameters. *Ann. Surg.* **243:** 108–114.

63. Hajnal, A. *et al.* 2010. Gastric bypass surgery alters behavioral and neural taste functions for sweet taste in obese rats. *Am. J. Physiol. Gastrointest. Liver Physiol.* **299:** G967–G979.

64. Tichansky, D.S. *et al.* 2011. Decrease in sweet taste in rats after gastric bypass surgery. *Surg. Endosc.* **25:** 1176–1181.

65. Zheng, H. *et al.* 2009. Meal patterns, satiety, and food choice in a rat model of Roux-en-Y gastric bypass surgery. *Am. J. Physiol. Regul. Integr. Comp. Physiol.* **297:** R1273–R1282.

66. Finlayson, G., N. King & J.E. Blundell. 2007. Liking vs. wanting food: importance for human appetite control and weight regulation. *Neurosci. Biobehav. Rev.* **31:** 987–1002.

67. Finlayson, G., N. King & J.E. Blundell. 2007. Is it possible to dissociate 'liking' and 'wanting' for foods in humans? A novel experimental procedure. *Physiol. Behav.* **90:** 36–42.

68. Finlayson, G., N. King & J. Blundell. 2008. The role of implicit wanting in relation to explicit liking and wanting for food: implications for appetite control. *Appetite* **50:** 120–127.

69. Finlayson, G. *et al.* 2011. Implicit wanting and explicit liking are markers for trait binge eating. A susceptible phenotype for overeating. *Appetite* **57:** 722–728.

70. Lemmens, S.G. *et al.* 2011. Stress augments food 'wanting' and energy intake in visceral overweight subjects in the absence of hunger. *Physiol. Behav.* **103:** 157–163.

71. Gearhardt, A.N. *et al.* 2011. Can food be addictive? Public health and policy implications. *Addiction* **106:** 1208–1212.

72. Avena, N.M., P. Rada & B.G. Hoebel. 2008. Evidence for sugar addiction: behavioral and neurochemical effects of intermittent, excessive sugar intake. *Neurosci. Biobehav. Rev.* **32:** 20–39.

73. Vendruscolo, L.F. *et al.* 2010. Sugar overconsumption during adolescence selectively alters motivation and reward function in adult rats. *PLoS One* **5:** e9296.

74. Frazier, C.R. *et al.* 2008. Sucrose exposure in early life alters adult motivation and weight gain. *PLoS One* **3:** e3221.

75. Davis, C.A. *et al.* 2009. Dopamine for "wanting" and opioids for "liking:" a comparison of obese adults with and without binge eating. *Obesity* **17:** 1220–1225.

76. Davis, C. *et al.* 2008. Reward sensitivity and the D2 dopamine receptor gene: a case-control study of binge eating disorder. *Prog. Neuropsychopharmacol. Biol. Psychiatr.* **32:** 620–628.

77. Stice, E. *et al.* 2008. Relation between obesity and blunted striatal response to food is moderated by TaqIA A1 allele. *Science* **322:** 449–452.

78. Felsted, J.A. *et al.* 2010. Genetically determined differences in brain response to a primary food reward. *J. Neurosci.* **30:** 2428–2432.

79. Stice, E. *et al.* 2009. Relation of obesity to consummatory and anticipatory food reward. *Physiol. Behav.* **97:** 551–560.

80. Geiger, B.M. *et al.* 2008. Evidence for defective mesolimbic dopamine exocytosis in obesity-prone rats. *FASEB J.* **22:** 2740–2746.

81. Davis, J.F. *et al.* 2008. Exposure to elevated levels of dietary fat attenuates psychostimulant reward and mesolimbic dopamine turnover in the rat. *Behav. Neurosci.* **122:** 1257–1263.

82. Pecina, S. *et al.* 2003. Hyperdopaminergic mutant mice have higher "wanting" but not "liking" for sweet rewards. *J. Neurosci.* **23:** 9395–9402.

83. Ettenberg, A. 2009. The runway model of drug self-administration. *Pharmacol. Biochem. Behav.* **91:** 271–277.

84. Steele, K.E. *et al.* 2010. Alterations of central dopamine receptors before and after gastric bypass surgery. *Obes. Surg.* **20:** 369–374.

85. Dunn, J.P. *et al.* 2010. Decreased dopamine type 2 receptor availability after bariatric surgery: preliminary findings. *Brain Res.* **1350:** 123–130.

86. Atalayer, D. *et al.* 2010. Food demand and meal size in mice with single or combined disruption of melanocortin type 3 and 4 receptors. *Am. J. Physiol. Regul. Integr. Comp. Physiol.* **298:** R1667–R1674.

87. Salamone, J.D. *et al.* 2009. Dopamine, behavioral economics, and effort. *Front. Behav. Neurosci.* **3:** 13.

88. Salamone, J.D. *et al.* 2003. Nucleus accumbens dopamine and the regulation of effort in food-seeking behavior: implications for studies of natural motivation, psychiatry, and drug abuse. *J. Pharmacol. Exp. Ther.* **305:** 1–8.

89. Ishiwari, K. *et al.* 2004. Accumbens dopamine and the regulation of effort in food-seeking behavior: modulation of work output by different ratio or force requirements. *Behav. Brain Res.* **151:** 83–91.

90. Farrar, A.M. *et al.* 2010. Nucleus accumbens and effort-related functions: behavioral and neural markers of the interactions between adenosine A2A and dopamine D2 receptors. *Neuroscience* **166:** 1056–1067.

91. Mingote, S. *et al.* 2005. Ratio and time requirements on operant schedules: effort-related effects of nucleus accumbens dopamine depletions. *Eur. J. Neurosci.* **21:** 1749–1757.

92. Huang, X.F. *et al.* 2005. Differential expression of dopamine D2 and D4 receptor and tyrosine hydroxylase mRNA in mice prone, or resistant, to chronic high-fat diet-induced obesity. *Brain Res. Mol. Brain Res.* **135:** 150–161.

93. Geiger, B.M. *et al.* 2009. Deficits of mesolimbic dopamine neurotransmission in rat dietary obesity. *Neuroscience* **159:** 1193–1199.

94. Hajnal, A. *et al.* 2007. Obese OLETF rats exhibit increased operant performance for palatable sucrose solutions and differential sensitivity to D2 receptor antagonism. *Am. J. Physiol. Regul. Integr. Comp. Physiol.* **293:** R1846–R1854.

95. Fulton, S. *et al.* 2006. Leptin regulation of the mesoaccumbens dopamine pathway. *Neuron* **51:** 811–822.

96. Thanos, P.K. *et al.* 2008. Leptin receptor deficiency is associated with upregulation of cannabinoid 1 receptors in limbic brain regions. *Synapse* **62:** 637–642.

97. Wang, G.J. *et al.* 2001. Brain dopamine and obesity. *Lancet* **357:** 354–357.

98. Volkow, N.D. *et al.* 2008. Overlapping neuronal circuits in addiction and obesity: evidence of systems pathology. *Philos. Trans. R. Soc. Lond. B. Biol. Sci.* **363:** 3191–3200.

99. Weller, R.E. *et al.* 2008. Obese women show greater delay discounting than healthy-weight women. *Appetite* **51:** 563–569.

100. Epstein, L.H. *et al.* 2010. Food reinforcement, delay discounting and obesity. *Physiol. Behav.* **100:** 438–445.

Ann. N.Y. Acad. Sci. ISSN 0077-8923

Brain-derived neurotrophic factor as a regulator of systemic and brain energy metabolism and cardiovascular health

Sarah M. Rothman, Kathleen J. Griffioen, Ruiqian Wan, and Mark P. Mattson

Laboratory of Neurosciences, National Institute on Aging Intramural Research Program, National Institutes of Health, Baltimore, Maryland

Address for correspondence: Sarah M. Rothman, Laboratory of Neurosciences, National Institute on Aging Intramural Research Program, National Institutes of Health, 251 Bayview Boulevard, Baltimore, MD 21224. rothmansm@mail.nih.gov

Overweight sedentary individuals are at increased risk for cardiovascular disease, diabetes, and some neurological disorders. Beneficial effects of dietary energy restriction (DER) and exercise on brain structural plasticity and behaviors have been demonstrated in animal models of aging and acute (stroke and trauma) and chronic (Alzheimer's and Parkinson's diseases) neurological disorders. The findings described later, and evolutionary considerations, suggest brain-derived neurotrophic factor (BDNF) plays a critical role in the integration and optimization of behavioral and metabolic responses to environments with limited energy resources and intense competition. In particular, BDNF signaling mediates adaptive responses of the central, autonomic, and peripheral nervous systems from exercise and DER. In the hypothalamus, BDNF inhibits food intake and increases energy expenditure. By promoting synaptic plasticity and neurogenesis in the hippocampus, BDNF mediates exercise- and DER-induced improvements in cognitive function and neuroprotection. DER improves cardiovascular stress adaptation by a mechanism involving enhancement of brainstem cholinergic activity. Collectively, findings reviewed in this paper provide a rationale for targeting BDNF signaling for novel therapeutic interventions in a range of metabolic and neurological disorders.

Keywords: autonomic nervous system; brain-derived neurotrophic factor; cognition; diabetes; exercise; neurogenesis; synaptic plasticity

Brain-derived neurotrophic factor and food intake

In the central nervous system (CNS), brain-derived neurotrophic factor (BDNF) and its high-affinity receptor TrkB are highly expressed in the hypothalamus, where this neurotrophic factor has major regulatory roles in the control of appetite and metabolism.[1] Mice that are heterozygous for targeted disruption of BDNF (BDNF[+/−] mice) show a 50% reduction in BDNF expression in the hypothalamus,[2] consume 47% more food than wild-type (WT) mice, and are obese.[3] Obesity in BDNF[+/−] mice is prevented by restricting food intake to match that of WT counterparts, implying that the loss of BDNF causes hyperphagia, which leads to obesity.[4]

A reduction in the expression of TrkB also leads to obesity in mice.[5] By six months of age, the average weight of male and female BDNF[+/−] mice are 44% and 33% increased over WT counterparts, respectively.[3] This age-related obesity and chronic hyperphagia is accompanied by hyperactivity, which is in contrast to lethargy normally associated with obesity.[3,6]

Diminished BDNF signaling results in hyperphagia and obesity, whereas an increase in BDNF signaling has the opposite effect. Intracerebroventricular (ICV) infusion of BDNF into normal mice results in a significant decrease in food consumption and a loss of body weight (Fig. 1). After seven days of intracerebroventricular (ICV) infusion of BDNF (1.2 μg over 24 h), WT mice

doi: 10.1111/j.1749-6632.2012.06525.x

49

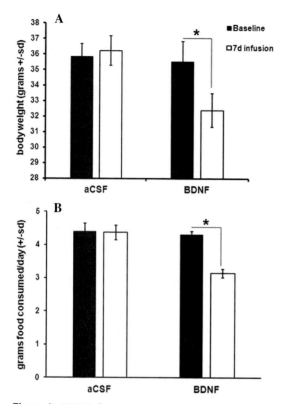

Figure 1. BDNF infusion causes a significant decrease in food consumption and a loss of body weight. ICV infusion of BDNF (1.2 μg over 24 h), causes a significant ($P = 0.005$) decrease in food consumption in WT mice as well as a significant ($P = 0.01$) decrease in body weight. Mice receiving infusion of artificial cerebrospinal fluid (aCSF) do not experience a significant change in either food consumption or body weight.

experience a highly significant ($P = 0.005$) decrease in food consumption and a significant ($P = 0.01$) decrease in body weight, whereas mice receiving infusion of artificial cerebrospinal fluid do not. This effect of BDNF also occurs in rats.[7] ICV infusion of BDNF into BDNF[+/−] mice normalizes food intake, body weight, and activity, further implying a physiological role for BDNF in regulating food intake.

The precise mechanism by which BDNF signaling suppresses appetite and food intake is not entirely known, although several hypotheses have emerged. The regions of the hypothalamus that regulate food intake include the paraventricular, arcuate (ARC), dorsomedial (DMH), and ventromedial nuclei (VMH). BDNF is expressed in some cells in the dorsomedial hypothalamus and at negligible levels in the arcuate; however, it is highly expressed in the VMH. Bilateral lesions of the VMH cause hypherphagia and obesity, implying an important role for this region of the hypothalamus in regulating food intake and energy metabolism.[8] Leptin, a polypeptide produced by adipocytes, targets neurons in the ARC; leptin positively regulates proopiomelanocortin neurons, which project to the VMH. In the VMH, expression of the melanocortin receptor 4 (MC4R) regulates expression of BDNF; a reduction in MC4R causes a downregulation of BDNF.[5] Further, administration of an MC4R agonist increased the level of BDNF mRNA in food-deprived mice, implying that MC4R signaling regulates alterations in BDNF in response to food deprivation.[5] Taken together, these data provide the beginning of a framework by which gut hormones target neurons in the hypothalamus that control expression of BDNF and therefore, food intake and metabolism.

Another region of the brain that may mediate the effects of BDNF signaling on food intake is the brainstem. Specifically, the dorsal vagal complex (DVC) contains insulin and leptin receptors as well as mechanisms that sense glucose levels.[9] Intraparenchymal infusion of BDNF to the DVC causes dose-dependent anorexia, implying that alterations in BDNF signaling in the DVC, without any alteration in signaling in the hypothalamus, are sufficient to alter feeding and potentially metabolism parameters.[10]

Experimental manipulations of BDNF levels affect feeding and metabolism, and environmental factors such as food deprivation and stress also cause changes in BDNF expression in the brain. Alternate day fasting stimulates the production of BDNF in neurons in different brain regions, including hippocampus, specifically the dentate gyrus.[11–13] It is noteworthy that dietary energy restriction (DER) restores BDNF levels in BDNF[+/−] mice and also reverses hyperphagia and obesity in those mice.[2] Conversely, food deprivation has an inhibitory effect on BDNF expression in the VMH of the hypothalamus that is partially reversed by administration of a MC4R antagonist.[5] In addition, food deprivation causes a reversible decrease in BDNF protein in the DVC, potentially indicating a role for DVC BDNF signaling in mediating metabolic feedback. The reason for the differential, brain region-specific changes in BDNF expression is likely that the neuronal circuits in the different regions are involved in coordinating complex behavioral and neuroendocrine responses to food intake or deprivation. Because food restriction is stressful, it can cause a rise in corticosterone (cortisol in humans), which has been shown to decrease the expression and production of BDNF in the brain. However, the brain stem may respond to food deprivation in a compensatory manner; food deprivation may cause an acute decrease in BDNF signaling, but it is possible that the brain stem upregulates BDNF expression in a compensatory response to a decrease in food intake.

BDNF regulates peripheral energy metabolism

BDNF regulation of food intake is also coupled to the ability of this neurotrophin to regulate peripheral energy metabolism. Low levels of circulating BDNF are noted in individuals with obesity and type 2 diabetes, implying a role for this neurotrophin in mediating obesity and metabolism.[14] However, patients with type 2 diabetes that are not obese display decreased levels of plasma BDNF, which could indicate that BDNF regulates obesity and metabolism via different mechanisms.[15] Mice with BDNF eliminated at birth are hypersensitive to stress, display elevated plasma glucose and insulin levels, and are obese.[6] Experimental studies show that ICV infusion of BDNF increases peripheral insulin sensitivity in normal rodents[16] and ameliorates diabetes in mice,[17] implying that BDNF signaling in the brain, particularly the hypothalamus, regulates peripheral energy metabolism. Further, BDNF[+/−] mice display a diabetic phenotype, including elevated levels of circulating glucose, insulin, and leptin.[2]

Obesity is associated with changes in the serum levels of leptin, insulin, glucose, and corticosterone/cortisol.[18] The interplay between these molecules is complex. In normal individuals, insulin is produced in response to increased glucose levels and stimulates glucose uptake by muscle and liver cells; corticosterone induces gluconeogenesis, and leptin production and secretion from adipocytes can be stimulated by insulin; and circulating leptin enters the brain and interacts with neurons in the hypothalamus to suppress appetite. Although the exact mechanisms by which BDNF regulates metabolism are not entirely known, several studies using leptin receptor mutant (*db/db*) mice demonstrate an important role of this hormone in BDNF-mediated

regulation of metabolism. The *db/db* mice are hyperglycemic and insulin resistant, and administration of BDNF normalizes glucose levels and insulin sensitivity.[16] Further, controlled studies in which both vehicle and BDNF-treated *db/db* mice consumed the same amount of food showed that BDNF administration significantly reduces blood glucose concentration, indicating that the effects of BDNF on energy metabolism are independent of its effects on appetite. Further proof of a role for leptin in mediating BDNF regulation of energy metabolism lies in data that shows that BDNF[+/−] mice display increased leptin levels and leptin resistance.[3] Conversely, *db/db* mice that cannot respond to leptin show reduced BDNF expression in the hippocampus and hypothalamus. Systemic BDNF infusion reduces food intake and blood glucose in *db/db* mice.[17,19] Lowering circulating corticosterone levels in *db/db* mice via adrenalectomy restores BDNF expression in the hippocampus but not the hypothalamus.[20]

Data show that some of the effects of leptin on energy metabolism may be mediated via changes in body temperature control. *Db/db* mice show reduced energy metabolism and lower body temperature and may be unable to activate thermogenesis to maintain body heat when their food intake is restricted.[16] In fact, administration of BDNF-raised body temperature in food-restricted *db/db* mice compared to *ad libitum* mice.[17]

Other models of diabetes in mice have been used to demonstrate a role for BDNF in regulating energy metabolism. Administration of streptozotocin (STZ) damages pancreatic β cells, leading to impaired insulin production and is therefore a model of type I diabetes. In STZ-treated mice, administration of BDNF causes a reduction in food intake but no change in circulating glucose levels, implying that the mechanism by which BDNF regulates food intake is independent of insulin signaling, but the mechanism by which BDNF regulates energy metabolism is insulin dependent.[16]

Involvement of BDNF in brain-stem control of the cardiovascular system

It is well known that neurotrophic factors such as BDNF are essential for the development of the autonomic nervous system (ANS), particularly in the formation of synaptic connectivity with peripheral targets. For example, BDNF is essential for the

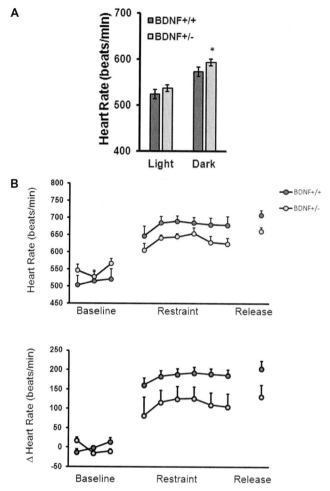

Figure 2. Evidence that BDNF signaling regulates heart rate. (A) BDNF$^{+/-}$ mice have significantly elevated basal heart rates compared to control (BDNF$^{+/+}$) mice during the dark cycle ($P < 0.05$). (B) BDNF$^{+/-}$ mice have attenuated heart rates during a 60-min restraint stress and release. Data points represent 10-min bins ($P < 0.001$). The change in heart rate in BDNF$^{+/-}$ mice is significantly lower than BDNF$^{+/+}$ mice during restraint and release ($P < 0.001$). Data are represented as mean ± SEM.

survival of arterial baroreceptors during vascular innervation.[21] However, recent findings indicate that BDNF also plays a major role in ANS control of cardiovascular function in adults. A role for BDNF in ANS regulation of heart rate is suggested by data showing exercise and DER, both of which increase BDNF levels in the CNS,[13,22,23] also decrease heart rate and increase heart rate variability.[24–26] The latter effects of exercise and DER are believed to be mediated by a relative increase in parasympathetic activity on the heart. Indeed, Yang *et al.*[27] showed that BDNF could alter the neurotransmitter release properties of sympathetic neurons innervating cardiac myocytes in culture, resulting in a shift from

excitatory to inhibitory cholinergic neurotransmission. The cardioprotective effects of DER may result, in part, from the relative increase in parasympathetic activity.[28]

Studies from our lab further suggest a role for BDNF in regulating parasympathetic and/or sympathetic inputs to the heart. BDNF$^{+/-}$ mice, which exhibit a 50% reduction in BDNF mRNA, have significantly elevated heart rates compared to WT mice (Fig. 2A). Further, when exposed to restraint stress, BDNF$^{+/-}$ mice fail to elevate their heart rate to WT levels (Fig. 2B), indicating an impaired cardiovascular stress response. Interestingly, a recent study revealed that humans with a BDNF polymorphism (Val66Met) that results in decreased activity-dependent BDNF secretion have altered sympathovagal balance leading to sympathetic dominance.[29] In addition, carriers of this mutation also exhibit attenuated heart rate responses to stress.[30] Taken together with our data, the latter findings suggest that reduced BDNF expression impairs heart rate stress responses and ANS function.

BDNF regulation of cardiovascular function likely occurs via signaling in central autonomic nuclei of the brain stem. BDNF is expressed in both baroreceptor and chemoafferents in the nodose and petrosal sensory ganglia, which terminate in the brainstem.[21,31] BDNF and TrkB are also produced in central autonomic nuclei of the brain stem,[32–35] as well as in higher cardiovascular control areas such as the hypothalamus, forebrain, and amygdala.[36,37] Local injection of BDNF into the rostral ventrolateral medulla increases blood pressure,[38] and BDNF acutely raises heart rate and blood pressure when applied to the third ventricle.[39] Further, BDNF is released in synchrony with baroreceptor activity from baroreceptor afferents onto second-order neurons in the nucleus tractus solitarius (NTS), the primary target of afferent cardiovascular input to the brainstem.[31,40] Injection of BDNF into the NTS of anesthetized rats increases blood pressure, heart rate, and lumbar sympathetic nerve activity.[41] Conversely, inhibition of tonic BDNF signaling in the NTS with a TrkB receptor antagonist decreases blood pressure and heart rate, indicating that BDNF signaling in the NTS tonically modulates cardiovascular regulation.[41]

BDNF mediates exercise- and energy restriction–induced neuroplasticity

The beneficial effects of exercise on health are unequivocal and, as reviewed previously,[42] include profound effects on the brain. Studies performed primarily in laboratory rodents have revealed highly reproducible effects of exercise, particularly aerobic endurance exercise, on the structure and function of the brain. Compared to their more sedentary counterparts, mice or rats that run voluntarily on running wheels (runners), exhibit increased numbers of dendritic spines (i.e., synapses) in hippocampal neurons, increased neurogenesis, and improved performance in some behavioral tests of cognitive function.[43–45] Mice and rats will typically run between 2 and 15 km in a 24-h period; there is consid-

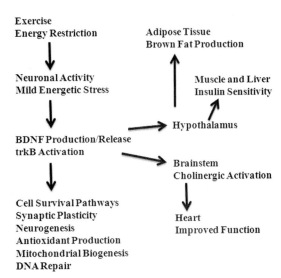

Figure 3. Central roles for BDNF as a mediator of beneficial effects of exercise and dietary energy restriction on neuroplasticity and overall health. Exercise and energy restriction (particular intermittent energy restriction) induce increased activity in neuronal circuits, as well as a mild energetic stress, throughout the nervous system. This mild cellular stress stimulates the production and release of BDNF. BDNF acts on neurons to promote their growth, enhance synaptic plasticity, and increase their resistance to injury and disease. Some of the genes induced in neurons in response to BDNF include those encoding antioxidant enzymes, DNA repair enzymes, and proteins involved in mitochondrial biogenesis. In addition, in the hippocampus, BDNF promotes the production and survival of new neurons from stem cells, and the integration of the new neurons into existing neuronal circuits. Acting in brain regions that control neuroendocrine pathways (e.g., hypothalamus) and the autonomic nervous system (e.g., brain-stem cholinergic neurons), BDNF mediates beneficial effects of exercise and energy restriction on glucose metabolism, body fat composition, and cardiovascular fitness.

erable interanimal variability in daily running distance, but less day-to-day variability in daily distance running for individual mice.[46–48] Where examined, there tends to be a graded positive effect of daily exercise intensity and duration on synaptic plasticity and neurogenesis. In addition, the positive effects of exercise on synaptic strength and learning and memory are not immediately apparent, and instead occur over periods of many days to weeks. The purpose of this section is to describe the evidence that BDNF plays a key role as a mediator of the effects of exercise on synaptic plasticity and neurogenesis (Fig. 3). In keeping with the main theme of this paper, we also consider recent evidence that BDNF mediates beneficial effects of intermittent energy restriction

(IER) on neuroplasticity and the vulnerability of neurons to injury and disease.[49,50]

Exercise and BDNF

Mice that are runners can perform better in tests of learning and memory, such as the radial arm maze, compared to nonrunners. Interestingly, when mice were provided running wheels for one week and then the running wheels were disabled, they learned the maze fastest when tested one week after cessation of running, but their memory retention was best when tested immediately after the one-week running period.[51] Hippocampal BDNF levels were elevated immediately after the one-week running period, remained elevated for two weeks after cessation of running and then returned to baseline by three weeks. There is also evidence that prior exercise can prime a BDNF response of hippocampal cells to a subsequent brief period of exercise that would be insufficient to significantly increase BDNF levels in previously sedentary animals.[52] The priming effect of exercise is maintained for at least two weeks after the cessation of exercise. The underlying molecular mechanism of the priming effect of exercise on BDNF production remains to be determined. In addition to improving performance in learning and memory tasks, running was shown to enhance the ability of mice to distinguish between objects located close to each other on a touch screen (spatial pattern separation), a process dependent upon hippocampal plasticity.[53] Moreover, a study of elderly human subjects showed that a daily aerobic exercise intervention can improve performance in memory tests, with an associated increase in the size of the hippocampus as measured by structural magnetic resonance imaging methods.[54]

The results of different behavioral tests have revealed that voluntary running and forced treadmill running can differentially affect performance on behavioral tasks. For example, whereas both voluntary and treadmill running improved performance of mice in the water maze, only treadmill running improved performance in a passive avoidance test.[55] The latter study further showed that both forced and voluntary running upregulated BDNF–TrkB signaling in the hippocampus, whereas only forced exercise increased BDNF signaling in the amygdala. In as much as forced exercise can be considered to be an aversive (or at least an annoying) stressor, it is un-

derstandable that it would engage neuronal circuits in the amygdala, a brain region heavily involved in fear-related learning.[56]

In rodents, running wheel exercise induces the expression of BDNF at the transcriptional level in multiple brain regions, and in particularly large amounts in dentate gyrus granule neurons and CA1 neurons in the hippocampus and in neurons in layers II and III of the cerebral cortex.[23,57] Elevations in BDNF transcripts can occur within minutes to hours of initiation of vigorous exercise. BDNF production can also be regulated at the translational level in response to exercise and other stimuli, with levels of BDNF protein typically increasing by a greater percentage over baseline compared to the increase in BDNF mRNA levels.[58] Compared to their more sedentary control counterparts, runner rats exhibit significantly more long-term potentiation (LTP) in response to theta rhythm-patterned stimulation, and this synaptic strengthening is associated with increased expression of the NR2B NMDA receptor subunit and BDNF.[44]

Activity-dependent production of nitric oxide may play an important role in the production of BDNF that occurs in response to exercise.[59,60] Likely of particular importance for the effects of exercise on synaptic plasticity is the stimulation of BDNF protein production locally from BDNF mRNA associated with ribosomes in dendrites.[61–63] In turn, BDNF induces the local (dendritic) translation of mRNAs encoding proteins critical for synaptic plasticity and learning and memory including Arc, *N*-methyl-D-aspartate (NMDA) receptor subunits, the postsynaptic density scaffolding protein Homer2, and CamKII.[64,65] In neurogenic niches, such as that in the dentate gyrus of the hippocampus and the subventricular zone, BDNF produced by neurons acts upon neural progenitor cells to promote their differentiation into neurons and the survival of those newly generated neurons.[13,59]

BDNF is synthesized as a longer protein called proBDNF that has little or no ability to activate TrkB. Biologically active BDNF is generated by enzymatic cleavage of proBDNF; plasmin and matrix metalloprotease 9 are two proteases that have been shown to cleave proBDNF.[66,67] Runner rats exhibit elevated activity of tissue-type plasminogen activator (tPA), an enzyme that converts plasminogen to plasmin, in the hippocampus, suggesting that exercise increases the generation of mature/active BDNF from proBDNF.[68] The latter study further showed that administration of a tPA inhibitor negates the effects of exercise on hippocampal synaptic plasticity. Another study showed that the antidepressant effect of exercise is associated with elevated levels of mature BDNF and increased expression of tPA in the hippocampus.[69]

As laboratory rodents age, decrements in performance in learning and memory tasks occur that are associated with reduced LTP and decreased expression of BDNF in the dentate gyrus. These age-related deficits in hippocampal neuroplasticity can be ameliorated by exercise and environmental enrichment.[70] In one study, the relative contributions of running and enrichment to the enhancement of neurogenesis in real world-like environments was determined by housing C57BL/6 mice under control, running, enrichment, or enrichment plus running conditions.[71] Progenitor cell proliferation and differentiation into neurons, as well as BDNF expression, were increased significantly only when running wheels were available, suggesting that exercise may be more effective than intellectual challenges alone in stimulating BDNF production and neurogenesis.

Collectively, the available data suggest that exercise enhances synaptic plasticity by increasing activity in neuronal circuits related to the exercise (sensory association cortices, entorhinal cortex, and hippocampus, as well as motor output-related circuits). Glutamate, the major excitatory neurotransmitter in all of the aforementioned neuronal circuits, activates postsynaptic α-amino-3-hydroxy-5-methyl-4-isoxazolepropionic acid (AMPA) and NMDA receptors, resulting in local Ca^{2+} influx. Ca^{2+} then binds calmodulin to activate CaMKII, resulting in the activation of nitric oxide synthase and the transcription factor cAMP response element-binding (CREB).[72] CREB then induces transcription of the *Bdnf* gene. In addition to glutamatergic signaling, activation of serotonergic and noradrenergic neurons, whose cell bodies are located in the brainstem, is necessary for a maximum effect of exercise on hippocampal BDNF production.[73] BDNF mediates effects of exercise on synaptic plasticity and neurogenesis by activating TrkB. As evidence, ablation of TrkB in hippocampal neural progenitor cells impairs their proliferation and their ability to differentiate into neurons under basal conditions and abolishes exercise-induced neurogenesis.[74] Down-

stream of TrkB, the PI3 kinase–Akt pathway plays a key role in effecting changes in synapses and neural progenitor cells. Thus, when a specific inhibitor of the PI3 kinase–Akt signaling pathway was infused into the brain, the ability of voluntary running to promote synaptic plasticity and the survival of newly generated neurons in the dentate gyrus of the hippocampus was compromised.[75]

Exercise has been shown to improve functional outcome in animal models of acute brain insults and progressive neurodegenerative disorders, and there is likely a role for BDNF signaling in the beneficial effects of exercise. For example, traumatic brain injury impairs cognitive function in rats, and postinjury exercise can enhance cognitive performance through a BDNF-mediated mechanism.[76] Neurogenesis and maze learning were suppressed in a model of systemic infection/encephalitis, and moderate running restored the proliferation and neuronal differentiation of hippocampal neural progenitor cells and also reversed the maze learning deficit.[77] In two different models of Alzheimer's disease (AD), running improved cognitive function, with notable reductions in markers of oxidative stress and inflammation in the brain.[78,79] Mice and rats that are runners exhibit increased resistance of their dopaminergic neurons to degeneration and improved motor performance, in dopaminergic toxin-induced experimental Parkinson's disease (PD).[80] In the latter study, it was found that levels of glial cell line–derived neurotrophic factor were greater in the striatum of runners compared to sedentary control animals. Exercise can also counteract adverse effects of a high-energy diet (a risk factor for stroke, and possibly AD and PD) on synaptic plasticity and cognitive function by a mechanism involving BDNF signaling.[81]

DER and BDNF

Dietary energy restriction, either by limited daily feeding (sustained caloric restriction) or intermittent energy restriction (IER) (e.g., alternate day fasting), significantly extends health span and life span in rats and mice.[82] DER can improve cognitive function and delay age-related cognitive impairment in rodents.[83,84] Limited daily feeding enhances learning consolidation and synaptic plasticity by a mechanism involving increased expression of the NR2B subunit of the NMDA receptor.[85] Recent findings have also demonstrated significant improvement in cognitive performance of nonhuman primates[86] and humans[87] maintained on caloric restriction diets. Hippocampal synapses in rats maintained on DER exhibit enhanced LTP.[88] Neurogenesis is increased in mice maintained on an alternate-day fasting diet, by a mechanism involving BDNF signaling, and increased survival of newly generated neurons.[13] However, whereas alternate day fasting increases expression of BDNF, limited daily feeding has less or no detectable effect on BDNF expression.[89] Further work is therefore required to determine how different types of DER affect BDNF signaling in different regions of the nervous system, and if and how the changes in BDNF signaling mediate behavioral and metabolic responses to DER.

DER is remarkable in its ability to protect neurons against dysfunction and degeneration in experimental models of neurological disorders. For example, neuroprotective and/or disease-modifying effects of IER have been demonstrated in animal models of severe epileptic seizures,[90] ischemic stroke,[91,92] Huntington's disease,[12,90] PD,[93,94] and AD.[95–98] Rats on a limited daily feeding (40% caloric restriction) diet exhibited recovery of spatial memory function, whereas rats fed *ad libitum* did not, in a model of cardiac arrest/global cerebral ischemia.[99] Rats maintained for four months on a 30% caloric restriction diet and then subjected to traumatic brain injury exhibited reduced neuronal degeneration and superior performance in a spatial learning test, compared to brain injured rats on a control (*ad libitum*) diet.[100] The improved outcome in the DER mice was associated with elevated levels of BDNF in the region of cerebral cortex around the site of injury and in the adjacent hippocampus. BDNF can protect neurons against oxidative, metabolic, and excitotoxic insults that are relevant to traumatic and ischemic brain injury by inducing the expression of antioxidant enzymes,[101] anti-apoptotic Bcl-2 family members,[102] and DNA repair enzymes (MPM, unpublished data).[103]

In a recent study, we maintained young, middle-aged, and old mice for three months on either IER (alternate day fasting) or *ad libitum* diets, and then subjected them to a focal ischemic stroke. The young and middle-aged mice in the IER group exhibited less brain damage and a better functional outcome, whereas IER had little or no protective effect in the old mice.[92] BDNF, the protein chaperones HSP-70

and GRP-78, and the antioxidant enzyme HO-1 were upregulated by IER in the striatum and cortex of young and middle-aged, but not old mice. Moreover, the ability of IER to suppress the production of proinflammatory cytokines was attenuated in old mice.[92] Thus, the ability of IER to enhance BDNF signaling and protect neurons against injury may be compromised with aging. The latter possibility leads to the conclusion that it is important to initiate DER (and exercise) in young adulthood or midlife to provide maximal protection of the brain against injury and disease late in life.

BDNF mediates effects of energy intake on cognitive function

The roles of BDNF in synaptic plasticity, learning, and memory have been well studied and documented.[104–107] BDNF has been shown to stimulate neurogenesis.[59,108] The results of physiological studies also demonstrate that BDNF plays an important role in LTP (see Refs. 109 and 110 for reviews).

Epidemiological studies in human populations suggest that obesity and diabetes may increase risk for developing cognitive impairments and dementia.[111–113] Multiple domains of cognitive function can be impaired in patients with diabetes.[114] Several studies have provided evidence that diabetes is a risk factor for age-related cognitive impairment and AD.[115,116] Studies with laboratory animal models have shown that overeating and obesity promote a range of major diseases including cardiovascular disease, diabetes, and many types of cancer.[117] Moreover, higher energy intake induces metabolic disorders and, together with a sedentary lifestyle, may further add risk for development of cognitive impairment.[118,119] A study conducted in rats indicated that animals maintained with food elevated in saturated fat or a diet with higher saturated fat and cholesterol committed memory errors compared to controls maintained with normal diets when they were tested in several different types of behavioral tasks.[120,121]

In a study from our lab using the triple-transgenic mouse model of AD (3xTgAD mice), the mice were maintained for one year (beginning at five months of age) on either *ad libitum* (control), 40% caloric restriction (CR), or IER (alternate-day fasting) diets. Behavioral testing of the 17-month-old 3xTgAD mice showed that those on the CR and IER diets exhibited higher levels of exploratory behavior and performed better in both the goal latency and probe trials of the water maze task compared to 3xTgAD mice on a control diet. 3xTgAD mice in the CR group also showed lower levels of Aβ1–40, Aβ1–42, and phospho-Tau in the hippocampus compared to the control diet group.[97] At least one other study has reported that long-term IER can enhance cognitive performance in rats and mice.[85]

Synaptic plasticity and BDNF

The mechanism(s) by which BDNF enables and enforces the changes in synaptic structure and function believed to underlie learning and memory are being elucidated. Although studies have shown that energy restriction affects BDNF expression in multiple brain areas, including cortical and subcortical regions, the hippocampus has received the most attention in various studies. The hippocampus has been a focus for studies of cognition-related synaptic plasticity because of the layered structural organization of the neurons and synaptic connections that provides the opportunity to readily perform electrophysiological recordings of synaptic transmission and long-term changes in synaptic strength. In addition, many of the currently well-characterized behavioral tasks used to evaluate memory, particularly for rodents, are hippocampus dependent. BDNF and LTP have received a great deal of attention in neuroscience research (see Refs. 109 and 110 for reviews). BDNF plays a critical role in activity-induced expression of proteins and in generating sustained structural and functional changes at hippocampal synapses. In particular, BDNF is sufficient to induce the transformation of early to late-phase LTP.[109] One of our recent studies showed that rats maintained on a high-fat, high-glucose diet exhibited impaired spatial learning ability, reduced hippocampal dendritic spine density, and reduced LTP at Schaffer collateral–CA1 synapses.[122] These changes were associated with reductions in levels of BDNF in the hippocampus. Therefore, a diabetogenic high-calorie diet reduces hippocampal synaptic plasticity and impairs cognitive function, possibly by impairing BDNF-mediated effects on dendritic spines. In a mouse model, we also found that excessive energy intake resulted in impaired leptin signaling and hippocampal neurogenesis, and was associated with deficits in synaptic plasticity at the perforant path—dentate granule neuron synapses

and impaired hippocampus-dependent memory.[45] Other studies reported a complementary result showing that a CR protocol prevents rats from developing age-related deficits in LTP and sustains NMDA and AMPA glutamate receptor levels in the hippocampus.[88,123]

BDNF and cellular energy metabolism

An additional mechanism by which BDNF may mediate beneficial effects of DER on cognition involves enhancing cellular energy metabolism.[124] In a recent study, it was shown that when 3xTgAD mice were fed a diet supplemented with 2-deoxyglucose (2DG) for seven weeks, levels of Aβ in their brains decreased, levels of BDNF increased, and neuronal bioenergetic capacity was increased.[125] Mice on the 2DG diet also exhibited elevated levels of ketone bodies, which are fatty acids known to affect neuronal excitability and to protect neurons against excitotoxic and metabolic insults.[126] BDNF has been shown to directly modify neuronal energy metabolism. For example, BDNF induces the expression of the monocarboxylate transporter that enables the use of lactate as an alternative energy source.[127] The ability of BDNF to induce the monocarboxylate transporter was blocked by inhibitors of PI3 kinase and p42/p44 MAP kinases. Another study showed that BDNF can stimulate glucose utilization and the upregulation of the glucose transporter, GLUT3 in cultured neurons.[128] BDNF also stimulates sodium-dependent amino acid transport and increases protein synthesis. The abilities of BDNF to increase cellular energy availability and enhance protein synthesis may play important roles in synaptic plasticity because energy substrates (ATP and NAD$^+$) and protein synthesis are critical for synaptic plasticity and learning and memory.[129,130]

The big picture: BDNF as an integrator of behavioral and neuroendocrine regulation of energy metabolism

Consider mammals, including our primate predecessors and ourselves, humans, in the natural environments in which they evolved. Further, consider two very different ends of the energy equation with regards to the availability/accessibility of energy (food) resources within the environment (Fig. 4). In one scenario (challenging environment), food is scarce, competition within and among species for the energy is high, and hazards such as

Challenging Environment
Energy (food) is scarce
Intra-species competition is high
Hazards (predators, climate, etc.)

Unchallenging Environment
Energy (food) is abundant
Competition is low
Hazards are few

Survival Advantages
Cognitive abilities
Agility and strength
Endurance
Energy conservation

Survival Advantages
Avoidance of hazards
Reproduction
Increased 'work' time

Mechanisms and Consequences
Neuronal activity
Neurotrophic signaling
Cognitive fitness
Insulin sensitivity
Cardiovascular fitness
Resistance to injury and disease

Mechanisms and Consequences
Neuronal inactivity
Reduced neurotrophic signaling
Cognitive deficits
Metabolic morbidity
Cardiovascular deconditioning
Vulnerability to a range of diseases

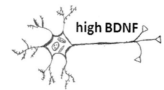
high BDNF

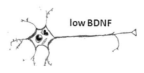
low BDNF

Figure 4. Characteristics of individuals living in natural or self-imposed environments at the two ends of the energy spectrum. (Left) An environment where food is scarce and competition and hazards are high poses major physical and cognitive challenges. From an evolutionary perspective, survival is favored by phenotypes that include superior cognitive abilities, strength and agility, endurance, and the ability to conserve energy. In this setting, neurons are active, and the production of neurotrophic factors such as BDNF are elevated. Emerging evidence suggests that BDNF mediates multiple responses of the individual to a Spartan environment, including enhanced neuronal plasticity, increased insulin sensitivity, improved cardiovascular fitness, and resistance of tissues to injury and disease. The different BDNF-mediated responses to a challenging environment can be considered a coordinated, adaptive cellular- and systems-level response to the reduced energy availability and increased energy expenditure required to survive. (Right) In an environment where food is abundant and the need for energy expenditure (exercise) minimal, individuals can increase their attention to reproduction and their occupation within the society. This environment results in reduced activation of adaptive stress response pathways, and thereby reduced BDNF signaling, insulin resistance, cardiovascular deconditioning, and susceptibility to a range of diseases, particularly age-related diseases.

predation also enter into the energy equation on both the acquisition and expenditure sides. Such challenging environments demand relatively higher levels of processing and retention (memory) of information from the environment (particularly sights, sounds, and odors). The underlying activity in neuronal circuits involves BDNF production and signaling, which promotes synaptic plasticity, the growth of dendrites, and neurogenesis. Individuals in physically challenging environments are at increased risk of injury, including damage to the nervous system. BDNF may provide a survival advantage to those suffering trauma to the nervous system by stimulating mitochondrial biogenesis and resistance of neurons to injury by upregulating the expression of genes encoding cytoprotective proteins. Recent findings suggest that BDNF signaling

in the brain mediates the positive effects of a cognitively and physically challenging environment on energy metabolism including increased insulin sensitivity,[17] increased genesis of brown fat,[131,132] and cardiovascular fitness.[133]

In the second scenario (unchallenging environment), energy is readily available and there are few hazards to obtaining the energy (Fig. 4). The time that would otherwise be used for food acquisition is used for reproduction and work to obtain other types of resources that provide an advantage to the individuals and their families. Because energy intake is increased and physical activity reduced in such an environment, BDNF signaling is reduced in neurons in the brain. Reduced BDNF signaling enhances food intake, reduces insulin sensitivity, deconditions the cardiovascular system, reduces cognitive abilities, and increases vulnerability to age-related diseases.

In summary, individuals will be lean, alert, and cognitively sharp when living in an environment in which there are major challenges for energy acquisition. This same (healthy) physical and cognitive phenotype can be achieved by self-imposed energy restriction and exercise routines in humans. BDNF signaling plays major roles in mediating adaptive responses of the nervous, cardiovascular, and energy-regulating organ systems in response to exercise and energy restriction. Efforts to develop pharmacological and molecular genetic methods for activating BDNF-signaling pathways may result in novel therapeutic treatments for a range of metabolic and neurological disorders.

Acknowledgment

This work was supported by the Intramural Research Program of the National Institute on Aging.

Conflicts of interest

The authors declare no conflicts of interest.

References

1. Nawa, H., J. Carnahan & C. Gall. 1995. BDNF protein measured by a novel enzyme immunoassay in normal brain and after seizure: partial disagreement with mRNA levels. *Eur. J. Neurosci.* **7:** 1527–1535.
2. Duan, W., Z. Guo, H. Jiang, *et al.* 2003b. Reversal of behavioral and metabolic abnormalities, and insulin resistance syndrome, by dietary restriction in mice deficient in brain-derived neurotrophic factor. *Endocrinol.* **144:** 2446–2453.
3. Kernie, S.G., D.J. Liebl & L.F. Parada. 2000. BDNF regulates eating behavior and locomotor activity in mice. *EMBO J.* **19:** 1290–1300.
4. Coppola, V. & L. Tessarollo. 2004. Control of hyperphagia prevents obesity in BDNF heterozygous mice. *Neuroreport* **15:** 2665–2668.
5. Xu, B., E.H. Goulding, K. Zang, *et al.* 2003. Brain-derived neurotrophic factor regulates energy balance downstream of melanocortin-4 receptor. *Nat. Neurosci.* **6:** 736–742.
6. Rios, M., G. Fan, C. Fekete, *et al.* 2001. Conditional deletion of brain-derived neurotrophic factor in the postnatal brain leads to obesity and hyperactivity. *Mol. Endocrinol.* **15:** 1748–1757.
7. Pelleymounter, M.A., M.J. Cullen & C.L. Wellman. 1995. Characteristics of BDNF-induced weight loss. *Exp. Neurol.* **131:** 229–238.
8. Shimizu, N., Y. Oomura, C.R. Plata-Salamán & M. Morimoto. 1987. Hyperphagia and obesity in rats with bilateral ibotenic acid-induced lesions of the ventromedial hypothalamic nucleus. *Brain Res.* **416:** 153–156.
9. Grill, H.J. & J.M. Kaplan. 2002. The neuroanatomical axis for control of energy balance. *Front. Neuroendocrinol.* **23:** 2–40.
10. Bariohay, B., B. Lebrun, E. Moyse & A. Jean. 2005. Brain-derived neurotrophic factor plays a role as an anorexigenic factor in the dorsal vagal complex. *Endocrinol.* **146:** 5612–5620.
11. Lee, J., W. Duan, J.M. Long, *et al.* 2000. Dietary restriction increases the number of newly generated neural cells, and induces BDNF expression, in the dentate gyrus of rats. *J. Mol. Neurosci.* **15:** 99–108.
12. Duan, W., Z. Guo, H. Jiang, *et al.* 2003a. Dietary restriction normalizes glucose metabolism and BDNF levels, slows disease progression, and increases survival in huntingtin mutant mice. *Proc. Natl. Acad. Sci. USA* **100:** 2911–2916.
13. Lee, J., W. Duan & M.P. Mattson. 2002. Evidence that brain-derived neurotrophic factor is required for basal neurogenesis and mediates, in part, the enhancement of neurogenesis by dietary restriction in the hippocampus of adult mice. *J. Neurochem.* **82:** 1367–1375.
14. Krabbe, K.S., A.R. Nielsen, R. Krogh-Madsen, *et al.* 2007. Brain-derived neurotrophic factor (BDNF) and type 2 diabetes. *Diabetologia.* **50:** 431–438.
15. Pedersen, B.K., M. Pedersen, K.S. Krabbe, *et al.* 2009. Role of exercise-induced brain-derived neurotrophic factor production in the regulation of energy homeostasis in mammals. *Exp. Physiol.* **94:** 1153–1160.
16. Nakagawa, T., M. Ono-Kishino, E. Sugaru, *et al.* 2002. Brain-derived neurotrophic factor (BDNF) regulates glucose and energy metabolism in diabetic mice. *Diabetes Metab. Res. Rev.* **18:** 185–191.
17. Nakagawa, T., A. Tsuchida, Y. Itakura, *et al.* 2000. Brain-derived neurotrophic factor regulates glucose metabolism by modulating energy balance in diabetic mice. *Diabetes* **49:** 436–444.
18. Spiegelman, B.M. & J.S. Flier. 1996. Adipogenesis and obesity: rounding out the big picture. *Cell* **87:** 377–389
19. Ono, M., Y. Itakura, T. Nonomura, *et al.* 2000. Intermittent administration of brain-derived neurotrophic factor ameliorates glucose metabolism in obese diabetic mice. *Metabolism* **49:** 129–133.
20. Stranahan, A.M., T.V. Arumugam & M.P. Mattson. 2011. Lowering corticosterone levels reinstates hippocampal brain-derived neurotropic factor and Trkb expression without influencing deficits in hypothalamic brain-derived neurotropic factor expression in leptin receptor-deficient mice. *Neuroendocrinol.* **93:** 58–64.
21. Brady, R., S.I. Zaidi, C. Mayer & D.M. Katz. 1999. BDNF is a target-derived survival factor for arterial baroreceptor and chemoafferent primary sensory neurons. *J. Neurosci.* **19:** 2131–2142.
22. Lee, J., K.B. Seroogy & M.P. Mattson. 2002b. Dietary restriction enhances neurotrophin expression and neurogenesis in the hippocampus of adult mice. *J. Neurochem.* **80:** 539–547.
23. Neeper, S.A., F. Gomez-Pinilla, J. Choi & C.W. Cotman. 1996. Physical activity increases mRNA for brain-derived neurotrophic factor and nerve growth factor in rat brain. *Brain Res.* **726:** 49–56.
24. Rosenwinkel, E.T., D.M. Bloomfield, M.A. Arwady & R.L. Goldsmith. 2001. Exercise and autonomic function in health and cardiovascular disease. *Cardiol. Clin.* **19:** 369–387.

25. Wan, R., S. Camandola & M.P. Mattson. 2003a. Intermittent fasting and dietary supplementation with 2-deoxy-D-glucose improve functional and metabolic cardiovascular risk factors in rats. *FASEB J.* **17:** 1133–1134.

26. Wan, R., S. Camandola & M.P. Mattson. 2003b. Intermittent food deprivation improves cardiovascular and neuroendocrine responses to stress in rats. *J. Nutr.* **133:** 1921–1929.

27. Yang, B., J.D. Slonimsky & S.J. Birren. 2002. A rapid switch in sympathetic neurotransmitter release properties mediated by the p75 receptor. *Nat. Neurosci.* **5:** 539–545.

28. Ahmet, I., R. Wan, M.P. Mattson, *et al.* 2005. Cardioprotection by intermittent fasting in rats. *Circulation* **112:** 3115–3121.

29. Yang, A.C., T.J. Chen, S.J. Tsai, *et al.* 2010. BDNF Val66Met polymorphism alters sympathovagal balance in healthy subjects. *Am. J. Med. Genet. B. Neuropsychiatr. Genet.* **153B:** 1024–1030.

30. Alexander, N., R. Osinsky, A. Schmitz, *et al.* 2010. The BDNF Val66Met polymorphism affects HPA-axis reactivity to acute stress. *Psychoneuroendocrinology* **35:** 949–953.

31. Martin, J.L., V.K. Jenkins, H.Y. Hsieh & A. Balkowiec. 2009. Brain-derived neurotrophic factor in arterial baroreceptor pathways: implications for activity-dependent plasticity at baroafferent synapses. *J. Neurochem.* **108:** 450–464.

32. Hofer, M., S.R. Pagliusi, A. Hohn, *et al.* 1990. Regional distribution of brain-derived neurotrophic factor mRNA in the adult mouse brain. *EMBO J.* **9:** 2459–2464.

33. Peiris, T.S., R. Machaalani & K.A. Waters. 2004. Brain-derived neurotrophic factor mRNA and protein in the piglet brainstem and effects of intermittent hypercapnic hypoxia. *Brain Res.* **1029:** 11–23.

34. Schober, A., N. Wolf, K. Huber, *et al.* 1998. TrkB and neurotrophin-4 are important for development and maintenance of sympathetic preganglionic neurons innervating the adrenal medulla. *J. Neurosci.* **18:** 7272–7284.

35. Yan, Q., M.J. Radeke, C.R. Matheson, *et al.* 1997. Immunocytochemical localization of TrkB in the central nervous system of the adult rat. *J. Comp. Neurol.* **378:** 135–157.

36. Altar, C.A., N. Cai, T. Bliven, *et al.* 1997. Anterograde transport of brain-derived neurotrophic factor and its role in the brain. *Nature* **389:** 856–860.

37. Helke, C.J., K.M. Adryan, J. Fedorowicz, *et al.* 1998. Axonal transport of neurotrophins by visceral afferent and efferent neurons of the vagus nerve of the rat. *J. Comp. Neurol.* **393:** 102–117.

38. Wang, H. & X.F. Zhou. 2002. Injection of brain-derived neurotrophic factor in the rostral ventrolateral medulla increases arterial blood pressure in anaesthetized rats. *Neuroscience* **112:** 967–975.

39. Nicholson, J.R., J.C. Peter, A.C. Lecourt, *et al.* 2007. Melanocortin-4 receptor activation stimulates hypothalamic brain-derived neurotrophic factor release to regulate food intake, body temperature and cardiovascular function. *J. Neuroendocrinol.* **19:** 974–982.

40. Andresen, M.C. & D.L. Kunze. 1994. Nucleus tractus solitarius–gateway to neural circulatory control. *Ann. Rev. Physiol.* **56:** 93–116.

41. Clark, C.G., E.M. Hasser, D.L. Kunze, *et al.* 2011. Endogenous brain-derived neurotrophic factor in the nucleus tractus solitarius tonically regulates synaptic and autonomic function. *J. Neurosci.* **31:** 12318–12329.

42. Cotman, C.W., N.C. Berchtold & L.A. Christie. 2007. Exercise builds brain health: key roles of growth factor cascades and inflammation. *Trends Neurosci.* **30:** 464–472.

43. van Praag, H., G. Kempermann & F.H. Gage. 1999. Running increases cell proliferation and neurogenesis in the adult mouse dentate gyrus. *Nat. Neurosci.* **2:** 266–270.

44. Farmer, J., X. Zhao, H. van Praag, *et al.* 2004. Effects of voluntary exercise on synaptic plasticity and gene expression in the dentate gyrus of adult male Sprague-Dawley rats in vivo. *Neuroscience* **124:** 71–79.

45. Stranahan, A.M., T.V. Arumugam, R.G. Cutler, *et al.* 2008. Diabetes impairs hippocampal function through glucocorticoid-mediated effects on new and mature neurons. *Nat. Neurosci.* **11:** 309–317.

46. Bruestle, D.A., R.G. Cutler, R.S. Telljohann & M.P. Mattson. 2009. Decline in daily running distance presages disease onset in a mouse model of ALS. *Neuromol. Med.* **11:** 58–62.

47. Knab, A.M., R.S. Bowen, T. Moore-Harrison, *et al.* 2009. Repeatability of exercise behaviors in mice. *Physiol. Behav.* **98:** 433–440.

48. Groves-Chapman, J.L., P.S. Murray, K.L. Stevens, *et al.* 2011. Changes in mRNA levels for brain-derived neurotrophic factor after wheel running in rats selectively bred for high- and low-aerobic capacity. *Brain Res.* **1425:** 90–97.

49. Mattson, M.P., W. Duan & Z. Guo. 2003. Meal size and frequency affect neuronal plasticity and vulnerability to disease: cellular and molecular mechanisms. *J. Neurochem.* **84:** 417–431.

50. Martin, B., M.P. Mattson & S. Maudsley. 2006. Caloric restriction and intermittent fasting: two potential diets for successful brain aging. *Ageing Res. Rev.* **5:** 332–353.

51. Berchtold, N.C., N. Castello & C.W. Cotman. 2010. Exercise and time-dependent benefits to learning and memory. *Neuroscience* **167:** 588–597.

52. Berchtold, N.C., G. Chinn, M. Chou, *et al.* 2005. Exercise primes a molecular memory for brain-derived neurotrophic factor protein induction in the rat hippocampus. *Neuroscience* **133:** 853–861.

53. Creer, D.J., C. Romberg, L.M. Saksida, *et al.* 2010. Running enhances spatial pattern separation in mice. *Proc. Natl. Acad. Sci. USA* **107:** 2367–2372.

54. Erickson, K.I., M.W. Voss, R.S. Prakash, *et al.* 2011. Exercise training increases size of hippocampus and improves memory. *Proc. Natl. Acad. Sci. USA* **108:** 3017–3022.

55. Liu, Y.F., H.I. Chen, C.L. Wu, *et al.* 2009. Differential effects of treadmill running and wheel running on spatial or aversive learning and memory: roles of amygdalar brain-derived neurotrophic factor and synaptotagmin I. *J. Physiol.* **587:** 3221–3231.

56. Lin, T.W., S.J. Chen, T.Y. Huang, *et al.* 2011. Different types of exercise induce differential effects on neuronal adaptations and memory performance. *Neurobiol. Learn. Mem.* **97:** 140–147.

57. Russo-Neustadt, A.A., R.C. Beard, Y.M. Huang & C.W. Cotman. 2000. Physical activity and antidepressant

treatment potentiate the expression of specific brain-derived neurotrophic factor transcripts in the rat hippocampus. *Neuroscience* **101**: 305–312.

58. Adlard, P.A., V.M. Perreau, C. Engesser-Cesar & C.W. Cotman. 2004. The timecourse of induction of brain-derived neurotrophic factor mRNA and protein in the rat hippocampus following voluntary exercise. *Neurosci. Lett.* **363**: 43–48.

59. Cheng, A., S. Wang, J. Cai, *et al.* 2003. Nitric oxide acts in a positive feedback loop with BDNF to regulate neural progenitor cell proliferation and differentiation in the mammalian brain. *Dev. Biol.* **258**: 319–333.

60. Chen, M.J., A.S. Ivy & A.A. Russo-Neustadt. 2006. Nitric oxide synthesis is required for exercise-induced increases in hippocampal BDNF and phosphatidylinositol 3' kinase expression. *Brain Res. Bull.* **68**: 257–268.

61. Jourdi, H., Y.T. Hsu, M. Zhou, *et al.* 2009. Positive AMPA receptor modulation rapidly stimulates BDNF release and increases dendritic mRNA translation. *J. Neurosci.* **29**: 8688–8697.

62. Stranahan, A.M., K. Lee, B. Martin, *et al.* 2009. Voluntary exercise and caloric restriction enhance hippocampal dendritic spine density and BDNF levels in diabetic mice. *Hippocampus* **19**: 951–961.

63. Wu, Y.C., R. Williamson, Z. Li, *et al.* 2011. Dendritic trafficking of brain-derived neurotrophic factor mRNA: regulation by translin-dependent and -independent mechanisms. *J. Neurochem.* **116**: 1112–1121.

64. Yin, Y., G.M. Edelman & P.W. Vanderklish. 2002. The brain-derived neurotrophic factor enhances synthesis of Arc in synaptoneurosomes. *Proc. Natl. Acad. Sci. USA* **99**: 2368–2373.

65. Schratt, G.M., E.A. Nigh, W.G. Chen, *et al.* 2004. BDNF regulates the translation of a select group of mRNAs by a mammalian target of rapamycin-phosphatidylinositol 3-kinase-dependent pathway during neuronal development. *J. Neurosci.* **24**: 7366–7377.

66. Pang, P.T., H.K. Teng, E. Zaitsev, *et al.* 2004. Cleavage of proBDNF by tPA/plasmin is essential for long-term hippocampal plasticity. *Science* **306**: 487–491.

67. Mizoguchi, H., J. Nakade, M. Tachibana, *et al.* 2011. Matrix metalloproteinase-9 contributes to kindled seizure development in pentylenetetrazole-treated mice by converting pro-BDNF to mature BDNF in the hippocampus. *J. Neurosci.* **31**: 12963–12971.

68. Ding, Q., Z. Ying & F. Gómez-Pinilla. 2011. Exercise influences hippocampal plasticity by modulating brain-derived neurotrophic factor processing. *Neuroscience* **192**: 773–780.

69. Sartori, C.R., A.S. Vieira, E.M. Ferrari, *et al.* 2011. The antidepressive effect of the physical exercise correlates with increased levels of mature BDNF, and proBDNF proteolytic cleavage-related genes, p11 and tPA. *Neuroscience* **180**: 9–18.

70. O'Callaghan, R.M., E.W. Griffin & A.M. Kelly. 2009. Long-term treadmill exposure protects against age-related neurodegenerative change in the rat hippocampus. *Hippocampus* **19**: 1019–1029.

71. Kobilo, T., Q.R. Liu, K. Gandhi, *et al.* 2011. Running is the neurogenic and neurotrophic stimulus in environmental enrichment. *Learn Mem.* **18**: 605–609.

72. Chen, M.J. & A.A. Russo-Neustadt. 2009. Running exercise-induced up-regulation of hippocampal brain-derived neurotrophic factor is CREB-dependent. *Hippocampus* **19**: 962–972.

73. Ivy, A.S., F.G. Rodriguez, C. Garcia, *et al.* 2003. Noradrenergic and serotonergic blockade inhibits BDNF mRNA activation following exercise and antidepressant. *Pharmacol. Biochem. Behav.* **75**: 81–88.

74. Li, Y., B.W. Luikart, S. Birnbaum, *et al.* 2008. TrkB regulates hippocampal neurogenesis and governs sensitivity to antidepressive treatment. *Neuron* **59**: 399–412.

75. Bruel-Jungerman, E., A. Veyrac, F. Dufour, *et al.* 2009. Inhibition of PI3K-Akt signaling blocks exercise-mediated enhancement of adult neurogenesis and synaptic plasticity in the dentate gyrus. *PLoS One.* **4**: e7901.

76. Griesbach, G.S., D.A. Hovda & F. Gomez-Pinilla. 2009. Exercise-induced improvement in cognitive performance after traumatic brain injury in rats is dependent on BDNF activation. *Brain Res.* **1288**: 105–115.

77. Wu, C.W., Y.C. Chen, L. Yu, *et al.* 2007. Treadmill exercise counteracts the suppressive effects of peripheral lipopolysaccharide on hippocampal neurogenesis and learning and memory. *J. Neurochem.* **103**: 2471–2481.

78. Parachikova, A., K.E. Nichol & C.W. Cotman. 2008. Short-term exercise in aged Tg2576 mice alters neuroinflammation and improves cognition. *Neurobiol. Dis.* **30**: 121–129.

79. García-Mesa, Y., J.C. López-Ramos, L. Giménez-Llort, *et al.* 2011. Physical exercise protects against Alzheimer's disease in 3xTg-AD mice. *J. Alzheimers Dis.* **24**: 421–454.

80. Zigmond, M.J., J.L. Cameron, R.K. Leak, *et al.* 2009. Triggering endogenous neuroprotective processes through exercise in models of dopamine deficiency. *Parkinsonism Relat. Disord.* **15**(Suppl. 3): S42–45.

81. Molteni, R., A. Wu, S. Vaynman, *et al.* 2004. Exercise reverses the harmful effects of consumption of a high-fat diet on synaptic and behavioral plasticity associated to the action of brain-derived neurotrophic factor. *Neuroscience* **123**: 429–440.

82. Masoro, E.J. 2005. Overview of caloric restriction and ageing. *Mech. Ageing Dev.* **126**: 913–922.

83. Means, L.W., J.L. Higgins & T.J. Fernandez. 1993. Mid-life onset of dietary restriction extends life and prolongs cognitive functioning. *Physiol. Behav.* **54**: 503–508.

84. Komatsu, T., T. Chiba, H. Yamaza, *et al.* 2008. Manipulation of caloric content but not diet composition, attenuates the deficit in learning and memory of senescence-accelerated mouse strain P8. *Exp. Gerontol.* **43**: 339–346.

85. Fontán-Lozano, A., J.L. Sáez-Cassanelli, M.C. Inda, *et al.* 2007. Caloric restriction increases learning consolidation and facilitates synaptic plasticity through mechanisms dependent on NR2B subunits of the NMDA receptor. *J. Neurosci.* **27**: 10185–10195.

86. Dal-Pan, A., F. Pifferi, J. Marchal, *et al.* 2011. Cognitive performances are selectively enhanced during chronic caloric restriction or resveratrol supplementation in a primate. *PLoS One* **6**: e16581.

87. Witte, A.V., M. Fobker, R. Gellner, *et al.* 2009. Caloric restriction improves memory in elderly humans. *Proc. Natl. Acad. Sci. USA* **106**: 1255–1260.

88. Eckles-Smith, K., D. Clayton, P. Bickford & M.D. Browning. 2000. Caloric restriction prevents age-related deficits in LTP and in NMDA receptor expression. *Mol. Brain Res.* **78:** 154–162.

89. Newton, I.G., M.E. Forbes, C. Legault, *et al.* 2005. Caloric restriction does not reverse aging-related changes in hippocampal BDNF. *Neurobiol. Aging* **26:** 683–688.

90. Bruce-Keller, A.J., G. Umberger, R. McFall & M.P. Mattson. 1999. Food restriction reduces brain damage and improves behavioral outcome following excitotoxic and metabolic insults. *Ann. Neurol.* **45:** 8–15.

91. Yu, Z.F. & M.P. Mattson. 1999. Dietary restriction and 2-deoxyglucose administration reduce focal ischemic brain damage and improve behavioral outcome: evidence for a preconditioning mechanism. *J. Neurosci. Res.* **57:** 830–839.

92. Arumugam, T.V., T.M. Phillips, A. Cheng, *et al.* 2010. Age and energy intake interact to modify cell stress pathways and stroke outcome. *Ann. Neurol.* **67:** 41–52.

93. Duan, W. & M.P. Mattson. 1999. Dietary restriction and 2-deoxyglucose administration improve behavioral outcome and reduce degeneration of dopaminergic neurons in models of Parkinson's disease. *J. Neurosci. Res.* **57:** 195–206.

94. Maswood, N., J. Young, E. Tilmont, *et al.* 2004. Caloric restriction increases neurotrophic factor levels and attenuates neurochemical and behavioral deficits in a primate model of Parkinson's disease. *Proc. Natl. Acad. Sci. USA* **101:** 18171–18176.

95. Patel, N.V., M.N. Gordon, K.E. Connor, *et al.* 2005. Caloric restriction attenuates Abeta-deposition in Alzheimer transgenic models. *Neurobiol. Aging* **26:** 995–1000.

96. Qin, W., M. Chachich, M. Lane, *et al.* 2006. Calorie restriction attenuates Alzheimer's disease type brain amyloidosis in Squirrel monkeys (Saimiri sciureus). *J. Alzheimers Dis.* **10:** 417–422.

97. Halagappa, V.K., Z. Guo, M. Pearson, *et al.* 2007. Intermittent fasting and caloric restriction ameliorate age-related behavioral deficits in the triple-transgenic mouse model of Alzheimer's disease. *Neurobiol. Dis.* **26:** 212–220.

98. Wu, P., Q. Shen, S. Dong, *et al.* 2008. Calorie restriction ameliorates neurodegenerative phenotypes in forebrain-specific presenilin-1 and presenilin-2 double knockout mice. *Neurobiol. Aging* **29:** 1502–1511.

99. Roberge, M.C., C. Messier, W.A. Staines & H. Plamondon. 2008. Food restriction induces long-lasting recovery of spatial memory deficits following global ischemia in delayed matching and non-matching-to-sample radial arm maze tasks. *Neuroscience* **156:** 11–29.

100. Rich, N.J., J.W. Van Landingham, S. Figueiroa, *et al.* 2010. Chronic caloric restriction reduces tissue damage and improves spatial memory in a rat model of traumatic brain injury. *J. Neurosci. Res.* **88:** 2933–2939.

101. Mattson, M.P., M.A. Lovell, K. Furukawa & W.R. Markesbery. 1995. Neurotrophic factors attenuate glutamate-induced accumulation of peroxides, elevation of intracellular Ca2 +concentration, and neurotoxicity and increase antioxidant enzyme activities in hippocampal neurons. *J. Neurochem.* **65:** 1740–1751.

102. Almeida, R.D., B.J. Manadas, C.V. Melo, *et al.* 2005. Neuroprotection by BDNF against glutamate-induced apoptotic cell death is mediated by ERK and PI3-kinase pathways. *Cell Death Differ.* **12:** 1329–1343.

103. Yang, J.L., T. Tadokoro, G. Keijzers, *et al.* 2010. Neurons efficiently repair glutamate-induced oxidative DNA damage by a process involving CREB-mediated up-regulation of apurinic endonuclease 1. *J. Biol. Chem.* **285:** 28191–28199.

104. Lu, B. & A. Figurov. 1997. Role of neurotrophins in synapse development and plasticity. *Rev. Neurosci.* **8:** 1–12.

105. Kuczewski, N., C. Porcher, V. Lessmann, *et al.* 2009. Activity-dependent dendritic release of BDNF and biological consequences. *Mol. Neurobiol.* **39:** 37–49.

106. Cowansage, K.K., J.E. LeDoux & M-H. Monfils. 2010. Brain-derived neurotrophic factor: a dynamic gatekeeper of neural plasticity. *Curr. Mol. Pharmacol.* **3:** 12–29.

107. Mattson, M.P. & R.Q. Wan. 2008. Neurotrophic factors in autonomic nervous system plasticity and dysfunction. *Neuromol. Med.* **10:** 157–168.

108. Schmidt, H.D. & R.S. Duman. 2007. The role of neurotrophic factors in adult hippocampal neurogenesis, antidepressant treatments and animal models of depressive-like behavior. *Behav. Pharmacol.* **18:** 391–418.

109. Lu, Y., K. Christian & B. Lu. 2008. BDNF: a key regulator for protein synthesis-dependent LTP and long-term memory? *Neurobiol. Learn Mem.* **89:** 312–323.

110. Minichiello, L. 2009. TrkB signalling pathways in LTP and learning. *Nat. Rev. Neurosci.* **10:** 850–860.

111. Castillo-Quan, J.I., D.J. Barrera-Buenfil, J.M. Pérez-Osorio & F.J. Alvarez-Cervera. 2010. Depression and diabetes: from epidemiology to neurobiology. *Rev. Neurol.* **51:** 347–59.

112. Nilsson, L.G. & E. Nilsson. 2009. Overweight and cognition. *Scand. J. Psychol.* **50:** 660–667.

113. Sabia, S., M. Kivimaki, M.J. Shipley, *et al.* 2009. Body mass index over the adult life course and cognation in late midlife: the Whitehall II Cohort Study. *Am. J. Clin. Nutr.* **89:** 601–607.

114. Kodl, C. & E.R. Seaquist. 2008. Cognitive dysfunction and diabetes mellitus. *Endocr. Rev.* **29:** 494–511.

115. Hölscher, C. 2011. Diabetes as a risk factor for Alzheimer's disease: insulin signalling impairment in the brain as an alternative model of Alzheimer's disease. *Biochem. Soc. Trans.* **39:** 891–897.

116. Kopf, D. & L. Frölich. 2009. Risk of incident Alzheimer's disease in diabetic patients: a systematic review of prospective trials. *J. Alzheimers Dis.* **16:** 677–685.

117. Muoio, D. M. & C.B. Newgard. 2006. Obesity-related derangements in metabolic regulation. *Ann. Rev. Biochem.* **75:** 367–401.

118. Qiu, C., D. De Ronchi & L. Fratiglioni. 2007. The epidemiology of the dementias: an update. *Curr. Opin. Psychiat.* **20:** 380–385.

119. Stranahan, A. M. & M.P. Mattson. 2008. Impact of energy intake and expenditure on neuronal plasticity. *Neuromol. Med.* **10:** 209–218.

120. Greenwood, C. E. & G. Winocur. 1996. Cognitive impairment in rats fed high-fat diets: a specific effect of saturated fatty-acid intake. *Behav. Neurosci.* **110:** 451–459.

121. Granholm, A. C., H.A. Bimonte-Nelson, A.B. Moore, *et al.* 2008. Effects of a saturated fat and high cholesterol diet on memory and hippocampal morphology in the middle-aged rat. *J. Alzheimers Dis.* **14:** 133–145.

122. Stranahan, A. M., E.D. Norman, K. Lee, *et al.* 2008a. Diet-induced insulin resistance impairs hippocampal synaptic plasticity and cognition in middle-aged rats. *Hippocampus* **18:** 1085–1088.

123. Shi, L., M.M. Adams, M.C. Linville, *et al.* 2007. Caloric restriction eliminates the aging-related decline in NMDA and AMPA receptor subunits in the rat hippocampus and induces homeostasis. *Exp. Neurol.* **206:** 70–79.

124. Gomez-Pinilla, F., S. Vaynman & Z. Ying. 2008. Brain-derived neurotrophic factor functions as a metabotrophin to mediate the effects of exercise on cognition. *Eur. J. Neurosci.* **28:** 2278–2287.

125. Yao, J., S. Chen, Z. Mao, *et al.* 2011. 2-Deoxy-D-glucose treatment induces ketogenesis, sustains mitochondrial function, and reduces pathology in female mouse model of Alzheimer's disease. *PLoS One* **6:** e21788. [Epub 2011 Jul 1].

126. Maalouf, M., J.M. Rho & M.P. Mattson. 2009. The neuroprotective properties of calorie restriction, the ketogenic diet, and ketone bodies. *Brain Res. Rev.* **59:** 293–315.

127. Robinet, C. & L. Pellerin. 2010. Brain-derived neurotrophic factor enhances the expression of the monocarboxylate transporter 2 through translational activation in mouse cultured cortical neurons. *J. Cereb. Blood Flow Metab.* **30:** 286–298.

128. Burkhalter, J., H. Fiumelli, I. Allaman, *et al.* 2003. Brain-derived neurotrophic factor stimulates energy metabolism in developing cortical neurons. *J. Neurosci.* **23:** 8212–8220.

129. Zhang, J., Y. Wang, Z. Chi, *et al.* 2011. The AAA+ AT-Pase Thorase regulates AMPA receptor-dependent synaptic plasticity and behavior. *Cell* **145:** 284–299.

130. Abraham, W.C. & J.M. Williams. 2008. LTP maintenance and its protein synthesis-dependence. *Neurobiol. Learn Mem.* **89:** 260–268.

131. Wang, C., E. Bomberg, C. Billington, *et al.* 2007. Brain-derived neurotrophic factor in the hypothalamic paraventricular nucleus increases energy expenditure by elevating metabolic rate. *Am. J. Physiol. Regul. Integr. Comp. Physiol.* **293:** R992–1002.

132. Cao, L., E.Y. Choi, X. Liu, *et al.* 2011. White to brown fat phenotypic switch induced by genetic and environmental activation of a hypothalamic-adipocyte axis. *Cell Metab.* **14:** 324–338.

133. Griffioen, K.J., R. Wan, T.R. Brown, *et al.* 2011. Aberrant heart rate and brainstem BDNF signaling in a mouse model of Huntington's disease. *Neurobiol. Aging.* In press.

Ann. N.Y. Acad. Sci. ISSN 0077-8923

ANNALS OF THE NEW YORK ACADEMY OF SCIENCES
Issue: *The Brain and Obesity*

Leptin action on nonneuronal cells in the CNS: potential clinical applications

Weihong Pan,[1] Hung Hsuchou,[1] Bhavaani Jayaram,[1] Reas S. Khan,[1] Eagle Yi-Kung Huang,[2] Xiaojun Wu,[1] Chu Chen,[3] and Abba J. Kastin[1]

[1]Blood-Brain Barrier Group, Pennington Biomedical Research Center, Baton Rouge, Lousiana. [2]Department of Pharmacology, National Defense Medical Center, Taipei, Taiwan, Republic of China. [3]Neuroscience Center of Excellence, Louisiana State University Health Science Center, New Orleans, Lousiana

Address for correspondence: Weihong Pan, M.D., Ph.D., Blood-Brain Barrier Group, Pennington Biomedical Research Center, 6400 Perkins Road, Baton Rouge, LA 70808. panw@pbrc.edu

Leptin, an adipocyte-derived cytokine, crosses the blood–brain barrier to act on many regions of the central nervous system (CNS). It participates in the regulation of energy balance, inflammatory processes, immune regulation, synaptic formation, memory condensation, and neurotrophic activities. This review focuses on the newly identified actions of leptin on astrocytes. We first summarize the distribution of leptin receptors in the brain, with a focus on the hypothalamus, where the leptin receptor is known to mediate essential feeding suppression activities, and on the hippocampus, where leptin facilitates memory, reduces neurodegeneration, and plays a dual role in seizures. We will then discuss regulation of the nonneuronal leptin system in obesity. Its relationship with neuronal leptin signaling is illustrated by *in vitro* assays in primary astrocyte culture and by *in vivo* studies on mice after pretreatment with a glial metabolic inhibitor or after cell-specific deletion of intracellular signaling leptin receptors. Overall, the glial leptin system shows robust regulation and plays an essential role in obesity. Strategies to manipulate this nonneuronal leptin signaling may have major clinical impact.

Keywords: leptin; CNS; obesity; astrocytes; blood–brain barrier

Introduction

Obesity is a global epidemic disorder. In the United States in 2010, no state had a prevalence of obesity less than 20%, and 12 states had a prevalence of 30% or more.[1] Leptin is a key adipocytokine that plays an essential role in the regulation of obesity. It is a 16 kDa polypeptide mainly produced by adipocytes in fat tissue. Blood concentrations of leptin correlate with the amount of adipocity.[2] Leptin levels are also increased in inflammatory situations. A product of the ob gene,[3] leptin has many cellular targets in different organs. One of its major targets is the central nervous system (CNS). Leptin acts to reduce feeding behavior and inhibit obesity. After administration of recombinant leptin, it acts directly on neuronal networks that control feeding and energy balance.[4]

Leptin exerts its actions on the CNS via regionally expressed leptin receptors (ObR or LepR). Belonging to the cytokine receptor class I superfamily, the ObRs consist of five alternatively spliced variants: a, b, c, d, and e. Among them, ObRb has the longest cytoplasmic tail and is responsible for phosphorylation and activation of signal transducer and activator for transcription (STAT)-3. To exert its functions in most CNS regions, leptin must cross the blood–brain barrier (BBB), which it does by a unique transport system;[5] ObRa is the most abundant isoform in the cerebral microvessels composing the BBB.[6] Given sufficient levels of expression, as shown in cultured cells overexpressing ObR isoforms, all of the membrane-bound forms can efficiently endocytose leptin.[7] By contrast, the soluble receptor ObRe may serve as an antagonist not only to leptin signaling but also to its transport across the BBB.[8] In this

doi: 10.1111/j.1749-6632.2012.06472.x

review, we focus on nonneuronal cellular effects of leptin in the CNS and address pathophysiological implications of these findings, not only in obesity but also in several CNS disorders. This extends beyond studies of the transport of leptin across the BBB.[5]

ObR distribution in the brain

In the initial study with nonradioactive *in situ* hybridization (ISH) of ObR mRNA in the brain of C57 mice, Huang *et al.* showed that the areas with the highest level of expression include the arcuate nucleus and median eminence of the hypothalamus. The dentate gyrus and CA1 region of the hippocampus are the second most abundant. Low levels are also seen in the piriform cortex and the medial margin of the medial habenular nucleus. Most of the ISH signals appear to be present in neurons, such as those producing neuropeptide Y.[9] This was confirmed by an independent study, which also showed the presence of ObR mRNA in the choroid plexus.[10] In the obese ob/ob mouse devoid of leptin production, ObR mRNA is increased in these hypothalamic regions, piriform and olfactory cortices, and the medial habenular nucleus in comparison with lean mice.[11] Although the control conditions may not be ideal, it appears that upregulation of ObR has regional specificity and may be dependent on the inducing factors, such as obesity, fasting, ischemia, or inflammation.

ISH with isoform-specific probes showed that ObRa and ObRb are abundant in the hypothalamus, whereas ObRa, ObRc, and ObRf, but not ObRb, are prominent in the choroid plexus. Besides brain regions described in the preceding paragraph, the thalamus, substantia nigra, and granular cell layer of the cerebellum also show intense ObR signals.[12] When relative levels of ObR isoforms are compared with those measured by reverse transcriptase–polymerase chain reaction (RT-PCR), the ObRb to ObRa mRNA ratio is highest in the hypothalamus of the normal adult mouse brain.[13] Shioda *et al.* first showed the protein expression of ObR by immunohistochemistry (IHC) and western blotting (WB) in rat brain. ObR immunoreactivity was seen in the arcuate, paraventricular, and ventromedial nuclei of the hypothalamus, lateral hypothalamus, olfactory bulb, neocortex, cerebellar cortex, dorsal raphe nucleus, inferior olive, nucleus of the solitary tract, and dorsal motor nucleus of the vagus nerve, with

WB showing a molecular weight corresponding to the 120 kDa major band.[14] The levels of protein and mRNA expression are congruous. In human autopsy samples, ObR immunoreactivity is seen in choroid plexus epithelium, ependymal lining, neurons of the hypothalamic nuclei (arcuate, suprachiasmatic, mamillary, paraventricular, dorsomedial, supraoptic, and posterior), nucleus basalis of Meynert, inferior olivary nuclei, and cerebellar Purkinje cells. The molecular weights are 97 kDa and 125 kDa. However, there was no significant regulatory change by obesity and diabetes in comparison with lean subjects.[15] More recently, advances in transgenic technology enabled a novel approach to generate a LepRb-IRES-Cre EYFP reporter mouse line. These mice have high levels of ObRb mRNA/EYFP coexpression, including in areas previously not shown to have abundant ObR, such as the dorsomedial nucleus of the hypothalamus, ventral premammillary nucleus, ventral tegmental area, parabrachial nucleus, the dorsal vagal complex, insular cortex, lateral septal nucleus, medial preoptic area, rostral linear nucleus, the Edinger–Westphal nucleus, and midbrain. The transgenic mice are thus useful to locate ObRb$^+$ cells and study their regulatory changes.[16]

With regard to the cellular distribution of ObR in the brain, choroid plexus epithelia show a heterogeneous distribution of strong immunoreactivity. This population is probably responsible for leptin transport across the blood–cerebrospinal fluid barrier.[17,18] ObR mRNA is also seen in meninges and blood vessels.[19] Among CNS parenchymal cells, neurons are the most studied target; indeed, leptin-induced immediate early gene c-Fos activation mainly resides in neurons, as seen in a study focusing on hypothalamus and brainstem.[20] However, ObR mRNA is also seen in astrocytes, as fluorescent ISH with double-labeling immunostaining of glial fibrillary acidic protein (GFAP) may be colocalized in normal rat hypothalamus.[21] In the arcuate nucleus of the hypothalamus, about 20% of ObR immunoreactivity is seen in GFAP$^+$ astrocytes, and this is further increased in mice with adult-onset obesity (Fig. 1, modified from Ref. 22), as will be further discussed in this review. Young showed that astrocytes have a specific anatomical relationship with leptin-sensitive neurons.[23] Later, the distribution and upregulation of astrocytic ObR was shown in mouse models of adult onset obesity.[22,24,25] Leptin

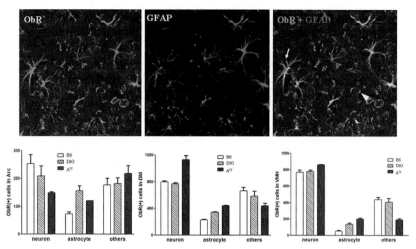

Figure 1. Obesity changes the number of ObR$^+$ cell types in hypothalamic regions. Upper panel: some ObR$^+$ cells are also GFAP$^+$, shown by confocal images in the arcuate nucleus of a B6 male mouse. Lower panel: counting of demarcated regions from independent sections from four mice showed differential changes in the number of ObR$^+$ neurons, astrocytes, and other types of cells in the arcuate nucleus (ArC), dorsomedial hypothalamus (DM), and ventromedial hypothalamus (VMH) of DIO and A^{vy} mice.

can attenuate oligodendrocyte development embryonically,[26] promote neurosphere self-renewal,[27,28] increase angiogenesis after stroke,[28] and show robust effects on microglial cytokine production.[29–31]

Extrahypothalamic actions of leptin in the brain

Besides its effect on neuroendorine regulation of feeding, the extrahypothalamic actions of leptin are becoming increasingly recognized. One major site of action is the hippocampus.[32,33] Leptin activates large conductance Ca^{2+}-activated K$^+$ channels in the hippocampus and promotes long-term depression of excitatory synaptic transmission.[34–38] Leptin may be neurotophic and promote neurogenesis,[39–42] and leptin deficient ob/ob mice show reduction of brain size and neuron numbers[43,44] as well as defective glial and synaptic proteins.[45] Leptin also acts through ObR and mitogen-activated protein kinase (MAPK) to facilitate N-methyl-D-aspartate (NMDA) receptor-mediated Ca^{2+} influx.[46] This suggests a potentially deleterious role of leptin signaling during excess excitation as seen in epilepsy.

Both anti- and proconvulsive roles of leptin have been shown by several groups. The antiepileptic effect of leptin is seen in ob/ob mice, which have increased seizure susceptibility, and in the effectiveness of intranasal leptin to decrease seizures.[47–49] The proepileptic effect is also well characterized. After intracerebroventricular injection, leptin has dose-dependent effects to potentiate penicillin-induced convulsion; lower doses (1 or 2 μg) are most effective, seen 90 min after delivery to the rat, whereas a high dose (10 μg) has no effect.[50] The proconvulsive effect of leptin appears to be mediated by the NMDA receptor.[51] The NMDA receptor mediates the leptin-induced excitatory effect by increasing intracellular calcium levels and synaptic transmission in rat hippocampal slices and cell culture.[34] Nonetheless, the same group of researchers showed that leptin can also act through PI3K and BK channels to inhibit epileptiform-like activity in neurons.[47] These contradictory findings for leptin may be related to different cellular effects upon activation of astrocytes, microglia, or neurons, as all three cell types express leptin receptors.[22,24]

Using a mouse model of epilepsy induced by pilocarpine, we detected robust upregulation of ObR in the reactive astrocytes (Fig. 2). By contrast, neuronal ObR in the hippocampus was only increased with severe seizures, as seen by high Racine scale scores. In the astrocyte-specific leptin receptor knockout (ALKO) mice generated in our laboratory,[52] high-dose pilocarpine-induced seizures had a milder presentation, and the acute mortality immediately after seizure induction was reduced in comparison with the wildtype littermates (Fig. 3). Better survival and less severe seizures in the ALKO mice in response to a high dose of pilocarpine suggest that astrocytic ObR is detrimental to epilepsy.

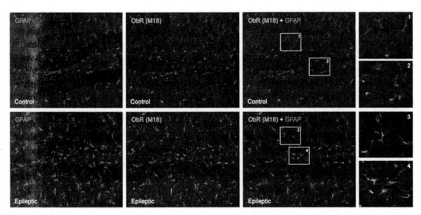

Figure 2. In the hippocampus of a mouse 6 weeks after pilocarpine-induced epilepsy (lower panel), there is increased GFAP (red) and ObR (green) immunoreactivity. Confocal colocalization shows that most ObR is present in astrocytes in control mice (upper panel) and upregulated by seizure.

With regard to spatial learning and cognitive functions, intrahippocampal leptin administered immediately after training improves subsequent retention of T-maze footshock avoidance and step-down inhibitory avoidance behavior in normal CD1 mice, and has a beneficial effect on memory processing of the SAMP8 mice with their spontaneously accelerated aging and elevated cerebral amyloid β protein.[33] This is consistent with a dose-dependent facilitatory role of leptin in hippocampal long-term potentiation.[32] It remains to be determined how astrocytic and neuronal leptin signaling interact with each other in the execution of normal and pathophysiological functions in the hippocampus.

Astrocytic leptin signaling in the brain regulates the development of obesity

It is now clear that adult-onset obesity is associated with region-specific upregulation of astrocytic ObR, shown in both agouti viable yellow (A^{vy}) mice that have a genetic mutation with constitutive production of a reverse melanocortin receptor antagonist and reduced apparent influx of leptin from blood to the brain,[22,53] and in control C57 mice with adult-onset obesity.[24] When astrocytic activity is inhibited by pretreatment of the A^{vy} mice with fluorocitrate, intracerebroventricular leptin-induced STAT3 activation is increased in neurons, concurrent with a higher signal intensity of fluorescently conjugated leptin taken up by periventricular neurons.[54] There are a few possibilities explaining how ObR$^+$ astrocytes may affect neuronal leptin signaling: (1) levels of ObR in astrocytes affect leptin permeation across the BBB; (2) once leptin crosses the BBB, it reaches astrocytes first but has a longer diffusion distance to neurons. Astrocytic ObR might compete with neuronal ObR for leptin endocytosis, and (3) astrocytes generate secondary signals in response to leptin, which, in turn, modulates neuronal responses (to leptin and to other stimuli, Fig. 4). Gliovascular coupling (between astrocytes and BBB microvessels) and metabolic coupling (between astrocytes and

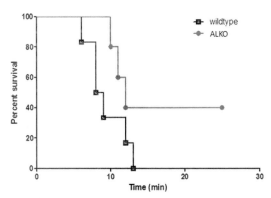

Figure 3. Seizures were induced in female ALKO and littermate controls with a high dose of pilocarpine (350 mg/kg i.p., sevenfold more than the usual dose), 30 min after pretreatment with scopolamine (1 mg/kg i.p.) to reduce peripheral cholinergic effects. The censor time was 25 min. In the control group, all mice died between 6 and 13 min, with a mean survival time of 8.5 min. In the ALKO group, three mice died at 10, 11, and 12 min, and the remaining two mice survived past the censored time of 25 minutes. The hazard ratio is 0.71. The log-rank test shows that the two survival curves had a trend toward significance ($P = 0.08$).

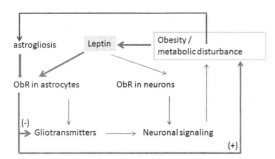

Figure 4. Possibilities of how obesity acts through the astrocytic leptin system to modulate neuronal leptin signaling. Metabolic factors, including leptin itself, may induce reactive astrogliosis and increase astrocytic ObR. Leptin uptake and/or signaling by astrocytes may thus be enhanced and, in turn, affects neuronal leptin signaling.

neurons) are dynamic processes responsible for regulation of cerebral blood flow and metabolism. Leptin is not a vasoactive polypeptide but probably can participate indirectly in these processes.

Astrocytes express several ObR splice variants, including ObRa, ObRb, ObRc, and ObRe. The presence of leptin signaling in astrocytes is shown by an increase of calcium influx in primary astrocytes after leptin superfusion during real-time calcium imaging.[24] To determine whether the level and isoforms of ObR in astrocytes affect leptin permeation across the BBB, C6 astrocytoma cells were transfected with different ObR isoforms before they were cocultured with hCMEC/D3 cerebral endothelial cells in a Transwell system. The apical-to-basolateral permeation of leptin in the *in vitro* BBB system was unchanged when the C6 cells overexpressed ObRa, but it was increased in C6 cells overexpressing ObRb or ObRe.[55] This appears to indicate that altered astrocytic leptin signaling facilitates leptin transport across the BBB. However, new preliminary data suggest that ObRa is the major form of upregulated astrocytic leptin receptor. Oppositely, removal of astrocytic leptin signaling does not have a significant effect on baseline transport of leptin across the BBB, since the newly generated ALKO mice do not show a reduction of leptin transport across the BBB in comparison with their wildtype littermates.[52] Of course, ALKO mice have an embryonic absence of astrocytic leptin receptors, so that compensatory mechanisms could have emerged.

It is yet to be determined whether astrocytic leptin receptor expression or signaling compromise the availability of leptin to ObR⁺ neurons, or whether secondary signals from leptin modulate neuronal leptin activities. This can be achieved by use of a gene knockdown approach before coculture of astrocytes and neurons, or by comparison of astrocytes from ALKO or wild-type mice on their effects on cocultured neurons. Nonetheless, *in vivo* inhibition of astrocytic metabolic activity by fluorocitrate has shown an increase of neuronal uptake of leptin after intracerebroventricular injection, accompanied by an elevation of STAT3 activation.[54]

The border between the area postrema and nucleus tractus solitarius and dorsal vagal complex in the caudal brainstem consists mainly of pallisading astrocytes. The columnar cells compose a continuous monolayer and are immunopositive for both the tight junction protein zona occludin-1 and the astrocyte marker GFAP.[56] These cells also express ObRa and ObRf as well as ObRb in rats; they not only constitute a diffusion barrier to fluorescent dyes after intravenous or intracerebroventricular injection, but also show upregulation of the ObR short forms.[57] This reflects regulatory changes of leptin transport from blood to the medulla and perhaps also other parts of the brainstem during energy insufficiency. This important set of data is complementary to the results identifying leptin receptors in selective brain regions in mice with adult-onset obesity,[22,24] and illustrates an important role of ObR⁺ astrocytes in the blood-to-CNS transport of leptin.

In the nucleus tractus solitarius, leptin plays an important role in regulating autonomic nervous system activities, including feeding, gastric motility,[58]

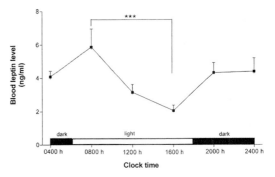

Figure 5. Leptin production and blood concentration show a 24-h rhythm that has a peak in the early morning and nadir in late afternoon in male CD1 mice that also show rhythmicity in blood–spinal cord barrier transport but not blood–brain barrier transport of leptin. Modified from Pan and Kastin.[64]

and the hypercapnic ventilator response.[59] Leptin is closely associated with obstructive sleep apnea syndrome,[60,61] whereas continuous positive airway pressure treatment for 8 weeks reduces the concentration of leptin as well as total cholesterol and low-density lipoprotein.[62] It is interesting to note that obesity attenuates the clock genes *Bmal1* and *Reverbalpha* and upregulates peroxisome proliferator-activated receptor alpha in this area.[63] Consistent with this, there is a circadian rhythm of leptin concentration and leptin transport across the BBB and blood–spinal cord barrier, despite partial saturation (Fig. 5).[64] This might be associated with a role of leptin in sleep regulation, as leptin-deficient mice have impaired sleep with more arousals and shorter sleep bouts despite an increase in the total amount of nonrapid eye movement sleep.[65]

Astrocytes are well positioned to regulate synaptic transmission and the neurovascular network. In astrocytes, glycogen is metabolized into lactate, which is subsequently transported into neurons to serve as a storehouse for glucose for neurons.[66] Cholesterol is also transported from astrocytes to neurons where it aids in synaptogenesis.[67] Astrocytic activation induces the accumulation of arachidonic acid and release of the gliotransmitters glutamate and adenosine-5′-triphosphate (ATP). The role of astrocytes in metabolic coupling has been reviewed extensively.[67,68] In the near future, we shall have a better understanding of how leptin signaling contributes to astrocyte–neuronal communication.

Summary

Though one of the best known functions of leptin is regulation of feeding behavior, leptin also plays essential roles in the regulation of cerebral blood flow and metabolism, cell differentiation, cognition and learning, and neurodegeneration, where it may play dual roles in stroke or epilepsy. The extrahypothalamic distribution and nonneuronal cellular distribution of leptin form a structural basis for its pleiotropic actions. By illustrating how astrocytic leptin signaling facilitates leptin transport across the *in vitro* BBB and attenuates the development of diet-induced obesity in mice, we have deduced a novel role for the astrocytic leptin system in gliovascular and metabolic coupling. The ObR+ astrocytes show dynamic changes in neurological and metabolic disorders, and are intricately linked to CNS functions.

Acknowledgments

We thank many past and current members of the Blood–Brain Barrier Group in the development of astrocytic and endothelial leptin signaling projects. This work was supported in part by NIH Grants DK54880, DK92245, and NS62291.

Conflicts of interest

The authors report no conflicts of interest.

References

1. Center for Disease Control. 2011. Retrieved from http://www.cdc.gov/ obesity/data/trends.html.
2. Frederich, R.C., A. Hamann, S. Anderson, *et al.* 1995. Leptin levels reflect body lipid content in mice: evidence for diet-induced resistance to leptin action. *Nat. Med.* **1:** 1311–1314.
3. Zhang, Y., R. Proenca, M. Maffei, *et al.* 1994. Positional cloning of the mouse obese gene and its human homologue. *Nature* **372:** 425–432.
4. Campfield, L.A., F.J. Smith, Y. Guisez, *et al.* 1995. Recombinant mouse OB protein: evidence for a peripheral signal linking adiposity and central neural networks. *Science* **269:** 546–549.
5. Banks, W.A., A.J. Kastin, W. Huang, *et al.* 1996. Leptin enters the brain by a saturable system independent of insulin. *Peptides* **17:** 305–311.
6. Bjørbæk,C., J.K. Elmquist, P. Michl, *et al.* 1998. Expression of leptin receptor isoforms in rat brain microvessels. *Endocrinology* **139:** 3485–3491.
7. Tu, H., W. Pan, L. Feucht & A.J. Kastin. 2007. Convergent trafficking pattern of leptin after endocytosis mediated by ObRa: ObRd. *J. Cell Physiol.* **212:** 215–222.
8. Tu, H., A.J. Kastin, H. Hsuchou & W. Pan. 2008. Soluble receptor inhibits leptin transport. *J. Cell. Physiol.* **214:** 301–305.
9. Huang, X.F., I. Koutcherov, S. Lin, *et al.* 1996. Localization of leptin receptor mRNA expression in mouse brain. *NeuroReport* **7:** 2635–2638.
10. Hakansson, M.L., A.L. Hulting & B. Meister. 1996. Expression of leptin receptor mRNA in the hypothalamic arcuate nucleus: relationship with NPY neurones *NeuroReport* **7:** 3087–3092.
11. Huang, X.F., S. Lin & R. Zhang. 1997. Upregulation of leptin receptor mRNA expression in obese mouse brain. *NeuroReport* **8:** 1035–1038.
12. Guan, X.M., J.F. Hess, H. Yu, *et al.* 1997. Differential expression of mRNA for leptin receptor isoforms in the rat brain. *Mol. Cell Endocrinol.* **133:** 1–7.
13. Lollmann, B., S. Gruninger, A. Stricker-Krongrad & M. Chiesi. 1997. Detection and quantification of the leptin receptor splice variants Ob-Ra, b, and, e in different mouse tissues. *Biochem. Biophys. Res. Commun.* **238:** 648–652.
14. Shioda, S., H. Funahashi, S. Nakajo, *et al.* 1998. Immunohistochemical localization of leptin receptor in the rat brain. *Neurosci. Lett.* **243:** 41–44.

15. Couce, M.E., B. Burguera, J.E. Parisi, *et al.* 1997. Localization of leptin receptor in the human brain. *Neuroendocrinology* **66**: 145–150.

16. Scott, M.M., J.L. Lachey, S.M. Sternson, *et al.* 2009. Leptin targets in the mouse brain. *J. Comp. Neurol.* **514**: 518–532.

17. Zlokovic, B.V., S. Jovanovic, W. Miao, *et al.* 2000. Differential regulation of leptin transport by the choroid plexus and blood-brain barrier and high affinity transport systems for entry into hypothalamus and across the blood-cerebrospinal fluid barrier. *Endocrinol.* **141**: 1434–1441.

18. Thomas, S.A., J.E. Preston, M.R. Wilson, *et al.* 2001. Leptin transport at the blood-cerebrospinal fluid barrier using the perfused sheep choroid plexus model. *Brain Res.* **895**: 283–290.

19. Elmquist, J.K., C. Bjorbaek, R.S. Ahima, *et al.* 1998. Distributions of leptin receptor mRNA isoforms in the rat brain. *J. Comp. Neurol.* **395**: 535–547.

20. Elias, C.F., J.F. Kelly, C.E. Lee, *et al.* 2000. Chemical characterization of leptin-activated neurons in the rat brain. *J. Comp. Neurol.* **423**: 261–281.

21. Hsuchou, H., W. Pan, M.J. Barnes & A.J. Kastin. 2009. Leptin receptor mRNA in rat brain astrocytes. *Peptides* **30**: 2275–2280.

22. Pan, W., H. Hsuchou, Y. He, *et al.* 2008. Astrocyte leptin receptor (ObR) and leptin transport in adult-onset obese mice. *Endocrinology* **149**: 2798–2806.

23. Young, J.K. 2002. Anatomical relationship between specialized astrocytes and leptin-sensitive neurones. *J. Anat.* **201**: 85–90.

24. Hsuchou, H., Y. He, A.J. Kastin, *et al.* 2009. Obesity induces functional astrocytic leptin receptors in hypothalamus. *Brain* **132**: 889–902.

25. Pan, W., H. Hsuchou, Y. He & A.J. Kastin. 2011. Glial leptin receptors and obesity. In *Modern Insights into Disease: from Molecules to Man: Adipokines.* V.R. Preedy, Ed.: 185–196. Science Publishers. Enfield, NH, USA.

26. Udagawa, J., M. Nimura & H. Otani. 2006. Leptin affects oligodendroglial development in the mouse embryonic cerebral cortex. *Neuro. Endocrinol. Lett.* **27**: 177–182.

27. Udagawa, J., A. Ono, M. Kawamoto & H. Otani. 2010. Leptin and its intracellular signaling pathway maintains the neurosphere. *NeuroReport* **21**: 1140–1145.

28. Avraham, Y., N. Davidi, V. Lassri, *et al.* 2011. Leptin induces neuroprotection neurogenesis and angiogenesis after stroke. *Curr. Neurovasc. Res.* **8**: 313–322.

29. Pinteaux, E., W. Inoue, L. Schmidt, *et al.* 2007. Leptin induces interleukin-1beta release from rat microglial cells through a caspase 1 independent mechanism. *J. Neurochem.* **102**: 826–833.

30. Tang, C.H., D.Y. Lu, R.S. Yang, *et al.* 2007. Leptin-induced IL-6 production is mediated by leptin receptor, insulin receptor substrate-1, phosphatidylinositol 3-kinase, Akt, NF-kappaB, and p300 pathway in microglia. *J. Immunol.* **179**: 1292–1302.

31. Lafrance, V., W. Inoue, B. Kan & G.N. Luheshi. 2010. Leptin modulates cell morphology and cytokine release in microglia. *Brain Behav. Immun.* **24**: 358–365.

32. Oomura, Y., N. Hori, T. Shiraishi, *et al.* 2006. Leptin facilitates learning and memory performance and enhances hippocampal CA1 long-term potentiation and CaMK II phosphorylation in rats. *Peptides* **27**: 2738–2749.

33. Farr, S.A., W.A. Banks & J.E. Morley. 2006. Effects of leptin on memory processing. *Peptides* **27**: 1420–1425.

34. Shanley, L.J., A.J. Irving & J. Harvey. 2001. Leptin enhances NMDA receptor function and modulates hippocampal synaptic plasticity. *J. Neurosci.* **21**: RC186.

35. Shanley, L.J., A.J. Irving, M.G. Rae, *et al.* 2002. Leptin inhibits rat hippocampal neurons via activation of large conductance calcium-activated K+ channels. *Nat. Neurosci.* **5**: 299–300.

36. Durakoglugil, M., A.J. Irving & J. Harvey. 2005. Leptin induces a novel form of NMDA receptor-dependent long-term depression. *J. Neurochem.* **95**: 396–405.

37. O'malley, D., N. Macdonald, S. Mizielinska, *et al.* 2007. Leptin promotes rapid dynamic changes in hippocampal dendritic morphology. *Mol. Cell. Neurosci.* **35**: 559–572.

38. Harvey, J. 2007. Leptin: a diverse regulator of neuronal function. *J. Neurochem.* **100**: 307–313.

39. Guo, Z., H. Jiang, X. Xu, *et al.* 2008. Leptin-mediated cell survival signaling in hippocampal neurons mediated by JAK STAT3 and mitochondrial stabilization. *J. Biol. Chem.* **283**: 1754–1763.

40. Garza, J.C., M. Guo, W. Zhang & X.Y. Lu. 2008. Leptin increases adult hippocampal neurogenesis in vivo and in vitro. *J. Biol. Chem.* **283**: 18238–18247.

41. Zhang, F., S. Wang, A.P. Signore & J. Chen. 2007. Neuroprotective effects of leptin against ischemic injury induced by oxygen-glucose deprivation and transient cerebral ischemia. *Stroke* **38**: 2329–2336.

42. Avraham, Y., N. Davidi, M. Porat, *et al.* 2010. Leptin reduces infarct size in association with enhanced expression of CB2, TRPV1, SIRT-1 and leptin receptor. *Curr. Neurovasc. Res.* **7**: 136–143.

43. Steppan, C.M. & A.G. Swick. 1999. A role for leptin in brain development. *Biochem. Biophys. Res. Commun.* **256**: 600–602.

44. Mobbs, C.V. 2006. Fathead: the gain in brain falls mainly with leptin wane. *Endocrinology* **147**: 645–646.

45. Ahima, R.S., C. Bjorbaek, S. Osei & J.S. Flier. 1999. Regulation of neuronal and glial proteins by leptin: implications for brain development. *Endocrinology* **140**: 2755–2762.

46. Irving, A.J., L. Wallace, D. Durakoglugil & J. Harvey. 2006. Leptin enhances NR2B-mediated N-methyl-D-aspartate responses via a mitogen-activated protein kinase-dependent process in cerebellar granule cells. *Neuroscience* **138**: 1137–1148.

47. Shanley, L.J., D. O'malley, A.J. Irving, *et al.* 2002. Leptin inhibits epileptiform-like activity in rat hippocampal neurones via PI 3-kinase-driven activation of BK channels. *J. Physiol.* **545**: 933–944.

48. Xu, L., N. Rensing, X.F. Yang, *et al.* 2008. Leptin inhibits 4-aminopyridine- and pentylenetetrazole-induced seizures and AMPAR-mediated synaptic transmission in rodents. *J. Clin. Invest.* **118**: 272–280.

49. Erbayat-Altay, E., K.A. Yamada, M. Wong & L.L. Thio. 2008. Increased severity of pentylenetetrazol induced seizures in leptin deficient ob/ob mice. *Neurosci. Lett.* **433**: 82–86.

50. Ayyildiz, M., M. Yildirim, E. Agar & A.K. Baltaci. 2006. The effect of leptin on penicillin-induced epileptiform activity in rats. *Brain Res. Bull.* **68:** 374–378.

51. Lynch, J.J., III, E.W. Shek, V. Castagne & S.W. Mittelstadt. 2010. The proconvulsant effects of leptin on glutamate receptor-mediated seizures in mice. *Brain Res. Bull.* **82:** 99–103.

52. Hsuchou, H., A.J. Kastin, H. Tu, *et al.* 2011. Effects of cell type-specific leptin receptor mutation on leptin transport across the BBB. *Peptides* **32:** 1392–1399.

53. Pan, W. & A.J. Kastin. 2007. Mahogany, blood-brain barrier, and fat mass surge in A^{vy} mice. *Int. J. Obes.* **31:** 1030–1032.

54. Pan, W., H. Hsuchou, C.L. Xu, *et al.* 2011. Astrocytes modulate distribution and neuronal signaling of leptin in the hypothalamus of obese A^{vy} mice. *J. Mol. Neurosci.* **43:** 478–484.

55. Hsuchou, H., A.J. Kastin, H. Tu, *et al.* 2010. Role of astrocytic leptin receptor subtypes on leptin permeation across hCMEC/D3 human brain endothelial cells. *J. Neurochem.* **115:** 1288–1298.

56. Wang, Q.P., J.L. Guan, W. Pan, *et al.* 2008. A diffusion barrier between the area postrema and nucleus tractus solitarius. *Neurochem. Res.* **33:** 2035–2043.

57. Dallaporta, M., E. Pecchi, J. Pio, *et al.* 2009. Expression of leptin receptors by glial cells of the nucleus tractus solitarius: possible involvement in energy homeostasis. *J. Neuroendocrinol.* **21:** 57–67.

58. Grill, H.J. & M.R. Hayes. 2009. The nucleus tractus solitarius: a portal for visceral afferent signal processing, energy status assessment and integration of their combined effects on food intake. *Int. J. Obes. (Lond)* **33**(Suppl 1): S11–S15.

59. Inyushkin, A.N., E.M. Inyushkina & N.A. Merkulova. 2009. Respiratory responses to microinjections of leptin into the solitary tract nucleus. *Neurosci. Behav. Physiol.* **39:** 231–240.

60. Malli, F., A.I. Papaioannou, K.I. Gourgoulianis & Z. Daniil. 2010. The role of leptin in the respiratory system: an overview. *Respir. Res.* **11:** 152.

61. Basoglu, O.K., F. Sarac, S. Sarac, *et al.* 2011. Metabolic syndrome, insulin resistance, fibrinogen, homocysteine, leptin, and C-reactive protein in obese patients with obstructive sleep apnea syndrome. *Ann. Thorac. Med.* **6:** 120–125.

62. Cuhadaroglu, C., A. Utkusavas, L. Ozturk, *et al.* 2009. Effects of nasal CPAP treatment on insulin resistance, lipid profile, and plasma leptin in sleep apnea. *Lung* **187:** 75–81.

63. Kaneko, K., T. Yamada, S. Tsukita, *et al.* 2009. Obesity alters circadian expressions of molecular clock genes in the brainstem. *Brain Res.* **1263:** 58–68.

64. Pan, W. & A.J. Kastin. 2001. Diurnal variation of leptin entry from blood to brain involving partial saturation of the transport system. *Life Sci.* **68:** 2705–2714.

65. Laposky, A.D., J. Shelton, J. Bass, *et al.* 2006. Altered sleep regulation in leptin-deficient mice. *Am. J. Physiol Regul. Integr. Comp Physiol.* **290:** R894–R903.

66. Wender, R., A.M. Brown, R. Fern, *et al.* 2000. Astrocytic glycogen influences axon function and survival during glucose deprivation in central white matter. *J. Neurosci.* **20:** 6804–6810.

67. Goritz, C., D.H. Mauch, K. Nagler & F.W. Pfrieger. 2002. Role of glia-derived cholesterol in synaptogenesis: new revelations in the synapse-glia affair. *J. Physiol. Paris* **96:** 257–263.

68. Tsacopoulos, M. & P.J. Magistretti. 1996. Metabolic coupling between glia and neurons. *J. Neurosci.* **16:** 877–885.

Ann. N.Y. Acad. Sci. ISSN 0077-8923

Brain orexin promotes obesity resistance

Catherine Kotz,[1,2,4] Joshua Nixon,[1,2] Tammy Butterick,[1,2] Claudio Perez-Leighton,[1,4] Jennifer Teske,[5,6] and Charles Billington[1,3]

[1]Department of Veterans Affairs, GRECC and Research Service, Minneapolis, Minnesota. [2]Food Science and Nutrition, University of Minnesota, Saint Paul, Minnesota. [3]Department of Medicine, University of Minnesota, Minneapolis, Minnesota. [4]Department of Neuroscience, University of Minnesota, Minneapolis, Minnesota. [5]Department of Nutritional Sciences, University of Arizona, Tucson, Arizona. [6]Research Service, Southern VA Healthcare System, Tucson, Arizona

Address for correspondence: Catherine Kotz, Ph.D., Department of Veterans Affairs--GRECC and Research Service, One Veterans Drive, Minneapolis, Minnesota 55417. Email: kotzx004@umn.edu

Resistance to obesity is becoming an exception rather than the norm, and understanding mechanisms that lead some to remain lean in spite of an obesigenic environment is critical if we are to find new ways to reverse this trend. Levels of energy intake and physical activity both contribute to body weight management, but it is challenging for most to adopt major long-term changes in either factor. Physical activity outside of formal exercise, also referred to as activity of daily living, and in stricter form, spontaneous physical activity (SPA), may be an attractive modifiable variable for obesity prevention. In this review, we discuss individual variability in SPA and NEAT (nonexercise thermogenesis, or the energy expended by SPA) and its relationship to obesity resistance. The hypothalamic neuropeptide orexin (hypocretin) may play a key role in regulating SPA and NEAT. We discuss how elevated orexin signaling capacity, in the context of a brain network modulating SPA, may play a major role in defining individual variability in SPA and NEAT. Greater activation of this SPA network leads to a lower propensity for fat mass gain and therefore may be an attractive target for obesity prevention and therapy.

Keywords: orexin; obesity; spontaneous physical activity; nonexercise activity thermogenesis; energy expenditure

Obesity and individual variability

Obesity is a condition defined by a chronic excess of body fat[1] and is positively correlated with shorter life expectancy, metabolic syndrome, type 2 diabetes, and coronary heart disease.[2] Obesity has become a public health issue, as its incidence in adults and children has increased in the last two decades across both developed and underdeveloped societies.[3–5]

Humans show large variation in their susceptibility to obesity, which is determined by both environmental and genetic factors.[6–9] A major factor in determining this variability is physical activity, and specifically a component of overall energy expenditure known as nonexercise induced thermogenesis (NEAT).[10–12] NEAT includes all forms of energy expenditure not associated with formal exercise, such as standing and fidgeting.[10,13–15] The complementary concept of spontaneous physical activity (SPA) is used to describe "any type of physical activity that does not qualify as voluntary exercise."[16,17] Both SPA and NEAT have a heredability component.[17,18]

SPA and NEAT are not interchangeable but are complementary concepts: NEAT refers to energy expenditure while SPA describes the types of physical activity that result in NEAT. Given the association between SPA and NEAT, it is not surprising that variability in SPA also contributes to variability in sensitivity to obesity. For example, lean people spend larger amounts of time standing (approximately two hours daily) than do obese people.[12] While the contribution of NEAT to weight gain resistance may seem small, it can be significant. For example, one study showed that after overfeeding humans with 1,000 kcal daily for eight weeks, fat mass gain was significantly and negatively correlated with the increase in SPA and NEAT. Importantly, there was no

doi: 10.1111/j.1749-6632.2012.06585.x
Ann. N.Y. Acad. Sci. 1264 (2012) 72–86 © 2012 New York Academy of Sciences.

change in volitional exercise, and no relationship between the observed change in fat mass and basal metabolism or postprandial thermogenesis.[11]

The neural mechanisms that underlie human variability in SPA are distributed processes involving multiple brain regions, neurotransmitters and neuropeptides, including cholecystokinin, corticotrophin releasing hormone, neuromedin, neuropeptide Y (NPY), leptin, and orexin (also known as hypocretin).[19] While all are important, in this review we focus on the biological role of central orexin peptides and their receptors with respect to their role in obesity and obesity resistance.

Orexin peptides and their receptors

The orexins are two closely related peptides, orexin A (OXA, hypocretin 1) and orexin B (OXB, hypocretin 2) that are produced by cleavage from a single propeptide.[20,21] In mammals, the majority of CNS orexin peptides are synthesized in neurons located in the lateral hypothalamus and perifornical area. The hypothalamic orexin neurons are glutamatergic neurons with tonic firing, low-threshold spike on recovery from hyperpolarization and little spike adaptation.[22,23] Recently, the existence of orexin neuronal subpopulations has been proposed based on morphological and electrophysiological evidence.[24]

The orexin peptides act through two G protein-coupled receptors, orexin receptor type 1 (OX1R, hypocretin receptor 1) and orexin receptor type 2 (OX2R, hypocretin receptor 2).[20,21] Both orexin receptor subtypes can bind to OXA and OXB, but with differential affinity: OX1R has a higher affinity for OXA, while OX2R has equal affinity for either orexin peptide.[20,25] Activation of both receptor subtypes leads to an increase in neuronal firing and an increase in intracellular calcium.[26–30] Preadministration of the OX1R antagonist SB334867 can block OXA-induced SPA and NEAT,[16,31–34] suggesting an important role for OX1R in mediating SPA and NEAT; however, OX2R involvement has not been ruled out.

An important characteristic of the orexin neurons are their projections to multiple brain regions.[35–39] Neuroanatomical studies have shown that the orexin neurons have collateral projections within the CNS,[40–42] transsynaptically collateral CNS efferents,[43] or collateral efferents to both CNS regions and brown adipose tissue.[44] The distribution pattern of the orexin receptors reflects the widespread projections of orexin neurons, as both orexin receptor subtypes are expressed throughout the brain. The orexin receptors show distinctive, yet overlapping patterns of expression, with a good agreement between mRNA and protein data.[45–50] However, studies addressing colocalization of the orexin receptor subtypes at a cellular level are lacking. The wide distribution of the orexin receptors and orexinergic fibers initially suggested the orexin system was involved in multiple physiological processes, and current research supports a role for orexin in the control of arousal and sleep, reward, stress, and energy homeostasis.[51–56]

The main contribution of the orexin peptides to energy metabolism is elegantly exemplified in a mouse model that exhibits postnatal loss of orexin neurons.[57] In these mice, the orexin promoter drives expression of the neurodegenerative gene ataxin-3, leading to progressive loss of the orexin neurons during development. These mice exhibit hypophagia, lower levels of SPA, and develop spontaneous onset obesity when fed a regular diet.[57,58] These results suggest that one primary function of the orexin peptides is to drive energy expenditure, although they can also modulate food intake. Additional support for this idea comes from another mouse model in which the β-actin cytomegalovirus promoter drives overexpression of the orexin peptides.[59] Consistent with the role of orexin in promoting energy expenditure, these mice show resistance to high-fat diet–induced obesity.[60]

Orexin-mediated signaling

As discussed above, the orexin peptides exert their effects by binding to two closely related G protein-coupled receptors. *In vitro* and *in vivo* models show that orexin signaling is of an excitatory nature at the cellular level. Increased intracellular Ca^{2+} influx has been accepted as the most immediate cellular response to orexin receptor activation in both overexpression and *in vivo* models.[25] Signaling responses for orexin receptors and their specific G-α subunit activation are currently under intense investigation. The activation of either orexin receptor can be coupled to G_q, $G_{i/o}$, or G_s G-α subunit proteins, which can modulate ion channels and exchangers to induce neuronal depolarization.[28,61–70] Thus far, increased cellular activity by either OX1R and OX2R can be mediated by modulation of nonselective cationic currents (NSCC), voltage-gated calcium channels, the Na^+/Ca^{2+} exchanger, and inwardly rectifying potassium channels.[27,28,30,71–79]

The type of intracellular mechanism triggered by activation of the receptors appears to be cell dependent. For example, in nucleus accumbens and nucleus of the solitary tract, OX1R/OX2R mediated depolarization requires a simultaneous decrease in K^+ conductance and increase in NSCC.[26,73] In GABAergic neurons from the arcuate nucleus, it occurs through a decrease in K^+ conductance and activation of the Na^+/Ca^{2+} exchanger.[80] Finally, there are differences in the temporal profile of intracellular Ca^{2+} increases after OX1R/OX2R activation between neurons from the dorsal raphe and laterodorsal tegmental areas.[29,81]

The specific mechanisms involved in orexin mediated second messenger cascades, and their physiological relevance to obesity, are relatively undefined. Homogeneous overexpression models with human OX1R have revealed alterations in adenylyl cyclase activity via $G_{i/o}$, G_s, and G_q subunits but differ in their potency.[25,61] OXA can also activate extracellular signal-regulated kinases (ERK1/2) and p38 mitogen-activated phosphate kinase (MAPK) in recombinant and adrenal cell culture models.[82] OXA activation of either OX1R/OX2R in cells overexpressing either receptor can elicit the activation of ERK1/2 and p38 MAPK via multiple G-α subunits.[70] Food deprivation in Wistar rats has also revealed differences in G-α subunit activation in hypothalamic tissue homogenates in response to OXA.[83] However, discrete hypothalamic OXA-induced G-α subunit signaling responses for either OX1R or OX2R have yet to be determined when coexpressed.

The relevance of OX1R/OX2R in obesity has been exemplified in the obesity resistant (OR) and obesity prone (OP) rats. OR rats have higher basal levels of both intrinsic SPA and OXA-induced SPA following injections into the rostral lateral hypothalamic area (rLH) than OP or Sprague Dawley rats.[34] Increased OXA sensitivity in OR rats appears to be due to an increase in receptor abundance compared to OP rats. While a difference in receptor density may address OXA sensitivity in OR rats, receptor functionality may also help explain the influence of orexin on SPA. Some aspects of orexin receptor sensitivity, distribution, and intracellular signaling mechanisms important in mediating OXA effects on SPA are currently under investigation using the OP or OR rat and other polygenic models of obesity. One such possibility under investigation is that that rLH orexin receptors in OR rats couple to G_s rather than $G_{i/o}$ proteins, while the opposite occurs in OP rats.

Orexin in an animal model of obesity resistance

Levin *et al.* showed that when fed a high-fat diet a tertile of outbred Sprague Dawley rats gained no more weight than chow-fed controls.[84] These diet-induced obese rats and their weight-gain resistant counterparts, referred to as diet resistant, were selectively bred by a commercial vendor for over ten years,[85,86] resulting in the current OP/OR polygenic model of obesity. The OP and OR rats have divergent weight gain profiles despite inconsistently observed differences in energy intake.[34,85,86] While early studies demonstrated that obese rats had a dampened feeding response to satiety-promoting agents such as leptin[87] and insulin,[88] it was clear that other neural modulators as well as differences in SPA likely contributed to the polygenic obesity observed in OR and OP rats. As the orexins modulate SPA,[31,32,89] these findings underscored the potential significance of orexin as a neural modulator regulating body weight in this rodent model.

Like the outbred diet-resistant rats, the selectively bred OR rats exhibit lower body weight and fat mass gain on a low-fat diet and gain less weight when fed high-fat diet relative to their obesity-prone counterparts.[33,34,85,90] These OR rats consume significantly fewer absolute calories, but they consume statistically *more* calories when calculated on a per gram body mass basis.[34] Differences in SPA between OP and OR rats suggested by an early study[91] were confirmed by tracking SPA levels in several groups of OP and OR rats at various ages using a chamber that tracks activity in the *x*, *y*, and *z* axes using infrared beams.[34,90,92,93] OR rats display more ambulatory and vertical movement independent of age or the presence of food,[34,90,92] and this finding has been consistent across different groups of OR and OP rats.[92] Subsequent studies revealed that this greater SPA was associated with greater energy expenditure,[92] a lower propensity to gain fat mass throughout development, maturation, and aging,[90] and maintenance of higher SPA levels after high-fat diet feeding.[33]

OXA-induced SPA is associated with a dose-dependent increase in energy expenditure.[31] Together with our previous studies showing OXA-induced hyperphagia following rLH infusion,[94,95] we hypothesized that heightened responses to SPA-promoting agents and a dampened response to feeding-stimulatory agents such as OXA would perpetuate the lean phenotype in OR rats. To test this, OR and OP rats with chronically implanted guide cannulae targeting the rLH were given graded doses of OXA. In separate experiments, SPA and food intake were measured postinjection in young and adult rats. As expected, OR rats had greater OXA-induced SPA independent of age,[34] but OR rats also had greater OXA-induced food intake per gram body mass than OP rats (also independent of age).[34] This increase in caloric intake fits with the above-described enhanced 24 hour basal caloric intake in OR versus OP rats, and the idea that any behavioral effect of OXA in OR rats may be heightened due to higher OXA signaling capacity in OR rats (described in more detail below). OR rats maintain a lean phenotype over time, suggesting that the negative caloric benefit of OXA-induced SPA appears to outweigh the positive calories due to OXA-induced hyperphagia. A potential contributing mechanism for this is the observed longer duration of OXA action on SPA relative to that on food intake.[89] Further supporting the idea that OR rats have higher endogenous SPA, we later showed that OR rats are also more sensitive to other SPA-promoting stimuli including caloric restriction[96] and appear to be intrinsically protected from treatments that lower SPA, such as high-fat diet feeding. We and others have shown that in contrast to OP rats, which display lower SPA levels after high-fat diet consumption, OR rats maintain high basal SPA levels and have greater OXA-induced SPA after high fat diet feeding.[97] Most importantly, we also showed that rLH-OXA increases energy expenditure,[93] and others found that daily OXA treatment reduces body weight by increasing SPA.[98] These findings support the hypothesis that elevated energy expenditure due to SPA-promoting agents such as OXA and defense from SPA-dampening treatments protects against excessive adiposity gain in OR rats.

To understand whether differences in responsivity to OXA are driven by greater orexin signaling at the level of the peptide or the receptor, we analyzed mRNA data for prepro-orexin, OX1R, and

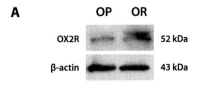

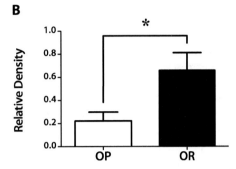

Figure 1. Difference in orexin 2 receptor (OX2R) protein levels between obesity-prone (OP) and obesity-resistant (OR) rats on chow diet. (A) Western blot analysis of caudal lateral hypothalamus (cLH) of OP and OR rats. The 52–53 kDA band is the predicted size of OXR2. Samples were normalized to β-actin. (B) Densitometric analysis of the Western blot data for all OP and OR rats. $N = 4$/group; $^*P < 0.05$.

OX2R from brain micropunches in OR and OP rats. Our data show that relative to OP rats, OR rats have greater orexin receptor mRNA in the rLH despite similar levels of preproorexin in the rLH or within whole hypothalamus.[34] This elevated receptor mRNA is mirrored by elevated receptor peptide levels in OR rats (Fig. 1). We later showed greater orexin receptor mRNA in the dorsal raphe, locus coeruleus, and ventrolateral preoptic area, in addition to better sleep quality in OR rats, which would be expected to contribute to the favorable weight status observed.[99] Together with our earlier work, these data suggest that elevated orexin receptor mRNA within distinct brain sites function to create a brain-wide orexin signaling network at the level of the receptor that perpetuates heightened basal and OXA-stimulated SPA levels, which attenuates adiposity gain in OR rats.

There are many animal models developed to amplify divergent locomotor activity, wheel running, or aerobic capacity.[17,100] These models help clarify the role of overall physical activity level in obesity, but they do not mirror models varying in SPA, as exercise has been shown to have a strong motivational component, whereas SPA may not. A recent series of studies undertaken by Novak *et al.*[101,102] in rodents selectively bred for high and low wheel running capacity (HCR and LCR, respectively)[103] have shown that HCR rats also have greater SPA, and basal and activity-induced energy expenditure.[102] Furthermore, like OR rats, HCR rats have heightened SPA following OXA infusion in the hypothalamic PVN.[101] However, in contrast to OR rats, greater OXA-induced SPA in HCR rats is accompanied by greater OXA content but not increased orexin receptor mRNA in the perifornical lateral hypothalamus.[101]

It is possible that the differences in orexin function between animal models bred for high exercise and the OP and OR rats arise from long-term effects of high fat consumption in this particular model. In itself, this makes the OP and OR rats a more appropiate model for studies of diet-induced obesity. The evidence from the OP and OR model suggests that the interaction between high fat consumption and the orexins is determined by individual susceptibility to high fat consumption, which in turn might be determined by baseline orexin signaling through particular brain regions, including the rLH. In summary, while animal models of high exercise may share some of the same orexin signaling characteristics as that of high SPA models, there are likely differences in the regulatory control of orexin between these models, which might be a consequence of the high-fat intake used to derive the OP and OR rats.

Orexin and sleep

As mentioned above, OR rats, with higher orexin signaling capacity, also exhibit more consolidated sleep relative to OP rats. The overall distribution of orexin fibers in the brain has suggested that the orexins play a role in a number of systems, including the maintenance of arousal.[104,105] Orexin fibers have been shown to project to several brain nuclei implicated in the control of sleep state.[35–39] Application of OXA in the locus coeruleus[104,106] and lateral preoptic area[107] of the rat increase wakefulness, primarily through a decrease in rapid eye movement (REM) sleep.[106] Activity in locus coeruleus (LC) neurons increases following application of OXA.[104–106] More recently, direct stimulation of electrical activity in orexin neurons using optogenetic techniques was shown to induce wakefulness in sleeping mice.[108]

In addition to projecting to sleep-wake nuclei, orexin cells receive input from brain systems involved in regulation of sleep-wakefulness. In mammals, circadian organization of activity including sleep-wake behavior is regulated by the endogenous clock located in the suprachiasmatic nucleus (SCN).[109] Orexin cell bodies receive both limited direct contact from the SCN,[110] as well as substantial indirect contact from the SCN via the medial preoptic area and the subparaventricular zone.[111,112] Orexin neurons show circadian patterns of activation,[113] and ablation of the SCN eliminates rhythmicity of orexin release.[114] Introduction of chemicals known to increase arousal in rats, such as methamphetamines or the anti narcoleptic drug modafinil, increase nuclear Fos expression in orexin cell bodies.[115,116] Furthermore, increasing the behavioral arousal of rats by sleep deprivation induced due to handling also increases the expression of nuclear Fos in OXA cells.[116] Finally, in a diurnal rodent model, Fos expression patterns in orexin neurons are correlated with individual variation in the timing of daily wheel running activity.[117] The orexins thus appear to be capable of both receiving information related to the arousal state of the animal, and relaying arousal information to other nuclei known to promote wakefulness.

The association between the orexins and arousal was strengthened by the discovery that sleep disorder narcolepsy is associated with a defect in the orexin system.[57,115] While it is clear from this evidence that orexin is not necessary for wakefulness, data suggest that orexin is important in maintaining high levels of arousal, and that one major function of the orexin system is to stabilize sleep–wake transitions. Furthermore, there is a recognition that orexin activity may be incompatible with sleep, as direct activation of orexin neurons causes wakefulness in rodents,[108] and silencing of orexin neuronal activity during the inactive period results in slow wave sleep.[118]

Two comorbidities associated with narcolepsy—cataplexy and obesity—help shed light on the importance of orexin in normal physiology. Early studies of orexin effects suggested that orexin results in the activation of motor activity,[119,120] and it is well established that physical activity is correlated with both activation of orexin neurons[116,117] and increases in OXA release.[121] In narcoleptic individuals, cataplexy (defined as a loss of muscle tone) is often triggered by emotional stress or physical exertion,[122] and is preceded by a reduction of neuronal firing in the LC.[123] Injection of OXA into the locus coeruleus activates LC neurons[104–106] and increases muscle tone.[124] Promotion of motor activity may thus be one important function of orexin, and this orexin-induced activity may be coupled strongly to behavioral state to maintain normal motor tone during periods of emotional or physical stress.

Obesity is a comorbidity of narcolepsy in both human and animal models.[57,122,125] Both human and animal subjects with narcolepsy eat less as would be expected given the association between orexins and feeding behavior.[57,125] Yet the effect of reduced caloric consumption is offset by decreases in physical activity, as narcoleptic individuals exhibit a significantly elevated body mass index relative to non-narcoleptic patients.[122] In human subjects, while the total time spent awake is not reduced, there is a decreased amplitude of circadian activity patterns, consistent with reduced overall physical activity.[126] In a mouse model of narcolepsy, in which orexin neurons are ablated postnatally, physical activity during the active (but not the resting) phase is reduced in affected animals,[57] and these animals subsequently become obese. These human and animal studies demonstrate that the effects of orexin on sleep–wake patterns and physical activity are consistent with the idea that orexin is a neuropeptide conveying resistance to obesity.

Instability of sleep patterns is known to contribute to weight gain.[127,128] In this light, it could be argued that weight gain in narcolepsy is more due to disturbance of sleep than to reductions in activity due to lack of orexin signaling. However, evidence from a diurnal rodent model and from laboratory rats suggests that orexin-associated activity can be altered without disturbing total sleep. Wheel running activity in some Nile grass rats can occur exclusively at night, during the inactive phase, while in others wheel running follows the normal diurnal pattern.[129] While Fos activation in orexin neurons occurs only during the light period in day-active animals, in night-active grass rats Fos is elevated both during nightly activity bouts and during the day.[117] Behavioral measures of sleep in these animals showed that while the timing of sleep was changed in night-active animals due to wheel running at night, neither total sleep nor duration of sleep bouts differed between groups.[130] Finally, pre-

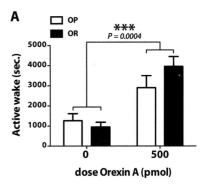

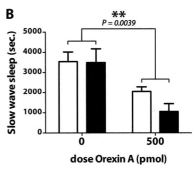

Figure 2. Total active wake (A) and slow wave sleep (B), in seconds, for obesity-prone (OP) and obesity-resistant (OR) rats for a 2-h period after treatment with vehicle or 500 pmol orexin A (OXA). In both panels, there is a significant overall effect of OXA on increasing active wake ($P = 0.0004$) and decreasing quiet sleep ($P = 0.0039$), respectively. However, there is no difference for either wake or sleep between OP and OR rats within treatment groups, despite the fact that OR rats show significantly higher levels of spontaneous physical activity following OXA treatment. $N = 6$/group; data are means ± SEM. $**P < 0.005; ***P < 0.0005$.

viously unpublished data from our laboratory show that in OP and OR rats the effects of orexin on activity and arousal are not inextricably linked. Application of exogenous orexin significantly increases physical activity in OR relative to OP rats; however, recordings of total sleep using implanted EEG/EMG transmitters showed that this increase in activity was not caused by increased arousal or reduced sleep time in OR rats (Fig. 2). Importantly, while both sets of rats had increased time spent in wakefulness, there was not a greater increase in OR rats relative to OP rats. This lack of difference between OP and OR rats in time spent awake after orexin treatment suggests that SPA following orexin is not merely due to increased wakefulness; the SPA effect in OR rats is in addition to the wake-promoting effect of orexin.

Translating SPA to energy expenditure

An important consideration is the relevance of SPA to overall energy expenditure. Energy expenditure (EE) comprises at least four main components, including basal metabolic rate (BMR; 60–70% of EE), which is the minimal energy required to maintain life, including heart beat, respiration, endocrine secretion, and kidney filtration. Diet-induced thermogenesis constitutes about 10% of total EE and includes energy related to the digestion, absorption, and metabolism of foods. Adaptive thermogenesis ranges from 10% to 15% of total EE and is related to adjustments in energy expenditure due to environment changes (e.g., shivering). Physical activity is the most variable of these components, ranging from 6% to 10% of total EE.[131]

SPA is neither a part of BMR nor a part of exercise physical activity. Therefore, it may be an attractive obesity target. SPA in humans was identified as early as 1954, and defined as a component of energy expenditure.[132] Ravussin *et al.*[15] showed that SPA in a human respiratory chamber averaged 348 kcal/day and, importantly, identified a large range in values: 100–700 kcal/day. Clearly, this wide range in values among humans suggests that there could be a large range in the weight gain response to overfeeding in humans, as demonstrated by Levine *et al.*[11] Zurlo *et al.*[133] showed that levels of SPA clustered in families, and could prospectively help explain propensity for weight gain in males, which suggests heritability of SPA. As discussed above, the idea that SPA level is an intrinsic heritable trait has recently been strengthened by Levine *et al.*, in his study showing that lean humans stand and ambulate for approximately two hours daily more than obese, which is not affected by weight loss or weight gain, in the obese and lean respectively.[11] That spontaneous physical activity levels differ between mouse strains[134] also suggests that SPA is an intrinsic, inherited trait that varies within animal species.

Despite large differences in body fat, energy intake, and body size, OR and OP rats expend a similar number of absolute kilocalories.[92] This suggests that OR rats are less efficient in their calorie use, as they are expending relatively large amounts of calories to support their relatively smaller energy needs related to their reduced body circumference and body fat, which affect levels of heat loss. This supports the idea that elevated SPA and the resultant NEAT in OR rats contributes to their obesity resistant phenotype.[34]

When considering the therapeutic potential of SPA, a practical consideration is the comparability of the SPA difference between OP and OR rats to that in obesity-prone versus obesity-resistant humans. Calories expended via spontaneous physical activity correlate with whole body energy expenditure in humans and in animals.[14,92] However, are the extra calories mediated by SPA in an OR rat, when compared to that in humans, enough to make meaningful changes in body weight? Based on indirect calorimetry studies, we estimate the energy flux in a rat to be about 100 kcal/day; SPA differences between OP and OR rats, when corrected for lean mass, amount to about 4 kcal/day or 4%.[92] In humans, a 100 kcal energy gap (that level of energy intake above the daily requirement to maintain a stable body weight) per day, or 5% for a person in balance at 2,000 kcal/day, can lead to a 10-lb weight differential over one year.[135] Based on this synthesis, we conclude that it is clear that SPA differences explaining obesity resistance in rodents can be relevant for human body weight control. Further, orexin-mediated mechanisms identified here may explain the differential body weight gain response to overfeeding in humans.[11] For example, a prior human study showed that gain or loss of 10–20 pounds resulted in linear changes in energy expenditure, and that the majority of the change was specifically in nonresting energy expenditure (i.e., NEAT).[136] We know that physical activity is correlated with both activation of orexin neurons[116,117] and increases in OXA release,[121] although the pathways through which this is effected are largely undefined. Thus increased SPA (either endogenous or artificial) could also lead to feedback mechanisms that maintain higher SPA levels in the future.

As discussed elsewhere in this review, orexin enhances feeding behavior and physical activity in a site-specific manner, with some sites conveying information regarding eating behavior, others activity, yet others both or neither. As our data and others show, the energetic consequence of these two behavioral outputs, when added up on a caloric basis, may result in negative energy balance and reduce body weight. In other words, the calories taken in by the effects of the orexin signal are outweighed by those expended via physical activity.

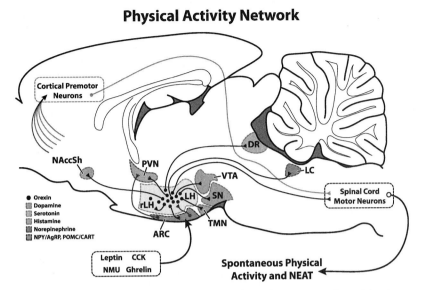

Figure 3. Overview of the involvement of orexin in a neural network regulating nonexercise activity thermogenesis (NEAT), indicating a sample of the brain areas, neuropeptides, and transmitters involved. Colors correspond to specific neuropeptide/hormone as follows: Black circles/projections: orexin; green, dopamine; orange, serotonin; pink, histamine; blue, norepinephrine; brown, neuropeptide Y/agouti-related protein (NPY/AgRP) and proopiomelanocortin/cocaine and amphetamine-related transcript (POMC/CART). Signals from all of these areas have the potential to influence cortical premotor neurons (indicated by arrows), and feedback from premotor neurons and orexinergic projections may interact to drive SPA. See the text for details. CCK, cholecystokinin; DR, dorsal raphe; LC, locus coeruleus; LH, lateral hypothalamus; NAccSH, shell of nucleus accumbens; NMU, neuromedin U; PVN, hypothalamic paraventricular nucleus; VTA, ventral tegmental area; rLH, rostral LH; SN, substantia nigra; TMN, tuberomammillary nucleus. Brain areas are not all to scale, and connections and neuropeptides/transmitters indicated are not all inclusive. Not all connections shown are discussed in this review. Figure modified from work by Kotz *et al.*[92]

Networks regulating spontaneous physical activity

Orexin plays a key role in an interdependent distributed brain network that regulates spontaneous physical activity. A significant number of brain sites that participate in this regulatory network in the forebrain and hindbrain have been identified, and there are a number of neurotransmitters that participate in this network, as depicted in Figure 3. The SPA network is distinct from the brain pathways that regulate purposeful activity, although the final common pathway leading to movement is clearly shared. Many of the brain sites that participate in the SPA network also participate in regulatory networks for food intake and other aspects of energy balance such as thermogenesis, but based both on distribution and on functional responses to stimulation, including orexin stimulation, it is clear that the SPA network is different from the networks that otherwise regulate energy balance.

Orexin is a unique contributor to this distributed SPA regulatory network by virtue of its position and projections. Orexin is made in one relatively small area of the hypothalamus, involving the caudal lateral hypothalamus and adjacent perifornical area.[20,21] From these sites, orexin projects throughout the brain. Orexin neurons project throughout the hypothalamus, including the paraventricular, arcuate, rostrolateral, perifornical, and ventromedial areas, as well as to several extrahypothalamic sites, including the septal nuclei, bed nucleus of the stria terminalis, paraventricular, and reunions nuclei of the thalamus, zona incerta, subthalamic nucleus, central gray, substantia nigra, dorsal raphe nuclei, parabrachial nucleus, locus coeruleus, medullary reticular formation, area postrema, and nucleus of the solitary tract.[92]

The sites receiving orexin signal vary considerably with respect to the primary function associated with that site.[53] Further, it is likely that in many sites the orexin effect on SPA is partially overlapping with other known orexin actions such as attention and wakefulness. While the functional outcomes of orexin action vary considerably from site to site, the production of SPA is common across many sites, and is widely distributed.[92] This organization, involving a focused site of origin with wide distribution of effect, greatly enhances the potential potency of orexin as a regulator of spontaneous physical activity.

Orexin neurons receive input from a number of sites throughout the brain that are thought to influence expression of the orexin signal. The lateral hypothalamus receives afferents from cortical structures, including the prefrontal/orbitofrontal, insular, and olfactory cortices; limbic sites, including the amygdala, the hippocampal formation, and the shell of the nucleus accumbens; and from brainstem sites, including the nucleus of the solitary tract.[137] Projections from other parts of hypothalamus include those from arcuate nucleus proopiomelanocortin (POMC)/cocaine and amphetamine-related transcript (CART) and NPY/agouti-related protein (AgRP) neurons.[138,139] In addition, there is connectivity within the lateral hypothalamic area, notably projections from anterior to posterior portions.[140] Whether all of these lateral hypothalamic projections play a significant role in regulating the activity of orexin neurons specifically is not yet determined, but a strong network of local lateral hypothalamic interneurons indicates the possibility of influences even by projections that do not directly synapse on orexin neurons themselves.[140] Tracing studies have specifically identified projections to lateral hypothalamic orexin neurons from several regions of the amygdala, nucleus accumbens shell, bed nucleus of the stria terminalis, laterodorsal tegmental area, basal forebrain cholinergic neurons, GABAergic neurons in the preoptic area, and serotonergic neurons in the median/paramedian raphe nuclei.[138]

Orexin neurons are in a baseline intrinsic state of depolarized activity[141] and are highly influenced by local conditions in an intralateral hypothalamic local network.[140] The functional effects of the many afferents, and the associated neural function and neurotransmitters to orexin neurons, are underexploration. Application of the cholinergic agent carbachol activates many orexin neurons, indicating that the cholinergic input to orexin neurons from basal forebrain is excitatory.[138] Cholinergic input from the laterodorsal tegmental area may also be

excitatory to orexin neurons. Input of an as yet unidentified chemical type from the amygdala and bed nucleus of stria terminalis may also stimulate orexin neurons.[138] CRF release from projections originating in the hypothalamic paraventricular nucleus also activates orexin neurons.[142] Inhibitory influences on orexin neurons come from the preoptic area and from the serotonergic neurons in the median raphe.[138]

The physiological conditions that are known to affect orexin neurons include suppression of activity by glucose,[143] along with prominent activation by hypoglycemia[144] and by food restriction.[145] There is evidence that low glucose states may be directly sensed by orexin neurons,[143] although the likelihood of low glucose signals originating in other glucose sensing neurons with projections to orexin neurons is substantial. Intracellular Foxa2 signaling from the insulin receptor may be a mechanism allowing orexin neurons to sense the state of glucose and possibly short-term nutrition.[146] Orexin neurons also receive local projections from leptin receptor bearing neurons, providing a means for translating the state of leptin signaling to the orexin neurons, and leptin action in lateral hypothalamus increases orexin action and decreases food intake.[147] Leptin and energy state sensing arcuate neurons that express POMC/CART and NPY/AgRP also project to lateral hypothalamic neurons.[138,139] A recent study has shown that amino acids, particularly nonessential amino acids, can stimulate orexin neurons directly through action on potassium channels and amino acid transporters.[148] The stimulation provided by amino acids may be potent enough to overcome inhibition by glucose.[148] Additional energy and nutrient related information may come to orexin neurons, perhaps directly, from the gut hormones ghrelin and glucagon-like peptide 1, both of which appear to activate orexin neurons in direct administration studies.[120,149] Orexin neurons also receive input about physiological stress–related information, likely through a corticotrophin releasing hormone pathway.[142] The integration of nutrition-related and other signals in orexin neurons, or in a network of which orexin is part, has not been defined.

Orexin signaling pathways can be further modified by the level of orexin-receptor expression in brain sites receiving orexin efferents. The conditions that lead to modulation of orexin receptor expression are incompletely defined. One example is the difference in orexin-receptor expression in a variety of brain sites associated with the difference between orexin-induced SPA response in polygenic OP and OR rat strains,[90,99] as described elsewhere in this review.

Many brain areas contribute to SPA, and all of these areas operate in a network; thus activity in one area, affected by environmental cues or physiology as discussed above, influences firing patterns in other areas. Behavioral studies of SPA can determine the output of specific brain activity, improving understanding of the brain sites and neurotransmitter systems that are most important. The existing literature, however, is not always interpretable in a straightforward way. Locomotor activity measured in a beam-break chamber (as has been done for SPA measures) has also been used to assess nonspecific drug effects, as in studies of drugs of abuse. Similarly, low locomotor activity has been used as a diagnostic criterion for depression or illness in rats and mice.[150] Thus, the data must be interpreted with care and in many cases repeated in a new context for full understanding.

Orexin A injected in almost all brain areas increases SPA, contrasting with feeding behavior that is stimulated after injection into only some of the same sites.[92] The time course of action is different for the feeding and activity effects of orexin A, so the presence of one behavior (feeding or SPA) does not depend upon the other.[95] Whether OXA-induced SPA is derived from orexin-enhanced wakefulness is not clear, but as discussed elsewhere in this review it appears likely that energy expenditure produced from SPA occurs after the initial waking event.

Many neurotransmitters have been shown to influence SPA in the network into which orexin action is projecting, although most of the available studies reporting on locomotor activity do not directly consider SPA itself, so at present some inferences must be made. Neurotransmitters that are likely to affect SPA include cholecystokinin, corticotrophin releasing hormone, neuromedin, NPY, leptin, and orexin.[19] The current state of the evidence does not permit straightforward interpretation of the direction of effect for these neurotransmitters since there is disparate evidence based in part on site and type of administration.[19] In general, it appears that in most situations each of these neurotransmitters can stimulate SPA, but orexin is the most consistent across all brain sites and types of stimulation. Little is presently known about the interactions of orexin with these other neurotransmitters with respect to the regulation of SPA.

Ultimately the output pathways for the SPA regulatory network must share engagement of the motor control pathways with voluntary movement brain mechanisms. The brain locations where these movement regulatory pathways begin to overlap have not yet been defined. It is likely that in part there are projections from forebrain structures, including the hypothalamus, to spinal motor neurons. An interesting possibility is that the SPA regulatory network engages cortical areas involved in voluntary motor control. The wide pattern of projections of the orexin neurons throughout the brain could mediate this function. In addition, a projection pathway from the accumbens through cortical premotor neurons and out to the spinal motor neurons has also been implicated.[92]

Conclusion

Brain mechanisms mediate SPA and NEAT (Fig. 3), and the understanding of this concept is beginning to shed light on new ways to target obesity prevention and treatment. Studies of orexin and its role in obesity resistance show that stimulation of orexin receptors may be an attractive therapy for altering the course of excess body weight gain with aging, and also demonstrate that modulating SPA and NEAT has important consequences for obesity resistance. The knowledge of this can guide human obesity therapy immediately, as the option to include more low-level activity throughout one's day is likely more feasible advice to prevent and treat obesity than standard approaches that repeatedly fail over time. The greater challenge is in how to use information on brain SPA and NEAT networks to provide better pharmaceutical and/or other therapeutic approaches for treatment of obesity. The rapid development of new neurochemical methods of altering brain neurophysiology and recent advances in computer/brain interface technologies provide confidence that knowledge of brain SPA and NEAT networks could be therapeutically applied in the near future.

Conflicts of interest

The authors declare no conflicts of interest.

References

1. Berthoud, H.R. & C. Morrison. 2008. The brain, appetite, and obesity. *Annu. Rev. Psychol.* **59:** 55–92.

2. Must, A. *et al.* 1999. The disease burden associated with overweight and obesity. *JAMA* **282:** 1523–1529.

3. Wang, Y. & T. Lobstein. 2006. Worldwide trends in childhood overweight and obesity. *Int. J. Pediatr. Obes.* **1:** 11–25.

4. Ogden, C.L. *et al.* 2006. Prevalence of overweight and obesity in the United States, 1999–2004. *JAMA* **295:** 1549–1555.

5. Flegal, K.M. *et al.* 2010. Prevalence and trends in obesity among US adults, 1999–2008. *JAMA* **303:** 235–241.

6. Bouchard, C. *et al.* 1990. The response to long-term overfeeding in identical twins. *N. Engl. J. Med.* **322:** 1477–1482.

7. Forbes, G.B. *et al.* 1986. Deliberate overfeeding in women and men: energy cost and composition of the weight gain. *Br. J. Nutr.* **56:** 1–9.

8. Mustelin, L. *et al.* 2009. Physical activity reduces the influence of genetic effects on BMI and waist circumference: a study in young adult twins. *Int. J. Obes.* **33:** 29–36.

9. Hamilton, M.T., D.G. Hamilton & T.W. Zderic. 2007. Role of low energy expenditure and sitting in obesity, metabolic syndrome, type 2 diabetes, and cardiovascular disease. *Diabetes* **56:** 2655–2667.

10. Levine, J.A. 2002. Non-exercise activity thermogenesis (NEAT). *Best Pract. Res. Clin. Endocrinol. Metab.* **16:** 679–702.

11. Levine, J.A., N.L. Eberhardt & M.D. Jensen. 1999. Role of nonexercise activity thermogenesis in resistance to fat gain in humans. *Science* **283:** 212–214.

12. Levine, J.A. *et al.* 2005. Interindividual variation in posture allocation: possible role in human obesity. *Science* **307:** 584–586.

13. Fruhbeck, G. 2005. Does a NEAT difference in energy expenditure lead to obesity? *Lancet* **366:** 615–616.

14. Ravussin, E. 2005. Physiology. A NEAT way to control weight? *Science* **307:** 530–531.

15. Snitker, S., P.A. Tataranni & E. Ravussin. 2001. Spontaneous physical activity in a respiratory chamber is correlated to habitual physical activity. *Int. J. Obes. Relat. Metab. Disord.* **25:** 1481–1486.

16. Kotz, C.M. *et al.* 2006. Orexin A mediation of time spent moving in rats: neural mechanisms. *Neuroscience* **142:** 29–36.

17. Garland, T., Jr. *et al.* 2011. The biological control of voluntary exercise, spontaneous physical activity and daily energy expenditure in relation to obesity: human and rodent perspectives. *J. Exp. Biol.* **214:** 206–229.

18. Dishman, R.K. 2008. Gene-physical activity interactions in the etiology of obesity: behavioral considerations. *Obesity* **16** Suppl 3: S60–S65.

19. Teske, J.A., C.J. Billington & C.M. Kotz. 2008. Neuropeptidergic mediators of spontaneous physical activity and non-exercise activity thermogenesis. *Neuroendocrinology* **87:** 71–90.

20. Sakurai, T. *et al.* 1998. Orexins and orexin receptors: a family of hypothalamic neuropeptides and G protein-coupled receptors that regulate feeding behavior. *Cell* **92:** 573–585.

21. de Lecea, L. *et al.* 1998. The hypocretins: hypothalamus-specific peptides with neuroexcitatory activity. *Proc. Natl. Acad. Sci. USA* **95:** 322–327.

22. Burdakov, D., O. Gerasimenko & A. Verkhratsky. 2005. Physiological changes in glucose differentially modulate the excitability of hypothalamic melanin-concentrating hormone and orexin neurons in situ. *J. Neurosci.* **25:** 2429–2433.

23. Li, Y. *et al.* 2002. Hypocretin/Orexin excites hypocretin neurons via a local glutamate neuron-A potential mechanism for orchestrating the hypothalamic arousal system. *Neuron* **36:** 1169–1181.

24. Schone, C. *et al.* 2011. Dichotomous cellular properties of mouse orexin/hypocretin neurons. *J. Physiol.* **589:** 2767–2779.

25. Ammoun, S. *et al.* 2003. Distinct recognition of OX1 and OX2 receptors by orexin peptides. *J. Pharmacol. Exp. Ther.* **305:** 507–514.

26. Yang, B. & A.V. Ferguson. 2003. Orexin-A depolarizes nucleus tractus solitarius neurons through effects on nonselective cationic and K+ conductances. *J. Neurophysiol.* **89:** 2167–2175.

27. van den Pol, A.N. et al. 1998. Presynaptic and postsynaptic actions and modulation of neuroendocrine neurons by a new hypothalamic peptide, hypocretin/orexin. *J. Neurosci.* **18:** 7962–7971.

28. Larsson, K.P. *et al.* 2005. Orexin-A-induced Ca2+ entry: evidence for involvement of trpc channels and protein kinase C regulation. *J. Biol. Chem.* **280:** 1771–1781.

29. Uramura, K. *et al.* 2001. Orexin-a activates phospholipase C- and protein kinase C-mediated Ca2 +signaling in dopamine neurons of the ventral tegmental area. *Neuroreport* **12:** 1885–1889.

30. Follwell, M.J. & A.V. Ferguson. 2002. Cellular mechanisms of orexin actions on paraventricular nucleus neurones in rat hypothalamus. *J. Physiol.* **545:** 855–867.

31. Kiwaki, K. *et al.* 2004. Orexin A (hypocretin 1) injected into hypothalamic paraventricular nucleus and spontaneous physical activity in rats. *Am. J. Physiol. Endocrinol. Metab.* **286:** E551–559.

32. Kotz, C.M. *et al.* 2002. Feeding and activity induced by orexin A in the lateral hypothalamus in rats. *Regul. Pept.* **104:** 27–32.

33. Novak, C.M., C.M. Kotz & J.A. Levine. 2006. Central orexin sensitivity, physical activity, and obesity in diet-induced obese and diet-resistant rats. *Am. J. Physiol. Endocrinol. Metab.* **290:** E396–403.

34. Teske, J.A. *et al.* 2006. Elevated hypothalamic orexin signaling, sensitivity to orexin A, and spontaneous physical activity in obesity-resistant rats. *Am. J. Physiol. Regul. Integr. Comp. Physiol.* **291:** R889–899.

35. Date, Y. *et al.* 1999. Orexins, orexigenic hypothalamic peptides, interact with autonomic, neuroendocrine and neuroregulatory systems. *Proc. Natl. Acad. Sci. USA* **96:** 748–753.

36. Mintz, E.M. *et al.* 2001. Distribution of hypocretin-(orexin) immunoreactivity in the central nervous system of Syrian hamsters (Mesocricetus auratus). *J. Chem. Neuroanat.* **21:** 225–238.

37. Moore, R.Y., E.A. Abrahamson & A. Van Den Pol. 2001. The hypocretin neuron system: an arousal system in the human brain. *Arch. Ital. Biol.* **139:** 195–205.

38. Nixon, J.P. & L. Smale. 2007. A comparative analysis of the distribution of immunoreactive orexin A and B in the brains of nocturnal and diurnal rodents. *Behav. Brain Funct.* **3:** 28.

39. Peyron, C. *et al.* 1998. Neurons containing hypocretin (orexin) project to multiple neuronal systems. *J. Neurosci.* **18:** 9996–10015.

40. Espana, R.A. *et al.* 2005. Organization of hypocretin/orexin efferents to locus coeruleus and basal forebrain arousal-related structures. *J. Comp. Neurol.* **481:** 160–178.

41. Ciriello, J. *et al.* 2003. Identification of neurons containing orexin-B (hypocretin-2) immunoreactivity in limbic structures. *Brain Res.* **967:** 123–131.

42. Krout, K.E., T.C. Mettenleiter & A.D. Loewy. 2003. Single CNS neurons link both central motor and cardiosympathetic systems: a double-virus tracing study. *Neuroscience* **118:** 853–866.

43. Geerling, J.C., T.C. Mettenleiter & A.D. Loewy. 2003. Orexin neurons project to diverse sympathetic outflow systems. *Neuroscience* **122:** 541–550.

44. Oldfield, B.J. *et al.* 2007. Lateral hypothalamic 'command neurons' with axonal projections to regions involved in both feeding and thermogenesis. *Eur. J. Neurosci.* **25:** 2404–2412.

45. Cluderay, J.E., D.C. Harrison & G.J. Hervieu. 2002. Protein distribution of the orexin-2 receptor in the rat central nervous system. *Regul. Pept.* **104:** 131–144.

46. Greco, M.A. & P.J. Shiromani. 2001. Hypocretin receptor protein and mRNA expression in the dorsolateral pons of rats. *Brain Res. Mol. Brain Res.* **88:** 176–182.

47. Hervieu, G.J. *et al.* 2001. Gene expression and protein distribution of the orexin-1 receptor in the rat brain and spinal cord. *Neuroscience* **103:** 777–797.

48. Marcus, J.N. *et al.* 2001. Differential expression of orexin receptors 1 and 2 in the rat brain. *J. Comp. Neurol.* **435:** 6–25.

49. Sunter, D. *et al.* 2001. Orexins: effects on behavior and localisation of orexin receptor 2 messenger ribonucleic acid in the rat brainstem. *Brain Res.* **907:** 27–34.

50. Trivedi, P. *et al.* 1998. Distribution of orexin receptor mRNA in the rat brain. *FEBS Lett.* **438:** 71–75.

51. Berridge, C.W., R.A. Espana & N.M. Vittoz. 2010. Hypocretin/orexin in arousal and stress. *Brain Res.* **1314:** 91–102.

52. Harris, G.C. & G. Aston-Jones. 2006. Arousal and reward: a dichotomy in orexin function. *Trends Neurosci.* **29:** 571–577.

53. Kotz, C.M. 2006. Integration of feeding and spontaneous physical activity: role for orexin. *Physiol. Behav.* **88:** 294–301.

54. Sakurai, T. 2005. Roles of orexin/hypocretin in regulation of sleep/wakefulness and energy homeostasis. *Sleep Med. Rev.* **9:** 231–241.

55. Sharf, R., M. Sarhan & R.J. Dileone. 2010. Role of orexin/hypocretin in dependence and addiction. *Brain Res.* **1314:** 130–138.

56. Tsujino, N. & T. Sakurai. 2009. Orexin/hypocretin: a neuropeptide at the interface of sleep, energy homeostasis, and reward system. *Pharmacol. Rev.* **61:** 162–176.

57. Hara, J. *et al.* 2001. Genetic ablation of orexin neurons in mice results in narcolepsy, hypophagia, and obesity. *Neuron* **30:** 345–354.

58. Akiyama, M. *et al.* 2004. Reduced food anticipatory activity in genetically orexin (hypocretin) neuron-ablated mice. *Eur. J. Neurosci.* **20:** 3054–3062.

59. Mieda, M. *et al.* 2004. Orexin peptides prevent cataplexy and improve wakefulness in an orexin neuron-ablated model of narcolepsy in mice. *Proc. Natl. Acad. Sci. USA* **101:** 4649–4654.

60. Funato, H. *et al.* 2009. Enhanced orexin receptor-2 signaling prevents diet-induced obesity and improves leptin sensitivity. *Cell Metab.* **9:** 64–76.

61. Ammoun, S. *et al.* 2006. OX1 orexin receptors activate extracellular signal-regulated kinase in Chinese hamster ovary cells via multiple mechanisms: the role of Ca2+ influx in OX1 receptor signaling. *Mol. Endocrinol.* **20:** 80–99.

62. Ekholm, M.E., L. Johansson & J.P. Kukkonen. 2007. IP3-independent signalling of OX1 orexin/hypocretin receptors to Ca2+ influx and ERK. *Biochem. Biophys. Res. Commun.* **353:** 475–480.

63. Johansson, L., M.E. Ekholm & J.P. Kukkonen. 2007. Regulation of OX1 orexin/hypocretin receptor-coupling to phospholipase C by Ca2 +influx. *Br. J. Pharmacol.* **150:** 97–104.

64. Johansson, L., M.E. Ekholm & J.P. Kukkonen. 2008. Multiple phospholipase activation by OX(1) orexin/hypocretin receptors. *Cell Mol. Life Sci.* **65:** 1948–1956.

65. Kukkonen, J.P. & K.E. Akerman. 2001. Orexin receptors couple to Ca2+ channels different from store-operated Ca2+ channels. *Neuroreport* **12:** 2017–2020.

66. Lund, P.E. *et al.* 2000. The orexin OX1 receptor activates a novel Ca2+ influx pathway necessary for coupling to phospholipase C. *J. Biol. Chem.* **275:** 30806–30812.

67. Magga, J. *et al.* 2006. Agonist potency differentiates G protein activation and Ca2+ signalling by the orexin receptor type 1. *Biochem. Pharmacol.* **71:** 827–836.

68. Nasman, J. *et al.* 2006. The orexin OX1 receptor regulates Ca2+ entry via diacylglycerol-activated channels in differentiated neuroblastoma cells. *J. Neurosci.* **26:** 10658–10666.

69. Peltonen, H.M. *et al.* 2009. Involvement of TRPC3 channels in calcium oscillations mediated by OX(1) orexin receptors. *Biochem. Biophys. Res. Commun.* **385:** 408–412.

70. Tang, J. *et al.* 2008. The signalling profile of recombinant human orexin-2 receptor. *Cell Signal.* **20:** 1651–1661.

71. Hoang, Q.V. *et al.* 2003. Effects of orexin (hypocretin) on GIRK channels. *J. Neurophysiol.* **90:** 693–702.

72. Hoang, Q.V. *et al.* 2004. Orexin (hypocretin) effects on constitutively active inward rectifier K +channels in cultured nucleus basalis neurons. *J. Neurophysiol.* **92:** 3183–3191.

73. Mukai, K. *et al.* 2009. Electrophysiological effects of orexin/hypocretin on nucleus accumbens shell neurons in rats: an in vitro study. *Peptides* **30:** 1487–1496.

74. van den Pol, A.N. *et al.* 2002. Hypocretin (orexin) enhances neuron activity and cell synchrony in developing mouse GFP-expressing locus coeruleus. *J. Physiol.* **541:** 169–185.

75. Samson, W.K. *et al.* 2002. Orexin actions in hypothalamic paraventricular nucleus: physiological consequences and cellular correlates. *Regul. Pept.* **104:** 97–103.

76. Burlet, S., C.J. Tyler & C.S. Leonard. 2002. Direct and indirect excitation of laterodorsal tegmental neurons by Hypocretin/Orexin peptides: implications for wakefulness and narcolepsy. *J. Neurosci.* **22:** 2862–2872.

77. Bayer, L. *et al.* 2002. Selective action of orexin (hypocretin) on nonspecific thalamocortical projection neurons. *J. Neurosci.* **22:** 7835–7839.

78. Bisetti, A. *et al.* 2006. Excitatory action of hypocretin/orexin on neurons of the central medial amygdala. *Neuroscience* **142:** 999–1004.

79. Liu, R.J., A. N. van den Pol & G.K. Aghajanian. 2002. Hypocretins (orexins) regulate serotonin neurons in the dorsal raphe nucleus by excitatory direct and inhibitory indirect actions. *J. Neurosci.* **22:** 9453–9464.

80. Burdakov, D., B. Liss & F.M. Ashcroft. 2003. Orexin excites GABAergic neurons of the arcuate nucleus by activating the sodium–calcium exchanger. *J. Neurosci.* **23:** 4951–4957.

81. Kohlmeier, K.A., T. Inoue & C.S. Leonard. 2004. Hypocretin/orexin peptide signaling in the ascending arousal system: elevation of intracellular calcium in the mouse dorsal raphe and laterodorsal tegmentum. *J. Neurophysiol.* **92:** 221–235.

82. Wenzel, J. *et al.* 2009. Hypocretin/orexin increases the expression of steroidogenic enzymes in human adrenocortical NCI H295R cells. *Am. J. Physiol. Regul. Integr. Comp. Physiol.* **297:** R1601–R1609.

83. Karteris, E. *et al.* 2005. Food deprivation differentially modulates orexin receptor expression and signaling in rat hypothalamus and adrenal cortex. *Am. J. Physiol. Endocrinol. Metab.* **288:** E1089–E1100.

84. Levin, B.E., S. Hogan & A.C. Sullivan. 1989. Initiation and perpetuation of obesity and obesity resistance in rats. *Am. J. Physiol.* **256:** R766–R771.

85. Ricci, M.R. & B.E. Levin. 2003. Ontogeny of diet-induced obesity in selectively bred Sprague-Dawley rats. *Am. J. Physiol. Regul. Integr. Comp. Physiol.* **285:** R610–R618.

86. Levin, B.E. *et al.* 1997. Selective breeding for diet-induced obesity and resistance in Sprague-Dawley rats. *Am.J. Physiol.* **273:** R725–R730.

87. Levin, B.E. & A.A. Dunn-Meynell. 2002. Reduced central leptin sensitivity in rats with diet-induced obesity. *Am. J. Physiol. Regul. Integr. Comp. Physiol.* **283:** R941–R948.

88. Clegg, D.J. *et al.* 2005. Reduced anorexic effects of insulin in obesity-prone rats fed a moderate-fat diet. *Am. J. Physiol. Regul. Integr Comp. Physiol.* **288:** R981–R986.

89. Thorpe, A.J. & C.M. Kotz. 2005. Orexin A in the nucleus accumbens stimulates feeding and locomotor activity. *Brain Res.* **1050:** 156–162.

90. Teske, J.A., C.J. Billington, M.A. Kuskowksi, & C.M. Kotz 2012. Spontaneous physical activity protects against fat mass gain. *Int. J. Obes.* (Lond). **36**(4): 603–613. doi:10.1038/ijo.2011.108.

91. Levin, B.E. 1991. Spontaneous motor activity during the development and maintenance of diet-induced obesity in the rat. *Physiol. Behav.* **50:** 573–581.

92. Kotz, C.M., J.A. Teske & C.J. Billington. 2008. Neuroregulation of nonexercise activity thermogenesis and obesity resistance. *Am. J. Physiol. Regul. Integr. Comp. Physiol.* **294:** R699–R710.

93. Teske, J.A., C.J. Billington & C.M. Kotz. 2010. Hypocretin/orexin and energy expenditure. *Acta Physiol. (Oxf).* **198:** 303–312.

94. Sweet, D.C. *et al.* 1999. Feeding response to central orexins. *Brain Res.* **821:** 535–538.

95. Thorpe, A.J. *et al.* 2003. Peptides that regulate food intake: regional, metabolic, and circadian specificity of lateral hypothalamic orexin A feeding stimulation. *Am. J. Physiol. Regul. Integr. Comp. Physiol.* **284:** R1409–1417.

96. Teske, J.A. & C.M. Kotz. 2009. Effect of acute and chronic caloric restriction and metabolic glucoprivation on spontaneous physical activity in obesity-prone and obesity-resistant rats. *Am. J. Physiol. Regul. Integr. Comp. Physiol.* **297:** R176–184.

97. Novak, C.M., M. Zhang & J.A. Levine. 2007. Sensitivity of the hypothalamic paraventricular nucleus to the locomotor-activating effects of neuromedin U in obesity. *Brain Res.* **1169:** 57–68.

98. Novak, C.M. & J.A. Levine. 2009. Daily intraparaventricular orexin—A treatment induces weight loss in rats. *Obesity* **17:** 1493–1498.

99. Mavanji, V. *et al.* 2010. Elevated sleep quality and orexin receptor mRNA in obesity-resistant rats. *Int. J. Obes* **34:** 1576–1588.

100. Feder, M.E. *et al.* 2010. Locomotion in response to shifting climate zones: not so fast. *Annu. Rev. Physiol.* **72:** 167–190.

101. Novak, C.M. *et al.* 2010. Spontaneous activity, economy of activity, and resistance to diet-induced obesity in rats bred for high intrinsic aerobic capacity. *Hormo. Behav.* **58:** 355–367.

102. Novak, C.M. *et al.* 2009. Endurance capacity, not body size, determines physical activity levels: role of skeletal muscle PEPCK. *PloS ONE* **4:** e5869.

103. Britton, S.L. & L.G. Koch. 2001. Animal genetic models for complex traits of physical capacity. *Exe. Sport Sci. Rev.* **29:** 7–14.

104. Hagan, J.J. *et al.* 1999. Orexin A activates locus coeruleus cell firing and increases arousal in the rat. *Proc. Natl. Acad. Sci. USA* **96:** 10911–10916.

105. Horvath, T.L. *et al.* 1999. Hypocretin (orexin) activation and synaptic innervation of the locus coeruleus noradrenergic system. *J. Comp. Neurol.* **415:** 145–159.

106. Bourgin, P. *et al.* 2000. Hypocretin-1 modulates rapid eye movement sleep through activation of locus coeruleus neurons. *J. Neurosci.* **20:** 7760–7765.

107. Methippara, M.M. *et al.* 2000. Effects of lateral preoptic area application of orexin-A on sleep-wakefulness. *Neuroreport* **11:** 3423–3426.

108. Adamantidis, A.R. *et al.* 2007. Neural substrates of awakening probed with optogenetic control of hypocretin neurons. *Nature* **450:** 420–424.

109. Weaver, D.R. 1998. The suprachiasmatic nucleus: a 25-year retrospective. *J. Biol. Rhythms* **13:** 100–112.

110. Abrahamson, E.E., R.K. Leak & R.Y. Moore. 2001. The suprachiasmatic nucleus projects to posterior hypothalamic arousal systems. *Neuroreport* **12:** 435–440.

111. Deurveilher, S. & K. Semba. 2005. Indirect projections from the suprachiasmatic nucleus to major arousal-promoting cell groups in rat: implications for the circadian control of behavioural state. *Neuroscience* **130:** 165–183.

112. Yoshida, K. *et al.* 2006. Afferents to the orexin neurons of the rat brain. *J. Comp. Neurol.* **494:** 845–861.

113. Martinez, G.S., L. Smale & A.A. Nunez. 2002. Diurnal and nocturnal rodents show rhythms in orexinergic neurons. *Brain Res.* **955:** 1–7.

114. Zhang, S. *et al.* 2004. Lesions of the suprachiasmatic nucleus eliminate the daily rhythm of hypocretin-1 release. *Sleep* **27:** 619–627.

115. Chemelli, R.M. *et al.* 1999. Narcolepsy in orexin knockout mice: molecular genetics of sleep regulation. *Cell* **98:** 437–451.

116. Estabrooke, I.V. *et al.* 2001. Fos expression in orexin neurons varies with behavioral state. *J. Neurosci.* **21:** 1656–1662.

117. Nixon, J.P. & L. Smale. 2004. Individual differences in wheel-running rhythms are related to temporal and spatial patterns of activation of orexin A and B cells in a diurnal rodent (Arvicanthis niloticus). *Neuroscience* **127:** 25–34.

118. Tsunematsu, T. *et al.* 2011. Acute optogenetic silencing of orexin/hypocretin neurons induces slow-wave sleep in mice. *J. Neurosci.* **31:** 10529–10539.

119. Ida, T. *et al.* 1999. Effect of lateral cerebroventricular injection of the appetite-stimulating neuropeptide, orexin and neuropeptide Y, on the various behavioral activities of rats. *Brain Res.* **821:** 526–529.

120. Yamanaka, A. *et al.* 2003. Hypothalamic orexin neurons regulate arousal according to energy balance in mice. *Neuron* **38:** 701–713.

121. Wu, M.F. *et al.* 2002. Hypocretin release in normal and narcoleptic dogs after food and sleep deprivation, eating, and movement. *Am. J. Physiol. Regul. Integr. Comp. Physiol.* **283:** R1079–1086.

122. Overeem, S. *et al.* 2001. Narcolepsy: clinical features, new pathophysiologic insights, and future perspectives. *J. Clin. Neurophysiol.* **18:** 78–105.

123. Siegel, J.M. 2004. Hypocretin (orexin): role in normal behavior and neuropathology. *Annu. Rev. Psychol.* **55:** 125–148.

124. Kiyashchenko, L.I. *et al.* 2002. Release of hypocretin (orexin) during waking and sleep states. *J. Neurosci.* **22:** 5282–5286.

125. Schuld, A. *et al.* 2000. Increased body-mass index in patients with narcolepsy. *Lancet* **355:** 1274–1275.

126. Middelkoop, H.A. *et al.* 1995. Circadian distribution of motor activity and immobility in narcolepsy: assessment with continuous motor activity monitoring. *Psychophysiology* **32:** 286–291.

127. Stamatakis, K.A. & N.M. Punjabi. 2010. Effects of sleep fragmentation on glucose metabolism in normal subjects. *Chest* **137:** 95–101.

128. Spiegel, K. *et al.* 2004. Brief communication: sleep curtailment in healthy young men is associated with decreased leptin levels, elevated ghrelin levels, and increased hunger and appetite. *Ann. Intern. Med.* **141:** 846–850.

129. Blanchong, J.A. *et al.* 1999. Nocturnal and diurnal rhythms in the unstriped Nile rat, Arvicanthis niloticus. *J. Biol. Rhythms* **14:** 364–377.

130. Schwartz, M.D. & L. Smale. 2005. Individual differences in rhythms of behavioral sleep and its neural substrates in Nile grass rats. *J. Biol. Rhythms* **20:** 526–537.

131. Donahoo, W.T., J.A. Levine & E.L. Melanson. 2004. Variability in energy expenditure and its components. *Curr. Opin. Clin. Nutr. Metab. Care* **7:** 599–605.

132. Widdowson, E.M., O.G. Edholm & C.R. Mc. 1954. The food intake and energy expenditure of cadets in training. *Br. J. Nutr.* **8:** 147–155.

133. Zurlo, F. *et al.* 1992. Spontaneous physical activity and obesity: cross-sectional and longitudinal studies in Pima Indians. *Am. J. Physiol.* **263:** E296–E300.

134. Tou, J.C. & C.E. Wade. 2002. Determinants affecting physical activity levels in animal models. *Exp. Biol. Med.* **227:** 587–600.

135. Hill, J.O. 2009. Can a small-changes approach help address the obesity epidemic? A report of the Joint Task Force of the American Society for Nutrition, Institute of Food Technologists, and International Food Information Council. *Am. J. Clin. Nutr.* **89:** 477–484.

136. Leibel, R.L., M. Rosenbaum & J. Hirsch. 1995. Changes in energy expenditure resulting from altered body weight. *N. Engl. J. Med.* **332:** 621–628.

137. Berthoud, H.R. & H. Munzberg. 2011. The lateral hypothalamus as integrator of metabolic and environmental needs: from electrical self-stimulation to opto-genetics. *Physiol. Behav.* **104:** 29–39.

138. Sakurai, T. *et al.* 2005. Input of orexin/hypocretin neurons revealed by a genetically encoded tracer in mice. *Neuron* **46:** 297–308.

139. Kampe, J. *et al.* 2009. An anatomic basis for the communication of hypothalamic, cortical and mesolimbic circuitry in the regulation of energy balance. *Eur. J. Neurosci.* **30:** 415–430.

140. Burt, J. *et al.* 2011. Local network regulation of orexin neurons in the lateral hypothalamus. *Am. J. Physiol. Regul. Integr. Comp. Physiol.* **301:** R572–R580.

141. Eggermann, E. *et al.* 2003. The wake-promoting hypocretin-orexin neurons are in an intrinsic state of membrane depolarization. *J. Neurosci.* **23:** 1557–1562.

142. Winsky-Sommerer, R. *et al.* 2004. Interaction between the corticotropin-releasing factor system and hypocretins (orexins): a novel circuit mediating stress response. *J. Neurosci.* **24:** 11439–11448.

143. Williams, R.H. *et al.* 2008. Adaptive sugar sensors in hypothalamic feeding circuits. *Proc. Natl. Acad. Sci. USA* **105:** 11975–11980.

144. Moriguchi, T. *et al.* 1999. Neurons containing orexin in the lateral hypothalamic area of the adult rat brain are activated by insulin-induced acute hypoglycemia. *Neurosci. Lett.* **264:** 101–104.

145. Cai, X.J. *et al.* 1999. Hypothalamic orexin expression: modulation by blood glucose and feeding. *Diabetes* **48:** 2132–2137.

146. Silva, J.P. *et al.* 2009. Regulation of adaptive behaviour during fasting by hypothalamic Foxa2. *Nature* **462:** 646–650.

147. Louis, G.W. *et al.* 2010. Direct innervation and modulation of orexin neurons by lateral hypothalamic LepRb neurons. *J. Neurosci.* **30:** 11278–11287.

148. Karnani, M.M. *et al.* 2011. Activation of central orexin/hypocretin neurons by dietary amino acids. *Neuron* **72:** 616–629.

149. Acuna-Goycolea, C. & A. van den Pol. 2004. Glucagon-like peptide 1 excites hypocretin/orexin neurons by direct and indirect mechanisms: implications for viscera-mediated arousal. *J. Neurosci.* **24:** 8141–8152.

150. Malisch, J.L. *et al.* 2009. Behavioral despair and home-cage activity in mice with chronically elevated baseline corticosterone concentrations. *Behav. Genet.* **39:** 192–201.

Ann. N.Y. Acad. Sci. ISSN 0077-8923

ANNALS OF THE NEW YORK ACADEMY OF SCIENCES

Issue: *The Brain and Obesity*

Visceral adipose tissue: emerging role of gluco- and mineralocorticoid hormones in the setting of cardiometabolic alterations

Marco Boscaro, Gilberta Giacchetti, and Vanessa Ronconi

Division of Endocrinology, Ospedali Riuniti "Umberto I-G.M. Lancisi-G. Salesi," Università Politecnica delle Marche, Ancona, Italy

Address for correspondence: Dr. Gilberta Giacchetti, Division of Endocrinology, Ospedali Riunti "Umberto I-G.M. Lancisi-G. Salesi," Univeristà Politecnica delle Marche, Via Conca 71, 60126 Ancona, Italy. g.giacchetti@ospedaliriuniti.marche.it

Several clinical and experimental lines of evidence have highlighted the detrimental effects of visceral adipose tissue excess on cardiometabolic parameters. Besides, recent findings have shown the effects of gluco-and mineralocorticoid hormones on adipose tissue and have also underscored the interplay existing between such adrenal steroids and their respective receptors in the modulation of adipose tissue biology. While the fundamental role played by glucocorticoids on adipocyte differentiation and storage was already well known, the relevance of the mineralocorticoids in the physiology of the adipose organ is of recent acquisition. The local and systemic renin–angiotensin–aldosterone system (RAAS) acting on adipose tissue seems to contribute to the development of the cardiometabolic phenotype so that its modulation can have deep impact on human health. A better understanding of the pathophysiology of the adipose organ is of crucial importance in order to identify possible therapeutic approaches that can avoid the development of such cardiovascular and metabolic sequelae.

Keywords: adipose tissue; glucocorticoids; mineralocorticoids

Introduction

Overweight and obesity have become a major public health problem in industrialized countries. Epidemiological studies have highlighted a rapid increase of such conditions worldwide, with a prevalence of about 30% and 20%, respectively, in the general population of the United States of America.[1] Along with the increase in obesity there is a parallel increase in the prevalence of obesity-related diseases, such as type 2 diabetes, impaired glucose tolerance,[2,3] and arterial hypertension.

The combination of excessive food intake and reduced physical activity—with consequent overweight/obesity—can lead, in subjects who are genetically predisposed, to insulin resistance in peripheral tissues, such as skeletal muscle, liver, and adipose tissue, which is the *primum movens* for other pathological conditions. Insulin resistance is indeed a pathogenetic element that plays a key role in the development of metabolic and hemodynamic alter-

ations and is responsible, in turn, for the onset of the so-called cardiometabolic syndrome.[4]

But what is the role of adipose tissue in this scenario? Adipose tissue, traditionally regarded simply as an inert energy storage organ, is appreciated increasingly as an endocrine organ and an important part of the innate immune system. In the last few years it has been recognized that adipose tissue can produce and secrete into the blood stream a wide variety of bioactive mediators named *adipokines*. The relevance of adrenal steroid excess, and in particular of cortisol—and more recently of aldosterone—has been proposed, on the basis of old and new clinical and experimental studies, to be involved in the pathogenesis of metabolic complications.

Adipose tissue

Histologically, two fundamentally different adipose tissue types can be differentiated: white adipose tissue (WAT) and brown adipose tissue (BAT).

doi: 10.1111/j.1749-6632.2012.06597.x

Ann. N.Y. Acad. Sci. 1264 (2012) 87–102 © 2012 New York Academy of Sciences.

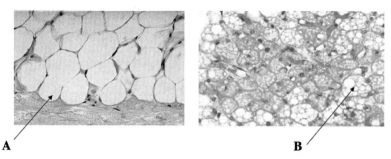

A　　　　　　　　　　　　　**B**

Figure 1. Morphological appearance of white (left) and brown (right) adipose tissue (WAT and BAT). (A) Unilocular adipocyte; (B) multilocular adipocyte.

Additionally, its anatomical distribution further classifies adipose tissue as either subcutaneous or visceral. Rodents have WAT and BAT in distinct depots, epididymal and interscapular, respectively, while the topographic distribution of BAT in humans is slightly different. Humans are born with BAT located mainly around the neck and large blood vessels of the thorax that is then partially replaced by WAT in adults.

Traditionally, the main functions ascribed to adipose tissue have been insulation, mechanical support, and storage of surplus fuel. In fact, in the presence of increased food intake and/or decreased energy expenditure, surplus energy is deposited through the action of lipogenic enzymes in the form of neutral triglycerides in adipocyte droplets. Conversely, when food is scarce and/or energy expenditure requirements increase, lipid reserves are released through lipolitic enzymes to provide fuel (i.e., free fatty acid, FFA) for energy generation in peripheral tissues and organs, as in the liver, muscle, and BAT.

From a morphological point of view, BAT and WAT are formed by different adipocytes. While white adipocytes form only single large vacuoles (unilocular cells) that contain triglycerides, brown adipocytes form numerous small vacuoles (multilocular cells) as "quick-access" fuel for heat production through mitochondrial uncoupling of oxidative phosphorylation of FFA[5] (Fig. 1). This thermogenic process is of vital importance in neonates exposed to the cold, and still persists in adult humans, in whom substantial amounts of metabolically active BAT have been detected in the paracervical and supraclavicular region.[6] The signal for the activation of BAT via activation of the sympathetic nervous system is a temperature below thermoneutrality (34 °C

for mice, 28 °C for rats, and 20–22 °C for humans). A second known stimulus for activation of BAT is food intake, thus suggesting the hypothesis of antiobesity properties of BAT.[7]

Adipose tissue is not a homogenous organ. It consists of a variety of different cell types: adipocytes, preadipocytes, stromal/vascular cells, and macrophages.[8] Macrophages are known to be crucial contributors to inflammation, and quite recently it has been shown that WAT inflammation, due to macrophage infiltration, is a relevant and early event in the development of obesity-related complications.[9–12] Cinti *et al.* have also shown that macrophage infiltration in both murine and human WAT of obese subjects is linked to adipocyte death.[13] The majority of such macrophages, likely attracted to phagocyte lipids and cellular debris, surround dead adipocytes, forming crown-like structures.

More recently, it has been recognized that adipocytes also demonstrate intrinsic inflammatory properties, and, like macrophages, they sense the presence of pathogens and inflammation and activate multiple inflammatory signal transduction cascades that result in the secretion of several inflammatory cytokines and acute-phase reactants.[14]

Adipose tissue as an endocrine organ

In 1994, the discovery of leptin—a satiety factor—as an adipocyte-secreted protein[15] led to the definition of WAT as an endocrine organ. Systemic analysis of the active genes in WAT, after constructing a 3′-directed complementary DNA library, revealed a high frequency of genes encoding secretory proteins.

As said, adipose tissue contains different cell types; and, importantly, each of these cell types presents its own secretion profile and specific

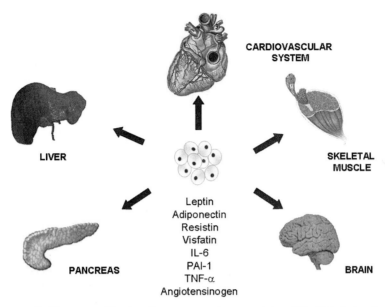

Figure 2. List of some adipokines secreted by the adipose organ and representation of the biological systems affected by such humoral products.

regulation. The additional cell types present in the adipose tissue, or its stromal-vascular fraction, include pericytes and endothelial cells, monocytes, macrophages, and pluripotent stem cells (including preadipocytes). Interestingly, these nonadipocyte cells may also be the main source of some secreted factors. The adipose organ expresses indeed more than 8,000 genes, including those for more than 120 receptors and 80 secreted proteins and hormones. Approximately 20–30% of all genes in WAT encode secretory proteins.[16] Such humoral products are involved in processes such as inflammation, lipid metabolism, energy balance, vascular tone, and atherosclerosis, but also glucose homeostasis and insulin sensitivity.[17] Leptin, adiponectin, plasminogen activator inhibitor (PAI 1), and components of the renin–angiotensin–aldosterone system are only a few of the substances produced by adipose tissue, which can act both with autocrine/paracrine mechanisms and in an endocrine manner—thus adipose tissue can be fully considered an adipose organ (Fig. 2).[18]

The relevance of adipocytes for health was shown through the use of the animal model of lipoatrophic diabetes,[19] in which mice had virtually no white fat tissue due to genetic manipulation. These animals displayed insulin resistance, hyperglycemia, hyperlipidemia, and fatty livers—all the characteristics of humans with severe lipoatrophic diabetes. This particular phenotype was completely reversed by the transplantation of adipose tissue from healthy mice, highlighting that the absence of adipocytes is metabolically detrimental. Conversely, WAT, especially the visceral fat, which is more hormonally active than the subcutaneous one, can also act as a "bad guy." It can potentially become one of the largest organs in the body, and when that happens, the total number of adipokines secreted from WAT can affect whole-body homeostasis. The massive increase in fat mass leads in fact to a dysregulation of circulating adipokine levels that may have pathogenic effects associated with obesity by triggering obesity-associated disorders, including systemic inflammation, insulin resistance, hypertension, hyperlipidemia, type 2 diabetes, and coronary heart disease.

In 1988, the American Heart Association identified obesity as a major, modifiable risk factor for coronary heart disease.[20] In particular, visceral obesity, that is, accumulation of adipose tissue within the abdomen, is associated with cardiovascular and metabolic complications.[21,22] Health problems are indeed more closely correlated to *android* obesity (visceral obesity or upper body obesity) than to *gynoid* obesity (lower body obesity). At a clinical level, visceral obesity is characterized by an increase in the waist-to-hip ratio, and this measurement is better than BMI as an indicator of cardiovascular and metabolic risk.[23]

Adipose tissue and the cardiometabolic syndrome

Cardiometabolic syndrome refers to the clustering of obesity-related metabolic disorders in one individual; in particular, it defines a syndrome characterized by the presence of glucose and lipid profile alterations, insulin resistance, hypertension, and, finally, cardiovascular diseases. This condition affects one in four adults, making it the leading public health issue associated with increased cardiovascular disease risk in the industrialized world.[24] The mechanisms that causally relate visceral obesity and metabolic syndrome—and therefore cardiovascular diseases—are not fully understood.

A hallmark of cardiometabolic syndrome is undoubtedly insulin resistance. Insulin resistance is defined as an inadequate response by insulin target tissues, such as skeletal muscle, liver, and adipose tissue, to the physiologic effects of circulating insulin. Impaired insulin sensitivity in these three tissues leads to reduced insulin-mediated glucose uptake by skeletal muscle, impaired insulin-mediated inhibition of endogenous glucose production in the liver, and a reduced ability of insulin to inhibit lipolysis in adipose tissue. Insulin resistance is a major predictor of the development of various metabolic complications, such as type 2 diabetes. It is indeed well established that in type 2 diabetes overt hyperglycemia is preceded by insulin resistance.[25] The causes of insulin resistance can be genetic and/or acquired, yet the genetic causes or predispositions toward insulin resistance in prediabetic populations are poorly understood. Although inherited defects in the basic insulin signaling cascade have been proposed,[26] it is likely that any genetic component must interact with environmental factors. In industrialized countries, the most common acquired factors causing insulin resistance are obesity, sedentary lifestyle, and aging—all of which are interrelated.[27] From a pathogenetic point, many lines of evidence have shown that chronic activation of proinflammatory pathways within insulin target cells can lead to obesity-related insulin resistance. Consistent with this, elevated levels of the proinflammatory cytokines TNF-α, IL-6, and C-reactive protein (CRP) have

been shown in individuals with insulin resistance and diabetes.[28–30]

Elevated plasma glucose is sensed by pancreatic beta cells, which increase insulin secretion to compensate for hyperglycemia, resulting in circulating hyperinsulinemia. However, over time, beta cells fail to secrete insulin normally and can no longer compensate for the decreased tissue insulin sensitivity, with consequent development of impaired glucose tolerance and eventually type 2 diabetes.[25] An increase of circulating FFA levels is observed before patients with insulin resistance develop glucose metabolism alterations. Impairment of insulin signaling in adipose tissue leads to increase of lipolysis and, possibly, defective storage of FFA in adipocytes.[31,32] Insulin resistance is also responsible for decreased activity of lipoprotein lipase and limited degradation of apoB. Together, these actions induce hypertriglyceridemia, characteristic of insulin resistance, and low aHDL phenotype, which are implicated in the development of atherosclerosis.[33]

Another relevant aspect of metabolic syndrome is its association with high blood pressure. Several mechanisms have been proposed to explain the pathogenesis of hypertension in relation to obesity and insulin resistance. Some of the physiological and tissue-specific consequences by which insulin resistance could result in hypertension include changes in vascular structure and function, alterations in cation flux, activation of the sympathetic nervous system, and enhanced renal sodium retention.[34] In this regard, the antinatriuretic action of insulin has been highlighted. The insulin-mediated reduction in Na^+ excretion appears to be mainly due to increased Na^+ reabsorption at the level of Henle's loop.[35,36] Chronic hyperinsulinemia can thus cause the rise in blood pressure through an increase in extracellular volume and cardiac output. Insulin can also enhance renal sodium retention through stimulation of the sympathetic nervous system and augmentation of angiotensin II–mediated aldosterone secretion.[37,38]

Moreover, the increase in FFA release by adipose tissue, which is observed in insulin resistant states, has a direct effect on peripheral resistance, as well as an effect mediated by the inhibition of NO synthase. In addition, insulin and insulin-like growth factors are mitogens capable of stimulating smooth muscle proliferation.[39] Therefore, hyperinsulinemia could result in vascular smooth muscle hypertrophy responsible for increased vascular resistance that, ultimately, leads to the development of high blood pressure. Another role for hyperinsulinemia in the etiology of hypertension related to insulin resistance is via upregulation of AT1R; this potentiates the physiologic actions of AngII, which include peripheral vasoconstriction and plasma volume expansion.[40,41]

Finally, metabolic syndrome is a condition characterized by proinflammatory and prothrombotic states. Elevation of cytokines (e.g., TNF-α and IL-6), as well as acute-phase reactants (CRP and fibrinogen), is indeed peculiar to the syndrome. Elevated CRP—defined as a risk factor for CVD[42]—seems to be linked to obesity, as excess adipose tissue releases inflammatory cytokines that may lead to higher CRP levels and to higher levels of the prothromotic factors plasminogen activator inhibitor (PAI)-1 and fibrinogen. Thus, prothrombotic and proinflammatory states may be metabolically interconnected.[43] But what is the interplay between visceral adipose tissue, the cardiometabolic syndrome, and adrenal steroids such as gluco- and mineralocorticoids?

Adipose tissue and glucocorticoids

Close phenotypic similarities exist between metabolic syndrome and conditions characterized by chronic exposure to glucocorticoid hormones, either exogenous or endogenous, such as in Cushing syndrome (CS). Common features are abdominal obesity, insulin resistance, hypertension, hyperglicemia, and dyslipidemia, which are all comorbidities strictly associated with the presence of obesity itself.

Obesity in CS is characterized by increased food intake with no changes in energy expenditure, reduced lean mass, and increased body weight with a redistribution of fat mass from peripheral toward central sites of the body, mainly in the truncal region and visceral depots.[44] This is not surprising if one considers that glucocorticoids (GCs) promote both the differentiation and the proliferation of human adipocytes through glucocorticoid receptors (GRs), which are more abundantly expressed in visceral than in subcutaneous adipose tissue.[45]

Although not univocal, several lines of evidence support the role of cortisol in the pathogenesis of obesity. Higher fasting cortisol levels have been found in patients with central obesity;[46] cortisol levels have been found to correlate with waist circumference,[47–48] while patients with abdominal obesity display an increased responsiveness of the hypothalamic–pituitary–adrenal (HPA) axis to a wide variety of stimuli, including food intake,[49] a low-dose ACTH test,[50] and CRH arginine vasopressin test.[51] In addition, in patients with central obesity, a loss of diurnal cortisol variation as well as a resistance to a low oral dexamethasone suppression test[52] have been described, suggesting hyperactivity of the HPA axis. Also increased urinary-free cortisol excretion has been found by different authors.[53–55]

Role of the glucocorticoid receptor polymorphisms

In the pathogenesis of GC-induced obesity, the relevance of individual sensitivity to GC exposure has been proposed and supported by genetic studies on the GR.[56] The GR is a ubiquitously expressed protein, encoded by exons 2–9 of the GR gene, located on chromosome 5 (5q31–32).[57,58] Alternative splicing of the GR precursor mRNA gives rise to 5 GR protein subtypes, termed GRα, GRβ, GRγ, GR-A, and GR-P. GRα is the functionally active protein, made up of 777 amino acids, and differs from GRβ only by its *C*-terminal residues.[59] Classically, GRβ has been considered a regulator of GRα via a dominant-negative effect. However, it has been recently demonstrated that GRβ also seems to have a role in cell signaling, regulating gene expression even in the absence of GRα.[60]

Polymorphisms in GR genes have been associated with variations in GR function[61,62] (Fig. 3). In particular, ER22/23EK carriers have a relative resistance to GCs and display a more favorable metabolic profile, including increased insulin sensitivity, and lower cholesterol levels and cardiovascular risk compared to wild-type carriers.[63] They also have beneficial effects on body composition (increased lean mass, lower fat mass, and same BMI) compared with noncarriers.[64]

On the other hand, N363S and *Bcl*I variations are associated with hypersensitivity to GCs, and may predispose to obesity.[65–67]

Although contrasting data have been reported on the association between *Bcl*I polymorphism and body composition,[68–70] it can be speculated that hypersensitivity to GCs has negative effects on abdominal fat mass early in life, while later in life GCs mainly affect lean mass by lowering it.[63] To confirm

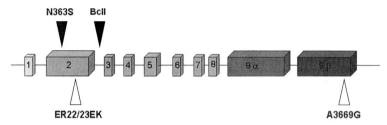

Figure 3. Schematic representation of the glucocorticoid receptor (GR) gene and its polymorphisms: N363S, codon 363 AAT→AGT asparagine (N) to serine (S) substitution; *Bcl*I, intron 2 C→G substitution, 646 nucelotides downstream from exon 2; ER22/23EK codon 22 GAG → GAA+ codon 23 AGG → AAG substitution with consequent arginine (R) to lysine (K) substitution. The black arrows indicate polymorphisms that increase the sensitivity to glucocorticoids, while the white arrows refer to polymorphisms that confer resistance to glucocorticoids.

such a hypothesis, a recent study has found an association between BMI, fat mass, and *Bcl*I variation in young carriers, while in older subjects the same variant is associated with lower BMI due to lower lean mass.[71]

As for the N363S variant, although a few studies found no association between it and BMI,[72,73] numerous other observations, conducted in different cohorts of subjects, have highlighted a contribution of the N363S variant to overweight and obesity.[67,74,75] Another studied GR polymorphism is the A3669G variant, which has been associated in Caucasian men with reduced central obesity and more favorable lipid profiles.[76] In support of a protective role by the A3669G polymorphism, a recent study found one in patients with CS for the development of diabetes mellitus.[77] Interestingly, in a subset of patients affected by eating disorders (anorexia nervosa, bulimia nervosa, and binge eating disorder) and obesity, the A3669G polymorphism was associated with binge eating disorder, while the N363S variant was associated with higher BMI independent of eating psychopathology.[78]

Role of 11β-hydroxysteroid dehydrogenase type 1

If one looks at GC action variability, one can distinguish between receptor level (i.e., at the GR) and prereceptor level action. Another fundamental variable that can modulate GC action in peripheral tissues is the activity of 11β–hydroxysteroid dehydrogenase type 1 (11β-HSD1), which determines GC availability at the prereceptor level in such tissues.

Circulating cortisol levels are not always increased in obese patients;[79–81] increased metabolic clearance has been claimed as a possible explanation for this,[82] but GC concentrations at a local level strictly linked to 11β-HSD1 activity is also of major importance. 11β-HSD1 is ubiquitously expressed and interconverts inactive cortisone and 11-dehydrocorticosterone to their active compounds cortisol and corticosterone. 11β-HSD1 is a bidirectional enzyme, although it acts predominantly as a reductase (converting inactive cortisone to active cortisol) rather than as a dehydrogenase, and it is highly expressed in adipose tissue, where it amplifies GC action independent of circulating cortisol levels.[83] Genetic studies in animal models have shown that 11β-HSD1 expression or activity plays a key role in determining a metabolic syndrome phenotype. Transgenic mice overexpressing 11β-HSD1 selectively in adipose tissue present visceral obesity, exaggerated by a high-fat diet, diabetes, insulin resistance, hyperlipemia, and hyperphagia.[84] Interestingly, overfed wild-type mice show reduced 11β-HSD1 expression in adipose tissue, indicating regulation of enzyme expression according to energy balance.[84] Hepatic overexpression of 11β-HSD1 in transgenic mice induces fatty liver, dyslipidemia, hypertension, and insulin resistance without obesity.[85] Knockout mice for 11β-HSD1 have a reduced risk for obesity and metabolic syndrome,[86] and data from our group support a tight correlation between obesity and 11β-HSD1, showing a significant positive correlation between BMI and 11β-HSD1 expression in different groups of subjects (CS and obese patients and control subjects).[87]

The lack of a correlation between circulating F levels and 11β-HSD1 expression in both obese subjects and Cushing's patients demonstrates that this enzyme is not directly regulated by plasma F concentrations. Surprisingly, we found 11β-HSD1 levels in visceral adipose tissue of patients with CS that are comparable to those observed in normal-weight control patients, which suggests that downregulation of the enzyme occurs as a result of long-term overstimulation.

Overall, data in the published literature indicate that 11β-HSD1 expression and activity are finely regulated tissue, specifically and strictly associated with energy balance status. This explains why 11β-HSD1 has been targeted therapeutically via the development of potent and specific inhibitors, including arylsulfonamidothiazoles,[88] adamantly triazoles,[89] anilinothiazolones,[90] and as further demonstrated in phase II studies recently published (inhibitor INCB13739) showing an improvement of glycemic control in DM2 patients, together with a modest reduction in body weight.[91] In adipose tissue, 11β-HSD1 inhibition is able to decrease lipolysis[92] and enhance lipogenesis, suggesting that the improvement of metabolic phenotype is likely mediated by decreased lipid mobilization.[83]

Role of AMP-activated protein kinase

AMP-activated protein kinase (AMPK) is a key element of energy metabolism, implicated in the regulation of several metabolic pathways resulting in the inhibition of anabolic pathways (fatty acid, triglycerides, cholesterol, and protein synthesis) and activation of catabolic pathways (glycolysis and fatty oxidation). AMPK switches off ATP-consuming processes and switches on catabolic processes that produce ATP, thus restoring the AMP:ATP ratio. AMPK phosphorylates and inactivates two rate-limiting enzymes in fatty acid and cholesterol synthesis, acetyl-CoA carboxylase (ACC) and HMG-CoA reductase.[93,94] Besides its lipid-related effects, AMPK has been implicated in carbohydrate and protein metabolism, cell-cycle regulation, and mitochondrisl biogenesis.

AMPK has a pivotal role in hypothalamic control of feeding behavior and is regulated by a wide variety of metabolic hormones, such as leptin, adiponectin, resistin, and ghrelin, and cannabinoids;[95] it also acts in peripheral tissues, such as skeletal muscle and liver, where it regulates glucose metabolism and decreases glycogen synthesis and gluconeogenesis. Decreased AMPK activity in visceral fat tissue can stimulste lipolysis and lipogenesis, although the effect on lipogenesis is predominant (Fig. 4).[96,97]

It has been proposed that many of the negative effects exerted by GCs, especially those related to

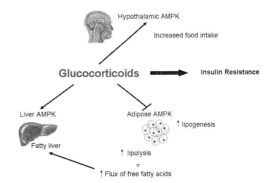

Figure 4. Summary of AMPK-mediated effects on different tissues and organs. AMPK switches off anabolic pathways and switches on catabolic pathways under the control of various metabolic hormones. Many metabolic effects exerted by glucocorticoids seem to be mediated by AMPK activation. In the central nervous system, AMPK activation favors increased appetite, while in the liver it leads to the development of hepatic steatosis through inhibition of gluconeogenesis and facilitation of lipid oxidation, associated with increased FFA availability due to AMPK inhibition in fat tissue, with consequent increased lipolysis and lipogenesis.

the development of metabolic alterations, could be mediated by GC-induced changes on AMPK activity, either directly or indirectly via stimulation of endocannabinoid synthesis. GCs can activate hypothalamic AMPK, which leads to a stimulation of appetite.[98] In adipose tissue, GC treatment reduces AMPK activity rather than AMPK expression, which leads to increased lipogenesis and fat storage.[83,98] GC's effects on adipocyte differentiation and lipid accumulation are more evident in visceral than in subcutaneous adipose tissue, which explains the development of central obesity observed during chronic GC exposure. Conversely, GCs were shown to increase AMPK activity in the rat liver, both *in vivo* and *in vitro*,[98,99] and, consequently, due to suppression of gluconeogenesis and facilitation of lipid oxidation with increased production of FFA, development of hepatic steatosis is observed. Increased circulating levels of FFAs facilitate the development of insulin resistance and concomitant impairment of insulin signaling via reduced content and phosphorylation of insulin receptor substrate 1 (IRS-1).[100–102] Finally, in the rat heart, GCs significantly decrease AMPK activity, suggesting that the observed detrimental effects of GC excess on the heart (i.e., left ventricular hypertrophy and myocardial ischemia), as is observed in patients with

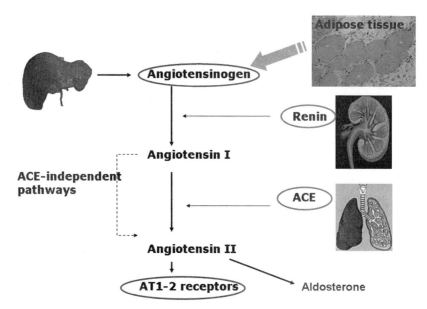

Figure 5. Schematic representation of the renin–angiotensin–aldosterone system (RAAS). Angiotensinogen (AGT) is secreted the liver and other tissues, such as adipose tissue. It is then converted by renin in angiotenin I, which is then transformed by the angiotensin-converting enzyme (ACE) in angiotensin II that, in turn, leads to adrenal production of aldosterone and has systemic effects on the cardiovascular system through its binding to AT1 and AT2 receptors.

chronic exposure to excess GCs, could be, at least in part, mediated by the decrease in AMPK activity.[98]

In summary, several features that characterize GC's excess exposure—as occurs in CS patients—can be ascribed to AMPK activity modification. A tissue-specific modulation of AMPK activity seems to be responsible for the metabolic alterations observed in patients with CS, including visceral obesity, insulin resistance, glucose metabolism, lipid profile alterations, hepatic steatosis, and cardiac changes, which together characterize cardiometabolic syndrome.

Adipose tissue and the renin–angiotensin–aldosterone system

Two distinct forms of the RAAS exist: a systemic or circulating form, and a local form that acts in peripheral tissues. Adipose tissue, especially the visceral type, possesses a local RAAS, which has paracrine as well as endocrine effects (Fig. 5).[103] In the condition of visceral obesity, both local and systemic RAAS activity are increased.[103] Such RAAS hyperactivity is responsible for the inhibition of preadipocyte differentiation and for the development of big insulin-resistant adipocytes able to secrete inflammatory adipokines, thus contributing to the cardiometabolic alterations associated with

insulin resistance and hyperinsulinemia.[104–105] In addition, in the presence of insulin resistance, it has been shown that increased FFA, through their effects on hepatic production of angiotensinogen (AGT), may stimulate aldosterone production, independent of renin.[103]

Pharmacological RAAS blockade seems to play an important positive role in insulin sensitivity. Numerous clinical trials have indicated that ACE inhibitors and angiotensin receptor blockers (ARBs) decrease the propensity to develop type 2 diabetes in high-risk patients,[106–108] likely due to insulin sensitivity improvement.[109]

These effects have been explained, in part, by experiments in animal models with obesity and diabetes mellitus type 2. Treatment with different ARBs (olmesartan, valsartan, telmisartan, or candesartan) has been shown to induce a significant reduction in adipocyte size,[110,111] body weight, and fat mass,[112] all associated with improvements in insulin sensitivity, partly due to increased release of insulin-sensitizing adipokines and concomitant reduction of diabetogenic adipokines.[112] In this regard, according to the hypothesis by Sharma *et al.*, while Ang II would inhibit preadipocyte differentiation, resulting in the formation of large insulin-resistant adipocytes,[113] RAS blockade would induce preadipocyte recruitment, thereby increasing the number of small insulin-sensitive adipocytes that secrete fewer inflammatory cytokines and more beneficial cytokines, including adiponectin. The effects of ARBs in adipocyte differentiation seem to be mediated by an AT1R-blocking effect and, at least in part, by AT2R stimulation.

RAAS and adipogenesis

It has been reported that some components of the RAAS (angiotensin II, angiotensin-converting enzyme (ACE), and plasma renin activity (PRA)) correlate with BMI.[105] Indeed, increased circulating AGT, renin, aldosterone, and ACE activity have been described in obese patients.[114,115] Moreover, increased RAAS gene expression has been found in adipose tissue of both obese animal models and obese humans.[116–118]

In animals, a trophic role for RAS on adipogenesis has been demonstrated using transgenic mice: the overexpression of AGT in adipose tissue is associated with increased fat mass, compared with wild-type mice, while AGT knockout mice present with reduced fat mass.[116] In humans, increasing AGT, ACE, and renin mRNA expression has been observed during preadipocyte differentiation,[119] thus indicating the relevance of local RAS on adipogenesis.

In addition, since the first studies on adipogenesis, which showed the ability of aldosterone to induce adipocyte differentiation,[120] further steps forward have been achieved, first with the identification of the mineralococrticoid receptor (MR) expression in both in BAT and WAT,[121,122] and then with the demonstration that such receptors are activated by aldosterone during adipose cell differentiation.[114] Finally, more recently, it has been shown that aldosterone is able to induce adipose conversion of mouse cell lines in a time- and dose-dependent manner, and that MR, but not GR, knockdown inhibits the glucocorticoid-induced adipose conversion of 3T3-L1 cells.[123]

In subsequent studies, the same group has shown that selective blockade of the MR exerts antiadipogenic effects through an alteration of transcriptional control, which suggests a novel therapeutic option for fat deposition and its related metabolic sequelae.[124]

Adipose tissue and aldosterone

A positive correlation between serum aldosterone levels and fat mass has been reported—especially in women—by several studies.[125–127] Aldosterone levels not only positively correlate with visceral obesity[125] and BMI[128] but also fall in serum after weight loss.[127] The mechanism by which weight loss can induce a reduction of aldosterone is not clear, although a reduced production of mineralocorticoid-releasing factors secreted by adipocytes[129] could be involved. Besides genetic factors, additional factors have been proposed to explain the positive correlation between serum aldosterone levels and fat mass, for example, the involvement of adipose tissue RAS,[130] the overactivity of the renal sympathetic nervous system, and stimulation of aldosterone secretion by an oxidized derivative of linoleic acid[131] or by unknown potent fat mineralocortcoid releasing factors.[126]

Fat cell–conditioned media collected from primary cultures of human adipocytes have been shown to stimulate aldosterone secretion from NCI-H295R adrenocortical cells and from primary human adrenocortical cells (from normal adrenals) in an Ang II–independent way.[129] Filtration of fat-conditioned media revealed the presence of two different fractions: an active one, with a molecular mass >50 kDa and representing 60% of the activity, and an inactive fraction with a mass of <3 kDa. Although the active molecules are still unidentified, the ability of adipocytes to secrete potent mineralocorticoid-releasing factors has been subsequently confirmed in a rat model of metabolic syndrome (SHR-cp).[132] Once again, aldosterone-releasing activity of fat cell–conditioned medium was not angiotensin II mediated, as it was not inhibited by ARB candesartan. In addition, the authors showed that treatment of adrenocortical cells with fat-derived medium upregulated mRNA expression of StAR protein. The production of such fat-derived mineralocorticoid-releasing substances could contribute to the hyperaldosteronism observed in obese subjects.

Aldosterone excess and glucose derangements in the clinical setting

The strict relationship linking aldosterone to adipose tissue has been underlined by several clinical and experimental studies conducted during the last few years. The deleterious effects exerted by aldosterone, via genomic and nongenomic actions in the heart, blood vessels, kidney, and brain have been extensively demonstrated both in experimental models and in humans.[133] Less is known about the negative effects of such a hormone on glucose metabolism and insulin sensitivity, although some recent papers have underlined the relevance of aldosterone and its excess on the development of glucose metabolism alterations.[134,135] In addition, a higher prevalence of glycemic abnormalities and of the metabolic syndrome has been demonstrated in patients with primary aldosteronism (PA), compared with patients with essential hypertension (EH).[136,137]

Conn *et al.*[138] first reported an increased incidence of impaired glucose tolerance in patients with primary aldosteronism. In subsequent studies on the potential mechanisms involved in this impaired glucose intolerance, both consistent and conflicting data have been generated. Decreased insulin-receptor expression and affinity in subcutaneous adipose tissue of a patient with primary aldosteronism have been reported.[139] Impaired pancreatic insulin has also been reported,[140] and other reports[141,142] have confirmed the finding of insulin resistance in primary aldosteronism patients.

As mentioned above, aldosterone levels positively correlate with visceral obesity and inversely with insulin sensitivity.[125,143] A dose-dependent increase of aldosterone caused a decrease in glucose uptake together with increased expression of proinflammatory cytokines such as leptin and MCP-1, but this study was carried out in mouse brown adipocytes.[144]

The prevalence of insulin resistance in primary aldosteronism—compared with essential hypertension—and its response to treatment are also areas in which different findings have been reported. In a small group of patients, Sindelka *et al.* found reduced insulin sensitivity in those with primary aldosteronism compared with healthy controls. In addition, they reported significant improvement of insulin sensitivity after adrenalectomy but not after medical treatment of bilateral adrenal hyperplasia. First Widimsky, and then other authors, similarly found primary aldosteronism patients to be insulin resistant;[141,142,145] on the other hand, they found no differences between patients with either primary aldosteronism or essential hypertension with regard to the prevalence of impaired glucose tolerance or diabetes.[146] The study by Catena *et al.*[142] produced findings different from those cited above in several ways. First, they found that although primary aldosteronism patients (with adrenal ade-

noma or bilateral adrenal hyperplasia) were more insulin resistant than age-, sex- and BMI–matched normal controls, the severity of the insulin resistance was less evident than in patients with EH. In addition, differently from our study, Catena *et al.* found an improvement in insulin sensitivity after surgical and medical treatment in patients with aldosterone-producing adenoma (APA) and those with idiopathic aldosteronism, respectively. Indeed, we have reported[147] a distinction between treatment results in the two groups; on the one hand, surgical treatment of adrenal adenoma improved glucose tolerance evaluated by the 2-h oral glucose tolerance test, despite the increase of BMI, while on the other hand medical treatment in patients with idiopathic aldosteronism blocked further progression of the metabolic complications rather than reversed them. An increased rate of diabetes mellitus in primary aldosteronism was also found in patients in the German Conn's Registry, having PA patients compared with EH with a prevalence of diabetes of 23% versus 10%.[148]

Although the exact relationship among aldosterone, glucose metabolism, insulin action, and development of multiple sclerosis remains mostly unresolved, several factors have been proposed as pathogenetic elements, including an indirect effect via hypokalemia, which seems responsible for a reduction in insulin secretion—even though the precise role of this ion deficiency remains unclear, and in the light of the persistence of insulin resistance in PA patients during potassium infusion.[149] In a previous study we showed that in patients with PA, homeostasis model assessment–estimated insulin resistance (HOMA IR) is higher in hypokalemic patients, thus indicating a possible effect of potassium on insulin sensitivity. It has also been suggested that nongenomic actions of aldosterone might result in an increase of collagen synthesis and fibrosis, not only in the heart but also in other tissues, such as the pancreas, liver, fat, and muscle, resulting in alterations of insulin release and insulin sensitivity.[150]

For pathogenetic mechanisms, great attention has been paid to adipose tissue and to the several recently discovered adipokines involved in glucose homeostasis. According to different studies, patients with PA present suppressed leptin levels,[151,152] reduced adiponectin levels,[153] and increased values of resistin.[154] In this light, adipose tissue seems to play a role in the pathogenesis of glucose metabolism alterations in PA patients.[155] To support this hypothesis, experimental studies with rodents have shown that aldosterone is able to increase the expression of proinflammatory adipokines responsible for a reduction in insulin receptor expression and impaired insulin-dependent glucose uptake.[156] However, it has been recently demonstrated that gene expression of insulin signaling/inflammatory molecules (PPAR-γ, insulin receptor, GLUT-4, IRS-1 and 2, leptin, adiponectin, IL-6, MCP-1, 11βHSD1, 11βHSD2, and GR) was similar in visceral adipose tissue of patients with APA compared with adipose tissue of patients with a nonfunctioning adrenal adenoma. These data do not support an effect of aldosterone excess on adipose tissue–mediated insulin sensitivity.[157]

Nevertheless, our unpublished data regarding mRNA expression studies on visceral adipose tissue samples of APA patients, compared with samples of healthy controls and patients with a nonfunctioning adrenal adenoma, showed significantly higher expression, in samples of APA patients, of IL-6, an important adipokine with proinflammatory properties and a marker of insulin resistance. The association of IL-6 and insulin resistance seems complex, and evidence suggests that IL-6 might act at multiple levels, both centrally and on peripheral tissues, to influence body weight, energy homeostasis, and insulin sensitivity.[158]

Finally, recent evidence indicates that the detrimental effects of aldosterone excess on insulin signaling are mediated by inflammatory/oxidative stress mechanisms of mineralocorticoids.[159] To support the involvement of the aldosterone stress mechanism in the TG(mRen2)22 rat, which has insulin resistance,[160] *in vivo* MR antagonism with spironolactone substantially improves *ex vivo* insulin-stimulated increases in glucose uptake in skeletal muscle.

Conclusions

A tight link between adipose tissue and the adrenal gland has been widely demonstrated. From a physiological point, GCs are essential for adipocyte biology, in particular for differentiation and lipid homeostasis.[80] From a clinical point of view, the dysregulation of both the hypothalamic–pituitary–adrenal axis (HPA) and of cortisol metabolism are implicated in the development of visceral obesity,

metabolic syndrome, and cardiovascular diseases.[41] Such characteristics are indeed typically observed in the presence of hypercortisolism, both endogenous, that is, CS, as well as in exogenous hypercortisolism. Similarly, the mineralocorticoid counterpart also has a role in adipocyte differentiation[120] and seems to be, at least in part, regulated by adipose-secreted factors, with reciprocal effects between adrenal gland and adipose organ.[126] Such interactions, in the presence of pathological condition, such as primary aldosteronism, result in the development of metabolic alterations. The picture is complicated by the interplay between cortisol and aldosterone and their respective receptors. It is indeed well known that the GR and MR display both structural and functional homology. The fact that cortisol is able to bind to the MR with an affinity higher than the affinity of aldosterone itself explains why it has been proposed that in the presence of chronic cortsiol excess this hormone could act via MR activation. In the same way, aldosterone action in particular conditions is mediated through GR activation. Such interplay is particularly evident in adipose tissue, and on adipogenesis and adipose biology.

Conflicts of interest

The authors declare no conflicts of interest.

References

1. Harris, M.I., K.M. Flegal, C.C. Cowie, *et al.* 1998. Prevalence of diabetes, impaired fasting glucose, and impaired glucose tolerance in U.S. adults. The Third National Health and Nutrition Examination Survey, 1988–1994. *Diabetes Care* **21:** 518–524.
2. Harris, M.I. 1998. Diabetes in America: epidemiology and scope of the problem. *Diabetes Care* **21:** C11–C14.
3. Ogden, C.L., M.D. Carroll, L.R. Curtin, *et al.* 2006. Prevalence of overweight and obesity in the United States, 1999–2004. *JAMA* **295:** 1549–1555.
4. Sowers, J.R., M. Epstein & E.D. Frohlich. 2001. Diabetes, hypertension, and cardiovascular disease: an update. *Hypertension* **37:** 1053–1059.
5. Cinti, S., C. Zancanaro, A. Sbarbati, *et al.* 1989. Immuno-electon microscopal identification of the uncoupling protein in brown adipose tissue mitochondria. *Biol. Cell.* **67:** 359–362.
6. Virtanen, K.A., M.E. Lidell, J. Orava, *et al.* 2009. Functional brown adipose tissue in healthy adults. *N. Engl. J. Med.* **360:** 1518–1525.
7. Rothwell, N.J. & M.J. Stock. 1979. A role for brown adipose tissue in diet-induced thermogenesis. *Nature* **281:** 31–35.

8. Wisse, B.E. 2004. The inflammatory syndrome: the role of adipose tissue cytokines in metabolic disorders linked to obesity. *J. Am. Soc. Nephrol.* **15:** 2792–2800.

9. Hotamisligil, G.S., N.S. Shargill & B.M. Spiegelman.1993. Adipose expression of tumor necrosis factor-alpha: direct role in obesity-link insulin resistance. *Science* **259:** 87–91.

10. Wellen, K.E., & G.S. Hotamisligil. 2003. Obesity-induced inflammation changes in adipose tissue. *J. Clin. Invest.* **112:** 1785–1788.

11. Weisberg, S.P., D. Mc Cann, M. Desai, *et al.* 2003. Obesity is associated with macrophage accumulation in adipose tissue. *J. Clin. Invest.* **112:** 1796–1808.

12. Xu, H., G.T. Barnes, Q. Yang, *et al.* 2003. Chronic inflammation in fat plays a crucial role in the development of obesity-related insulin resistance. *J. Clin. Invest.* **112:** 1821–1830.

13. Cinti, S., G. Mitchell, G. Barbatelli, *et al.* 2005. Adipocyte death defines macrophage localization and function in adipose tissue of obese mice and humans. *J. Lipid. Res.* **46:** 2347–2355.

14. Maury, E., K. Ehala-Aleksejev, Y. Guiot, *et al.* Adipokines oversecreted by omental adipose tissue in human obesity. *Am. J. Physiol. Endocrinol. Metab.* **293:** E656–E665.

15. Zhang, Y., R. Proenca, M. Maffei, *et al.* 1994. Positional cloning of the mouse obese gene and its human homologue. *Nature* **372:** 425–432.

16. Maeda, K., K. Okubo, I. Shimomura, *et al.* 1996. cDNA cloning and expression of a novel adipose specific collagen-like factor, apM1 (AdiPose most abundant gene transcript 1). *Biochem. Biophys. Res. Commun.* **221:** 286–289.

17. Sethi, J.K. & A.J. Vidal-Puig. 2007. Adipose tissue function and plasticity orchestrate nutritional adaptation. *J. Lipid Res.* **48:** 1253–1262.

18. Ahima, R.S. & J.S. Flier. 2000. Adipose tissue as an endocrine organ. *Trends Endocrinol. Metab.* **11:** 327–332.

19. Shimomura, I., R.E. Hammer, S. Ikemoto, *et al.* 1999. Leptin reverses insulin resistance and diabetes mellitus in mice with congenital lipodystrophy. *Nature* **401:** 73–76.

20. Eckel, R.H. & R.M. Krauss. 1998. American Heart Association call to action: obesity as a major risk factor for coronary heart disease. AHA Nutrition Committee. *Circulation* **97:** 2099–2100.

21. Donahue, R.P., R.D. Abbott, E. Bloom, *et al.* 1987. Central obesity and coronary heart disease in men. *Lancet* **1:** 821–824.

22. Després, J.P., I. Lemieux, J. Bergeron, *et al.* 2008. Abdominal obesity and the metabolic syndrome: contribution to global cardiometabolic risk. *Arterioscler. Thromb. Vasc. Biol.* **28:** 1039–1049.

23. Lean, M.E., T.S. Han & J.C. Seidell. 1998. Impairment of health and quality of life in people with large waist circumference. *Lancet* **351:** 853–856.

24. Ford, E.S.G. & W.H. Dietz. 2002. Prevalence of the metabolic syndrome among U.S. adults: findings from the Third National Health and Nutrition Examination Survey. *JAMA* **287:** 356–359.

25. Stumvoll, M., B.J. Goldstein & T.W. van Haeften. 2005. Type 2 diabetes: principles of pathogenesis and therapy. *Lancet* **365:** 1333–1346.

26. Zierath, J.R. & H. Wallberg-Henriksson. 2002. From receptor to effector: insulin signal transduction in skeletal muscle from type II diabetic patients. *Ann. N.Y. Acad. Sci.* **967:** 120–134.

27. Mokdad, A.H., E.S. Ford, B.A. Bowman, *et al.* 2003. Prevalence of obesity, diabetes, and obesity-related health risk factors. 2001. *JAMA* **289:** 76–79.

28. De Luca, C. & J.M. Olefsky. 2008. Inflammation and insulin resistance. *FEBS Lett.* **582:** 97–105.

29. Wang, P., E. Mariman, J. Renes, *et al.* 2008. The secretory function of adipocytes in the physiology of white adipose tissue. *J. Cell. Physiol.* **216:** 3–13.

30. Shoelson, S.E., L. Herrero & A. Naaz. 2007. Obesity, inflammation, and insulin resistance. *Gastroenterology* **132:** 2169–2180.

31. Villena, J.A., S. Roy, E. Sardaki-Nagy, *et al.* 2004. Desnutrin, an adipocyte gene encoding a novel patratin domain-containing protein, is induced by fasting and glucocorticoids: ectopic expression of desnutrin increase triglyceride hydrolysis. *J. Biol. Chem.* **279:** 47066–47075.

32. Foley, J.E. 1992. Rationale and application of fatty acid oxidation inhibitors in treatment of diabetes mellitus. *Diabetes Care* **15:** 773–784.

33. Ginsberg, H.N. 2006. Review: efficacy and mechanisms of action of statins in the treatment of diabetic dyslipidemia. *J. Clin. Endocrinol. Metab.* **91:** 383–392.

34. Cheung, B.M. & C. Li. 2012. Diabetes and hypertension: is there a common metabolic pathway? *Curr. Atheroscler. Rep.* **14:** 160–166.

35. Baum, M. 1987. Insulin stimulates volume absorption in the proximal convoluted tubule. *J. Clin. Invest.* **79:** 1104–1109.

36. Vierhapper, H., W. Waldhausl & P. Nowontny. 1983. The effect of insulin on the rise in blood pressure and plasma aldosterone after angiotensin II in normal man. *Clin. Sci.* **64:** 383–386.

37. Deibert, D.C. & R.A. DeFronzo. 1980 Epinephrine-induced insulin resistance in man. *J. Clin. Invest.* **65:** 717–721.

38. Rocchini, A.P., J. Key, D. Bondie, *et al.* 1989. The effect of weight loss on the sensitivity of blood pressure to sodium in obese adolescents. *N. Engl. J. Med.* **21:** 580–585.

39. Stout, R.W., E. Bierman & R. Ross. 1975. Effect of insulin on the proliferation of cultured pnmate arterial smooth muscle cells. *Circ. Res.* **36:** 319–327.

40. Sowers, J.R. 2004. Insulin resistance and hypertension. *Am. J. Physiol. Heart. Circ. Physiol.* **286:** H1597–H1602.

41. McFarlane, S.I., M. Banerji & J.R. Sowers. 2001. Insulin resistance and cardiovascular disease. *J. Clin. Endocrinol. Metab.* **86:** 713–718.

42. Third report of the National Cholesterol Education Program (NCEP) expert panel on detection, evaluation, and treatment of high blood cholesterol in adults (Adult Treatment Panel III): final report. 2002. *Circulation* **106:** 3143–3421.

43. Grundy, S.M., H.B. Brewer, J.I. Cleeman, *et al.* 2004. Definition of Metabolic Syndrome Report of the National Heart, Lung, and Blood Institute/American Heart Association Conference on Scientific Issues Related to Definition. *Circulation* **109:** 433–438.

44. Pasquali, R., V. Vicennati, M. Cacciari, *et al.* 2006. The Hypothalamic-pituitary-adrenal axis activity in obesity and the metabolic syndrome. *Ann. N.Y. Acad. Sci.* **1083:** 111–128.

45. Rebuffè-Scrive, M., U.A. Walsh, B. McEwen, *et al.* 1992. Effect of chronic stress and exogenous glucocorticoids on regional fat distribution and metabolism. *Physiol. Behav.* **52:** 583–590.

46. Brunner, E.J., H. Hemingway, B.R. Walker, *et al.* 2002. Adrenocortical, autonomic, and inflammatory causes of the metabolic syndrome: nested case–control study. *Circulation* **106:** 2659–2665.

47. Epel, E.S., B. McEwen, T. Seeman, *et al.* 2000. Stress and body shape: stress-induced cortisol secretion is consistently greater among women with central fat. *Psychosom. Med.* **62:** 623–632.

48. Pasquali, R. & V. Vicennati. 2000. Activity of the hypothalamic-pituitary-adrenal axis in different obesity phenotypes. *Int. J. Obes. Relat. Metab. Disord.* **24**(Suppl): S47–S49.

49. Ward, A.M., C.H. Fall, C.E. Stein, *et al.* 2003. Cortisol and the metabolic syndrome in South Asians. *Clin. Endocrinol.* **58:** 500–505.

50. Korbonits, M., P.J. Trainer, M.L. Nelson, *et al.* 1996. Differential stimulation of cortisol and dehydroepiandrosterone levels by food in obese and normal subjects: relation to body fat distribution. *Clin. Endocrinol.* **45:** 699–706.

51. Pasquali, R., B. Anconetani, R. Chattat, *et al.* 1996. Hypothalamic-pituitary-adrenal axis activity and its relationship to the autonomic nervous system in women with visceral and subcutaneous obesity: effects of the corticotropin-releasing factor/arginine-vasopressin test and of stress. *Metabolism* **45:** 351–356.

52. Pasquali, R., B. Ambrosi, D. Armanini, *et al.* 2002. Study Group on Obesity of the Italian Society of Endocrinology. Cortisol and ACTH response to oral dexamethasone in obesity and effects of sex, body fat distribution, and dexamethasone concentrations: a dose–response study. *J. Clin. Endocrinol. Metab.* **87:** 166–175.

53. Misra, M., M.A. Bredella, P. Tsai, *et al.* 2008. Lower growth hormone and higher cortisol are associated with greater visceral adiposity, intramyocellular lipids, and insulin resistance in overweight girls. *Am. J. Physiol. Endocrinol. Metab.* **295:** E385–E392.

54. Marin, P., N. Darin, T. Amemiya, *et al.* 1992. Cortisol secretion in relation to body fat distribution in obese premenopausal women. *Metabolism* **41:** 882–886.

55. Pasquali, R., S. Cantobelli, F. Casimirri, *et al.* 1993. The hypothalamic-pituitary-adrenal axis in obese women with different patterns of body fat distribution. *J. Clin. Endocrinol. Metab.* **77:** 341–346.

56. Manenschijn, L., E.L. van den Akker, S.W. Lamberts, *et al.* 2009. Clinical features associated with glucocorticoid receptor polymorphisms. An overview. *Ann. N.Y. Acad. Sci.* **1179:** 179–198.

57. Francke, U. & B.E. Foellmer. 1989. The glucocorticoid receptor gene is in 5q31-q32. *Genomics* **4:** 610–612.

58. Theriault, A., E. Boyd, S.B. Harrap, *et al.* 1989. Regional chromosomal assignement of the human glucocorticoid receptor gene to 5q31. *Hum. Genet.* **83:** 289–291.

59. Hollenberg, S.M., C. Weinberger, E.S. Ong, *et al.* 1985. Primary structure and expression of functional human glucocorticoid receptor cDNA. *Nature* **318:** 635–641.

60. Lewis-Tuffin, L.J., C.M. Jewell, R.J., Bienstock, *et al.* 2007. Human glucocorticoid receptor beta binds RU-486 and is transcriptionally active. *Mol. Cell. Biol.* **27:** 2266–2282.

61. Lamberts, S.W., A.T. Huizenga, P. de Lange, *et al.* 1996. Clinical aspects of glucocorticoid sensitivity. *Steroids* **61:** 157–160.

62. Russcher, H., P. Smit, E.L. van den Akker, *et al.* 2005. Two polymorphisms in the glucocorticoid receptor gene directly affect glucocorticoid-regulated gene expression. *J. Clin. Endocrinol. Metab.* **90:** 5804–5810.

63. van Rossum, E.F. & S.W. Lamberts. 2004. Polymorphisms in the glucocorticoid receptor gene and their associations with metabolic parameters and body composition. *Recent Prog. Horm. Res.* **59:** 333–357.

64. van Rossum, E.F.C. & S.W.J. Lamberts. 2004. Polymorphisms in the glucocorticoid receptor gene and their associations with metabolic parameters and body composition. *Recent Prog. Horm. Res.* **59:** 333–357.

65. Dobson, M.G., C.P. Redfern, N. Unwin, *et al.* 2001. The N363S polymorphism of the glucocorticoid receptor: potential contribution to central obesity in men and lack of association with other risk factors for coronary heart disease and diabetes mellitus. *J. Clin. Endocrinol. Metab.* **86:** 2270–2274.

66. Di Blasio, A.M., E.F. van Rossum, S. Maestrini, *et al.* 2003. The relation between two polymorphisms in the glucocorticoid receptor gene and body mass index, blood pressure and cholesterol in obese patients. *Clin. Endocrinol.* **59:** 68–74.

67. Lin, R.C., X.L.Wang, B. Dalziel, *et al.* 2003. Association of obesity, but not diabetes or hypertension, with glucocorticoid receptor N363S variant. *Obes. Res.* **11:** 802–818.

68. Buemann, B., M.C. Vohl, M. Chagnon, *et al.* 1997. Abdominal visceral fat is associated with a BclI restriction fragment length polymorphism at the glucocorticoid receptor gene locus. *Obes. Res.* **5:** 186–192.

69. Ukkola, O., L. Perusse, Y.C. Chagnon, *et al.* 2001. Interactions among the glucocorticoid receptor, lipoprotein lipase and adrenergic receptor genes and abdominal fat in the Quebec Family Study. *Int. J. Obes. Relat. Metab. Disord.* **25:** 1332–1339.

70. Rutters, F., A.G. Nieuwenhuizen, S.G. Lemmens, *et al.* 2011. Associations between anthropometrical measurements, body composition, single-nucleotide polymorphisms of the hypothalamus/pituitary/adrenal (HPA) axis and HPA axis functioning. *Clin. Endocrinol.* **74:** 679–686.

71. Voorhoeve, P.G., E.L. van den Akker, E.F. van Rossum, *et al.* 2009. Glucocorticoid receptor gene variant is associated with increased body fatness in youngsters. *Clin. Endocrinol.* **71:** 518–523.

72. Rosmond, R., C. Bouchard & P. Bjorntorp. 2001. Tsp509I polymorphism in exon 2 of the glucocorticoid receptor gene in relation to obesity and cortisol secretion: cohort study. *Br. Med. J.* **322:** 652–653.

73. Echwald, S.M., T.I. Sorensen, T. Andersen, *et al.* 2001. The Asn363Ser variant of the glucocorticoid receptor gene is not associated with obesity or weight gain in Danish men. *Intl. J. Obes. Relat. Metab. Disord.* **25:** 1563–1565.

74. Lin, R.C.Y., W.Y.S. Wang & B.J. Morris. 1999. High penetrance, overweight, and glucocorticoid receptor variant: case–control study. *Br. Med. J.* **349:** 1337–1338.

75. Huizenga, N.A., J.W. Koper, P. De Lange, *et al.* 1998. A polymorphism in the glucocorticoid receptor gene may be associated with and increased sensitivity to glucocorticoids in vivo. *J. Clin. Endocrinol. Metab.* **83:** 144–151.

76. Syed, A.A., J.A. Irving, C.P. Redfern, *et al.* 2006. Association of glucocorticoid receptor polymorphism A3669G in exon 9beta with reduced central adiposity in women. *Obesity* **14:** 759–764.

77. Trementino, L., G. Appolloni, C. Concettoni, *et al.* 2012. Association of glucocorticoid receptor polymorphism A3669G with decreased risk of developing diabetes in patients with Cushing's syndrome. *Eur. J. Endocrinol.* **166:** 35–42.

78. Cellini, E., G. Castellini, V. Ricca, *et al.* 2010. Glucocorticoid receptor gene polymorphisms in Italian patients with eating disorders and obesity. *Psychiatr. Genet* **20:** 282–288.

79. Walker, B.R. 2001. Tissue-specific dysregulation of cortisol metabolism in human obesity. *J. Clin. Endocrinol. Metab.* **86:** 1418–1421.

80. Walker, B.R., S. Soderberg, B. Lindahl, *et al.* 2000. Independent effects of obesity and cortisol in predicting cardiovascular risk factors in men and women. *J. Intern. Med.* **247:** 198–204.

81. Travison, T.G., A.B. O'Donnell, A.B. Araujo, *et al.* 2007. Cortisol levels and measures of body composition in middle-aged and older men. *Clin. Endocrinol.* **67:** 71–77.

82. Duclos, M., P. Marquez Pereira, P. Barat, *et al.* 2005. Increased cortisol bioavailability, abdominal obesity, and the metabolic syndrome in obese women. *Obes. Res.* **13:** 1157–1166.

83. Gathercole, L.L., S.A. Morgan, I.J. Bujalska, *et al.* 2011. Regulation of lipogenesis by glucocorticoids and insulin in human adipose tissue. *PLoS One* **6:** e26223.

84. Masuzaki, H., J. Paterson, H. Shinyama, *et al.* 2001. A transgenic model of visceral obesity and the metabolic syndrome. *Science* **294:** 2166–2170.

85. Paterson, J.M., N.M. Morton, C. Fievet, *et al.* 2004. Metabolic syndrome without obesity: hepatic overexpression of 11beta-hydroxysteroid dehydrogenase type 1 in transgenic mice. *Proc. Natl. Acad. Sci. USA* **101:** 7088–7093.

86. Morton, N.M., J.M. Paterson, H. Masuzaki, *et al.* 2004. Novel adipose tissue-mediated resistance to diet-induced visceral obesity in 11 β-hydroxysteroid dehydrogenase type 1-deficient mice. *Diabetes* **53:** 931–938.

87. Mariniello, B., V. Ronconi, S. Rilli, *et al.* 2006. Adipose tissue 11beta-hydroxysteroid dehydrogenase type 1 expression in obesity and Cushing's syndrome. *Eur. J. Endocrinol.* **155:** 435–441.

88. Barf, T., J. Vallgårda, R. Emond, *et al.* 2002. Arylsulfonamidothiazoles as a new class of potential antidiabetic drugs. Discovery of potent inhibitors of the 11β-hydroxysteroid dehydrogenase type I. *J. Med. Chem.* **45:** 3813–3815.

89. Olson, S., S.D. Aster, K. Brown, *et al.* 2005. Adamantyl triazoles as selective inhibitors of 11β-hydroxysteroid dehydrogenase type 1. *Bioorg. Med. Chem. Lett.* **15:** 4359–4362.

90. Yuan, C., D.J. St Jean Jr., Q. Liu, *et al.* 2007. The discovery of 2-anilinothiazolones as 11β-HSD1 inhibitors. *Bioorg. Med. Chem. Lett.* **17:** 6056–6061.

91. Rosenstock, J., S. Banarer, V.A. Fonseca, *et al.* 2010. The 11-beta-hydroxysteroid dehydrogenase type 1 inhibitor INCB13739 improves hyperglycemia in patients with type 2 diabetes inadequately controlled by metformin monotherapy. *Diabetes Care* **33:** 1516–1522.

92. Tomlinson, J.W., M. Sherlock, B. Hughes, *et al.* 2007. Inhibition of 11{beta}-HSD1 activity in vivo limits glucocorticoid exposure to human adipose tissue and decreases lipolysis. *J. Clin. Endocrinol. Metab.* **92:** 857–864.

93. Carling, D., V.A. Zammit & D.G. Hardie. 1987. A common biclycic protein kinase cascade inactivates the regulatory enzymes of fatty acid and cholesterol biosynthesis. *FEBS Lett.* **223:** 217–222.

94. Hardie, D.G., D. Carling & A.T.R Sim. 1989. The AMP-activated protein kinase: a multisubstrate regulator of lipid metabolism. *Trends Biochem. Sci.* **14:** 20–23.

95. Lim, C.T., B. Kola & M. Korbonits. 2010. AMPK as mediator of hormonal signalling. *J. Mol. Endocrinol.* **44:** 87–97.

96. Divertie, G.D., M.D. Jensen & J.M. Miles. 1991. Stimulation of lipolysis in humans by physiological hypercortisolemia. *Diabetes* **40:** 1228–1232.

97. Djurhuus, C.B., C.H. Gravholt, S. Nielsen, *et al.* 2002. Effects of cortisol on lipolysis and regional interstitial glycerol levels in human. *Am. J. Physiol.Endocrinol. Metab.* **283:** E172–E177.

98. Christ-Crain, M., B. Kola, F. Lolli, *et al.* 2008. AMP-activated protein kinase mediates glucocorticoid-induced metabolic changes: a novel mechanism in Cushing's syndrome. *FASEB J.* **22:** 1672–1683.

99. Viana, A.Y., H. Sakoda, M. Anai, *et al.* 2006. Role of hepatic AMPK activation in glucose metabolism and dexamethasone-induced regulation of AMPK expression. *Diab. Res. Clin. Pract.* **73:** 135–142.

100. Brown, P.D., S. Badal, S. Morrison, *et al.* 2007. Acute impairment of insulin signalling by dexamethasone in primary cultured rat skeletal myocytes. *Mol. Cell. Biochem.* **297:** 171–177.

101. Lundgren, M., J. Buren, T. Ruge, *et al.* 2004. Glucocorticoids down-regulate glucose uptake capacity and insulin-signaling proteins in omental but not subcutaneous human adipocytes. *J. Clin. Endocrinol. Metab.* **89:** 2989–2997.

102. Koricanac, G., E. Isenovic, V. Stojanovic-Susulic, *et al.* 2006. Time dependent effects of dexamethasone on serum insulin level and insulin receptors in rat liver and erythrocytes. *Gen. Physiol. Biophys.* **25:** 11–24.

103. El-Atat, F., A. Aneja, S. Mcfarlane, *et al.* 2003. Obesity and hypertension. *Endocrinol. Metab. Clin. North Am.* **32:** 823–854.

104. Zhang, R. & E. Reisin. 2000. Obesity-hypertension: the effects on cardiovascular and renal systems. *Am. J. Hypertens.* **13:** 1308–1314.

105. Reaven, G.M., H. Lithell & L. Landsberg. 1996. Hypertension and associated metabolic abnormalities—the role of insulin resistance and the sympathoadrenal system. *N. Engl. J. Med.* **334:** 374–381.

106. Hansson, L., L.H. Lindholm, L. Niskanen, *et al.* 1999. Effect of angiotensin-converting-enzyme inhibition compared with conventional therapy on cardiovascular morbidity and mortality in hypertension: the Captopril Prevention Project (CAPPP) randomized trial. *Lancet* **353:** 611–616.

107. Yusuf, S., P. Sleight, B. Pogue, *et al.* 2000. Effects of an angiotensin-converting-enzyme inhibitor, ramipril, on cardiovascular events in high-risk patients. The Heart Outcomes Prevention Evaluation Study Investigators. *N. Engl. J. Med.* **342:** 145–153.

108. Lindholm, L.H., H. Ibsen, K. Borch-Johnsen, *et al.* 2002. Risk of new-onset diabetes in the Losartan Intervention for Endpoint reduction in hypertension study. *J. Hypertens.* **20:** 1879–1886.

109. Jandeleit-Dahm, K.A.M., C. Tikellis, C.M. Reid, *et al.* 2005. Why blockade of the renin–angiotensin system reduces the incidence of new-onset diabetes. *J. Hypertens.* **23:** 463–473.

110. Furuhashi, M., N. Ura, H. Takizawa, *et al.* 2004. Blockade of the renin–angiotensin systemdecreases adipocyte size with improvement in insulin sensitivity. *J. Hypertens.* **22:** 1977–1982.

111. Mori, Y., Y. Itoh & N. Tajima. 2007. Angiotensin II receptor blockers downsize adipocytes in spontaneously type 2 diabetic rats with visceral fat obesity. *Am. J. Hypertens.* **20:** 431–436.

112. Zorad, S., J.T. Dou, J. Benicky, *et al.* 2006. Long-term angiotensin II AT1 receptor inhibition produces adipose tissue hypotrophy accompanied by increased expression of adiponectin and PPARγ. *Europ. J. Pharmacol.* **552:** 112–122.

113. Sharma, A.M., J. Janke, K. Gorzelniak, *et al.* 2002. Angiotensin blockade prevents type 2 diabetes by formation of fat cells. *Hypertension* **40:** 609–611.

114. Cooper, R., N. McFarlane-Anderson, F.I. Bennett, *et al.* 1997. ACE, angiotensinogen and obesity: a potential pathway leading to hypertension. *J. Hum. Hypertens.* **11:** 107–111.

115. Messerli, F.H., B. Christie, J.G. DeCarvalho, *et al.* 1981. Obesity and essential hypertension. Hemodinamycs, intravascular volume, sodium excretion, and plasma renin activity. *Arch. Intern. Med.* **141:** 81–85.

116. Massiera, F., M. Bloch-Faure, D. Ceiler, *et al.* 2001. Adipose angiotensinogen is involved in adipose tissue growth and blood pressure regulation. *FASEB J.* **15:** 2727–2729.

117. Faloia, E., C. Gatti, M.A. Camilloni, *et al.* 2002. Comparison of circulating and local adipose tissue renin-angiotensin system in normotensive and hypertensive obese subjects. *J. Endocrinol. Invest.* **25:** 309–314.

118. Giacchetti, G., E. Faloia, B. Mariniello, *et al.* 2002. Overexpression of the renin-angiotensin system in human visceral adipose tissue in normal and overweight subjects. *Am. J. Hypertens.* **15:** 381–388.

119. Janke, J., S. Engeli, K. Gorzelniak, *et al.* 2002. Mature adipocytes inhibit in vitro differentiation of human preadipocytes via angiotensin type 1 receptors. *Diabetes* **51:** 1699–1707.

120. Rondinone, C.M., D. Rodbard & M.E. Baker. 1993. Aldosterone stimulates differentiation of mouse 3T3-L1 cells into adipocytes. *Endocrinology* **132:** 2421–2426.

121. Zennaro, M.C., D. Le Menuet, S. Viengchareun, *et al.* 1998. Hibernoma development in transgenic mice identifies brown adipose tissue as a novel target of aldosterone action. *J. Clin. Invest.* **101:** 1254–1260.

122. Fu, M., T. Sun, A.L. Bookout, *et al.* 2005. A nuclear receptor atlas: 3T3-L1 adipogenesis. *Mol. Endocrinol.* **19:** 2437–2450.

123. Caprio, M., B. Feve, A. Claes, *et al.* 2007. Pivotal role of the mineralocorticoid receptor in corticosteroid-induced adipogenesis. *FASEB J.* **21:** 2185–2194.

124. Caprio, M., A. Antelmi, G. Chetrite, *et al.* 2011. Antiadipogenic effects of the mineralocorticoid receptor antagonist drospirenone: potential implications for the treatment of metabolic syndrome. *Endocrinology* **152:** 113–125.

125. Goodfriend, T.L., D.E. Kelley, B.H. Goodpaster, *et al.* 1999. Visceral obesity and insulin resistance are associated with plasma aldosterone levels in women. *Obes. Res.* **7:** 355–362.

126. El Gharbawy, A.H., V.S. Nadig, J.M. Kotchen, *et al.* 2001. Arterial pressure, left ventricular mass, and aldosterone in essential hypertension. *Hypertension* **37:** 845–850.

127. Engeli, S., J. Bohne, K. Gorzelniak, *et al.* 2005. Weight loss and the renin-angiotensin-aldosterone system. *Hypertension* **45:** 356–362.

128. Rossi, G.P., A. Belfiore, G. Bernini, *et al.* 2008. Body mass index predicts plasma aldosterone concentrations in overweight-obese primary hypertensive patients. *J. Clin. Endocrinol. Metab.* **93:** 2566–2571.

129. Ehrhart-Bornstein, M., V. Lamounier-Zepter, V. Schraven, *et al.* 2003. Human adipocytes secrete mineralocorticoid-releasing factors. *Proc. Natl. Acad. Sci. USA* **100:** 14211–14216.

130. Strazzullo, P., R. Iacone, L. Iacoviello, *et al.* 2003. Genetic variation in the renin-angiotensin system and abdominal adiposity in men: the Olivetti Heart Study. *Ann. Intern. Med.* **138:** 17–23.

131. Goodfriend, T.L., D.L. Ball, B.M. Egan, *et al.* 2004. Epoxyketo derivative of linoleic acid stimulates aldosterone secretion. *Hypertension* **43:** 358–363.

132. Nagase, M., S. Yoshida, S. Shibata, *et al.* 2006. Enhanced aldosterone signaling in the early nephropathy of rats with metabolic syndrome: possible contribution of fat-derived factors. *J. Am. Soc. Nephrol.* **17:** 3438–3446.

133. Funder J.W. 2005. The nongenomic actions of aldosterone. *Endocr. Rev.* **26:** 313–321.

134. Rossi, G.P., L.A. Sechi, G. Giacchetti, *et al.* 2008. Primary aldosteronism: cardiovascular, renal and metabolic implications. *Trends Endocrinol. Metab.* **19:** 88–90.

135. Giacchetti, G., L.A. Sechi, S. Rilli, *et al.* 2005. The renin–angiotensin–aldosterone system, glucose metabolism and diabetes. *Trends Endocrinol. Metab.* **16:** 120–126.

136. Fallo, F., F. Veglio, C. Bertello, *et al.* 2006. Prevalence and characteristics of the metabolic syndrome in primary aldosteronism. *J. Clin. Endocrinol. Metab.* **91:** 454–459.

137. Ronconi, V., F. Turchi, S. Rilli, *et al.* 2010. Metabolic syndrome in primary aldosteronism and essential

hypertension: relationship to adiponectin gene variants. *Nutr. Metab. Cardiovasc. Dis.* **20:** 93–100.

138. Conn, J.W., R.F. Knopf & R.M. Nesbit. 1964. Clinical characteristics of primary aldosteronism from an analysis of 145 cases. *Am. J. Surg.* **107:** 159–172.

139. Carranza, M.C., A. Torres & C. Calle. 1991. Decreased insulin receptor number and affinity in subcutaneous adipose tissue in a patient with primary hyperaldosteronism. *Rev. Clin. Exp.* **188:** 414–417.

140. Shimamoto, K., M. Shiiki, T. Ise, *et al.* 1994. Does insulin resistance participate in an impaired glucose tolerance in primary aldosteronism. *J. Hum. Hypertens.* **10:** 755–759.

141. Widimsky, J., G. Sindelka, T. Haas, *et al.* 2000. Impaired insulin action in primary aldosteronism. *Physiol. Res.* **49:** 241–244.

142. Catena, C., R. Lapenna, S. Baroselli, *et al.* 2006. Insulin sensitivity in patients with primary aldosteronism: a follow-up study. *J. Clin. Endocrinol. Metab.* **91:** 3457–3463.

143. Devenport, L.D., K.G. Goodwin & P.M. Hopkins. 1985. Continuous infusion of aldosterone: correlates of body weight gain. *Pharmacol. Biochem. Behav.* **22:** 707–709.

144. Kraus, D., J. Jager, B. Meier, *et al.* 2005. Aldosterone inhibits uncoupling protein-1, induces insulin resistance, and stimulates proinflammatory adipokines in adipocytes. *Horm. Metab. Res.* **37:** 455–459.

145. Skrha, J., T. Haas, G. Sindelka, *et al.* 2004. Comparison of the insulin action parameters from hyperinsulinemic clamps with homeostasis model assessment and QUICKI indexes in subjects with different endocrine disorders. *J. Clin. Endocrinol. Metab.* **89:** 135–141.

146. Widimsky, J., B. Strauch, G. Sindelka, *et al.* 2001. Can primary hyperaldosteronism be considered as a specific form of diabetes mellitus? *Physiol. Res.* **50:** 603–607.

147. Giacchetti, G., V. Ronconi, F. Turchi, *et al.* 2007. Aldosterone as a key mediator of the cardiometabolic syndrome in primary aldosteronism: an observational study. *J. Hypertens.* **25:** 177–186.

148. Reincke, M., C. Meisinger, R. Holle, *et al.* 2010. Is primary aldosteronism associated with diabetes mellitus? Results of the German Conn's Registry. *Horm. Metab. Res.* **42:** 435–439.

149. Sindelka, G., J. Widimský, T. Haas, *et al.* 2000. Insulin action in primary aldosteronism before and after surgical or pharmacological treatment. *Exp. Clin. Endocrinol. Diabetes* **108:** 21–25.

150. Corry, D.B. & M.L. Tuck. 2003. The effect of aldosterone on glucose metabolism. *Curr. Hypertens. Rep.* **5:** 106–109.

151. Haluzik, M., G. Sindelka, J. Widimsky Jr., *et al.* 2002. Serum leptin levels in patients with primary hyperaldosteronism before and after treatment: relationship to insulin sensitivity. *J. Hum. Hypertens.* **16:** 41–45.

152. Torpy, D.J., S.R. Bornstein, W. Taylor, *et al.* 1999. Leptin levels are suppressed in primary aldosteronism. *Horm. Metab. Res.* **31:** 533–536.

153. Fallo, F., P. Della Mea, N. Sonino, *et al.* 2007. Adiponectin and insulin sensitivity in primary aldosteronism. *Am. J. Hypertens.* **20:** 855–861.

154. Iacobellis, G., L. Petramala, D. Cotesta, *et al.* 2010. Adipokines and cardiometabolic profile in primary hyperaldosteronism. *J. Clin. Endocrinol. Metab.* **95:** 2391–2398.

155. Ronconi, V., F. Turchi, G. Appolloni, *et al.* 2012. Aldosterone, mineralocorticoid receptor and the metabolic syndrome: role of the mineralocorticoid receptor antagonists. *Curr. Vasc. Pharmacol.* **10:** 238–246.

156. Ehrhart-Bornstein, M., K. Arakelyan, A.W. Krug, *et al.* 2004. Fat cells may be the obesity-hypertension link: human adipogenic factors stimulate aldosterone secretion from adrenocortical cells. *Endocr. Res.* **30:** 865–870.

157. Urbanet, R., C. Pilon, A. Calcagno, *et al.* 2010. Analysis of insulin sensitivity in adipose tissue of patients with primary aldosteronism. *J. Clin. Endocrinol. Metab.* **95:** 4037–4042.

158. Salmenniemi, U., E. Ruotsalainen, J. Pihlajamaki, *et al.* 2004. Multiple abnormalities in glucose and energy metabolism and coordinated changes in levels of adiponectin, cytokines, and adhesion molecules in subjects with metabolic syndrome. *Circulation* **110:** 3842–3848.

159. Hitomi, H., H. Kiyomoto, A. Nishiyama, *et al.* 2007. Aldosterone suppresses insulin signaling via the downregulation of insulin receptor substrate-1 in vascular smooth muscle cells. *Hypertension* **50:** 750–755.

160. Lastra, G., A. Whaley-Connell, C. Manrique, *et al.* 2008. Low-dose spironolactone reduces reactive oxygen species generation and improves insulin-stimulated glucose transport in skeletal muscle in the TG(mRen2)27 rat. *Am. J. Physiol. Endocrinol. Metab.* **295:** E110-E116.

Ann. N.Y. Acad. Sci. ISSN 0077-8923

The circadian clock transcriptional complex: metabolic feedback intersects with epigenetic control

Selma Masri, Loredana Zocchi, Sayako Katada, Eugenio Mora, and Paolo Sassone-Corsi

Center for Metabolism and Epigenetics, U904 Inserm "Epigenetics and Neuronal Plasticity," School of Medicine, University of California, Irvine, California

Address for correspondence: Paolo Sassone-Corsi, Center for Metabolism and Epigenetics, U904 Inserm "Epigenetics and Neuronal Plasticity," School of Medicine, University of California, Irvine, California 92697, USA. psc@uci.edu

Chromatin remodeling is a prerequisite for most nuclear functions, including transcription, silencing, and DNA replication. Accumulating evidence shows that many physiological processes require highly sophisticated events of chromatin remodeling. Recent findings have linked cellular metabolism, epigenetic state, and the circadian clock. The control of a large variety of neuronal, behavioral, and physiological responses follows diurnal rhythms. This is possible through a transcriptional regulatory network that governs a significant portion of the genome. The harmonic oscillation of gene expression is paralleled by critical events of chromatin remodeling that appear to provide specificity and plasticity in circadian regulation. Accumulating evidence shows that the circadian epigenome appears to share intimate links with cellular metabolic processes. These notions indicate that the circadian epigenome might integrate tissue specificity within biological pacemakers, bridging systems physiology to metabolic control. This review highlights several advances related to the circadian epigenome, the contribution of NAD^+ as a critical signaling metabolite, and its effects on epigenetic state, followed by more recent reports on circadian metabolomics analyses.

Keywords: circadian clock; epigenetics; metabolism

Circadian rhythms: systems biology

A wide variety of physiological functions, including sleep–wake cycles, body temperature, hormone secretion, locomotor activity, and feeding behavior depend on the circadian clock—a highly conserved system that enables organisms to adapt to common daily changes, such as the day–night cycle and food availability.[1] Based on evidence accumulated during several decades, it is safe to conclude that circadian rhythms represent one of the most clear examples of systems biology.[2] Our understanding of circadian rhythms indicates that these cyclic events are self-sustained and centrally controlled, suggesting a complex and intricate biological timing mechanism that governs our daily behavior. Disruption of circadian rhythms has been linked to numerous diseases, including sleep disorders, depression, metabolic syndrome, and more recently an emerging role in tumorigenesis.[3,4]

In mammals, the anatomical structure in the brain that governs circadian rhythms is a small area consisting of $\sim$15,000 neurons localized in the anterior hypothalamus, called the suprachiasmatic nucleus (SCN).[5,6] This "central pacemaker" in the SCN receives signals from the environment and coordinates the oscillating activity of peripheral clocks that are located in almost all tissues.[1,7–9] One important feature of the circadian clocks is that they are self-sustained: circadian oscillations intrinsic to each cell can occur autonomously, without any environmental signals. However, because the period of oscillation is not exactly 24 h, the endogenous clock needs to be synchronized by external cues, a process called *entrainment*. External cues (also known as *zeitgebers*) reset the system daily and thereby prevent the endogenous clock from free-running out of phase. The predominant external cue of the central clock is light.[10] In mammals, specialized cells in the retina detect the light signal that is then transmitted to the

SCN via the retinohypothalamic tract (RHT).[11–13] At the level of SCN neurons, the light signal stimulates a cascade of signaling pathways that lead to the activation of a transcriptional program that involves immediate early genes and clock-controlled genes (CCGs). These gene expression events are associated with specific histone modifications leading to chromatin remodeling.[14] Peripheral tissues also contain functional circadian oscillators that are self-sustained at the single-cell level, but they do not respond to light–dark cycles and appear to require other physiological stimuli in order to sustain their circadian rhythms.

The systemic control of the central SCN clock over peripheral clocks necessitates a hierarchical network to maintain proper biological timing events, and several studies have elegantly demonstrated this idea. Lesions of the rodent SCN disrupt the circadian periodicity in peripheral tissues, whereas SCN transplantation into SCN-ablated arrhythmic animals restores this disfunction.[5,15] Additional experiments in which the transplantation approach was applied to peripheral tissues demonstrated a hierarchical dominance of the SCN over clocks in peripheral tissues.[16] To date, however, the means by which the SCN communicates with peripheral tissues to sustain and synchronize their cycles is still not clear. Several observations support the idea that communication may be exerted by a combination of neuronal signals through the autonomic nervous system and humoral factors, of which glucocorticoids, and retinoic acid are the most likely candidates.[3,17] In addition, expression of the SCN-secreted protein prokineticin 2 (PK2) is light sensitive, and levels of this protein are likely to regulate behavior and locomotor activity in mice, presumably through PK2 receptors (PKR2) found in surrounding regions of the brain.[18] Similarly, transforming growth factor alpha (TGF-α) is another output signal of the SCN that has been implicated in sleep and locomotor activity by binding epidermal growth factor receptors found in the hypothalamic subparaventricular zone.[19] Furthermore, peripheral rhythms in mammals are affected by other SCN-independent stimuli.[9] Although light is the main stimulus that entrains the central pacemaker, peripheral clocks can themselves be entrained by food,[20] probably through modifications of hormonal secretion or metabolite availability. Restricted access to food can reset the phase of peripheral oscillators, with little if any effects on the SCN central pacemaker.[21] These notions underscore the intimate links between the circadian clock and cellular metabolism.[3,22]

Another important environmental cue is temperature.[23] Temperature compensation is one of the most prominent features of the circadian system as it allows the integration of moderate variations in ambient temperature that do not affect the period length of circadian oscillation. Nevertheless, low-amplitude temperature cycles can synchronize the circadian clocks in peripheral tissues in mammals, independently of the central clock.[24]

The circadian transcriptome

At the heart of the molecular network that constitutes the circadian clock are the core transcription factors CLOCK and BMAL1 that heterodimerize and direct transcriptional activation of CCGs, by binding to E-box sites within their promoters. Among these CCGs, CLOCK and BMAL1 also direct transcription of their own repressors, period (PER), and cryptochrome (CRY) family members, creating a tightly self-regulated system.[4] During the day, transcription of PER and CRY is high, leading to protein translation of the circadian repressors, and resulting in formation of the inhibitory complex with CLOCK and BMAL1 that abolishes transcription of CCGs. The degradation of PER and CRY alleviates transcriptional repression and allows CLOCK:BMAL1-mediated transcription to again proceed, establishing an oscillatory rhythm in circadian gene expression. An additional level of circadian regulation exists with the orphan nuclear receptors RORα and REV-ERBα that activate and repress transcription of the *Bmal1* gene, respectively.[25,26] Furthermore, the possibility that the clock protein may be regulated in a posttranslational manner, as in the case of SUMOylation of BMAL1,[27] adds an additional level of regulation of the clock machinery.

While the basic molecular organization and conceptual design of these autoregulatory loops are common to both SCN and peripheral tissues, it is intuitive that circadian function and output of SCN, liver, or skeletal muscle are vastly divergent, begging the question on how the pacemakers intrinsic to these tissues may differ. Indeed, the property of circadian synchronicity in culture is unique to SCN neurons: cultured cells from peripheral tissues, although each has a sustained circadian cycle, do not display concerted oscillations.[28] On the other hand, it is reasonable to speculate that tissue-specific transcriptional regulators may contribute or intersect with the clock machinery. Several genome-wide array analyses have been centered on determining the proportion and specificity of cycling transcripts.[29] The first remarkable finding indicated that ~10% of all expressed genes in any tissue are under circadian regulation.[1,28,30,31] This unexpectedly high proportion of circadian transcripts suggests that the clock machinery may direct widespread events of cyclic chromatin remodeling and consequent transcriptional activation/repression. Furthermore, genome-wide studies comparing the central SCN pacemaker and peripheral tissues, such as the liver, revealed that between 5% and 10% of cycling genes were identical in both tissue types.[32,33] A recent analysis covering 14 mouse tissues identified ~10,000 known genes showing circadian oscillations in at least one tissue. The number of common genes showing circadian oscillation in multiple tissues decreased drastically as the number of tissues included in the comparative analysis increased, with only 41 genes displaying circadian oscillation in at least 8 out of 14 tissues.[34] These findings underscore the presence of molecular interplay between the core clockwork, which can be assumed to be common to all tissues, and cell-specific transcriptional systems. Taking into consideration the recent view of the mammalian circadian clock as a transcriptional network,[2,35] through which the oscillator acquires plasticity and robustness, it is reasonable to speculate that the clock network contributes to physiological responses by intersecting with cell-specific transcriptional pathways. This notion has been demonstrated in the way the circadian machinery interplays with other signaling-responsive transcription factors, such as CREB.[36]

Chromatin remodeling and epigenetic control of circadian expression

How does the complex organization of chromatin cope with the cyclic regulation of circadian genes? Several histone modifications contribute to chromatin remodeling and thereby to the control of a large array of nuclear processes.[37,38] A number of histone modifications have been associated with distinct chromatin-based outputs. For example, position-specific modifications of the histone H3 N-terminal tail have been coupled to transcrip-

tional regulation (Lys4 and Lys9/Lys14 acetylation, Ser10 phosphorylation), transcriptional silencing (Lys9 methylation), histone deposition (Lys9 acetylation), and chromosome condensation/segregation (Ser10/Ser28 phosphorylation). It is believed that specific signaling pathways lead to distinct histone modifications,[39] suggesting that various physiological stimuli translate into differential chromatin remodeling events.[40]

Histone acetylation has been shown to play a pivotal role in the modulation of chromatin structure associated with transcriptional activation.[41–45] In support of this notion, a wide variety of nuclear proteins involved in transcriptional control possess intrinsic histone acetyltransferase (HAT) activity. We have found that one of these proteins is the master regulator CLOCK, whose HAT function is essential for circadian control.[46] We have shown that chromatin remodeling is coupled to circadian clock function[14] and that the protein CLOCK functions as an enzyme, which induces chromatin remodeling.[46] This previously unforeseen activity of a core clock factor has several, far-reaching biological implications. CLOCK is a HAT, which preferentially modifies histone H3 in position Lys14, a site where addition of an acetyl group results in stimulation of gene expression (Fig. 1). Thereby, CLOCK acts as an enzyme that globally modifies genome functions, by inducing the opening of chromatin structure and allowing transcriptional activation. In addition, the enzymatic activity of CLOCK is not restricted to histones.[47] Our findings indicate that CLOCK acetylates its own transcriptional partner, BMAL1. This modification occurs at one unique lysine residue in position 537 of the protein and is essential for circadian rhythmicity.[47,48] Recent data indicate that the histone methyltransferase MLL1 directs the cyclic trimethylation of histone H3 Lys4 on circadian promoters, which subsequently mediates the recruitment of the CLOCK:BMAL1 complex to chromatin.[49] Moreover, the demethylase Jarid1a is recruited with CLOCK/BMAL1 to circadian gene promoters and is involved in modulating the acetylation of histone H3 by inhibiting HDAC1, and subsequently enhancing transcription of the clock complex.[50]

The recent discovery that the activity of SIRT1, a longevity-associated protein belonging to a family of nicotinamide adenine dinucleotide (NAD$^+$) activated histone deacetylases,[51] oscillates in a

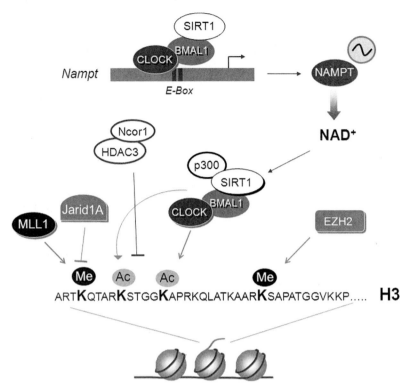

Figure 1. Linking cellular metabolism with the circadian clock transcriptional complex. A series of studies demonstrated that the circadian clock machinery controls the cyclic synthesis of NAD[+] through control of the NAD[+] salvage pathway.[55,56] The gene encoding the enzyme NAMPT, the rate-limiting step in the NAD[+] salvage pathway, contains E-boxes and is controlled by CLOCK-BMAL1. A crucial step in the NAD[+] salvage pathway is controlled by SIRT1, which also contributes to the regulation of the *Nampt* promoter by associating with CLOCK-BMAL1 in the CLOCK chromatin complex.[29] Oscillating levels of NAD[+] also regulate chromatin remodeling events through SIRT1 and ultimately connect clock-dependent transcriptional control with a cellular metabolic pathway. MLL1 also directs the cyclic histone H3 Lys4 trimethylation event that is responsible for CLOCK:BMAL1 recruitment and permits circadian gene expression.[62] Also included is the HDAC3 complex, which targets H3K9 acetylation in a circadian manner,[53] EZH2,[63] and Jarid1a.[50] NAD[+], nicotinamide adenine dinucleotide; ∼ indicates oscillation; Me, methylation; Ac, acetylation.

circadian fashion broadens our knowledge about the communication between the circadian clock and metabolism. Yet this finding also reveals a void in our understanding about the interplay between the metabolic state of the cell and circadian control on the molecular level. SIRT1 counterbalances the HAT function of CLOCK by deacetylating both H3 Lys9/14 and BMAL1,[48] as well as the deacetylation of the circadian regulatory protein PER2.[52] SIRT1 demonstrates an oscillation in activity, impinging back on the circadian clock by altering BMAL1 acetylation and CLOCK:BMAL1-induced gene transcription.[48,52] In addition to the deacetylase activity of SIRT1, HDAC3 has been reported to modulate chromatin marks in a circadian manner. HDAC3 is recruited to the genome in a rhythmic manner in mouse liver, and histone acetylation at H3K9 is inversely correlated with HDAC3 recruitment.[53] Moreover, it was found by genome-wide ChIP-seq analysis that HDAC3 and Rev-erbα were co-localized at a number of common genes involved in lipid metabolism, amino acid metabolism, and carbohydrate metabolism in the mouse liver.[53] A number of different lines of evidence suggest that an intricate relationship exists between chromatin state and cellular metabolism that is under the control of the circadian clock. What is quite intriguing is the bidirectional regulation between metabolism and epigenetics, suggesting that the circadian clock may control a complex network of feedback signals that we are only beginning to understand.

NAD⁺ as a central circadian regulator

The discovery of metabolite oscillations during the yeast metabolic cycle,[54] combined with evidence of circadian-directed sirtuin activity, allows speculation as to whether metabolites such as NAD^+ themselves serve a preponderant role in the cellular link between metabolism and the circadian clock. Indeed, NAD^+ itself is a critical signaling metabolite that is under the control of the circadian clock. Using accurate mass spectrometry/liquid chromatography measurements, our laboratory and others have confirmed this notion by demonstrating that NAD^+ levels oscillate in serum-entrained MEFs and in liver.[55,56] The circadian clock controls the expression of nicotinamide phosphoribosyltransferase (NAMPT), a key rate-limiting enzyme in the salvage pathway of NAD^+ biosynthesis. CLOCK, BMAL1, and SIRT1 are recruited to the *Nampt* promoter in a time-dependent manner. The oscillatory expression of NAMPT is abolished in *clock/clock* mice, which results in drastically reduced levels of NAD^+ in MEFs derived from these mice.[55] These results make a compelling case for the existence of an interlocking classical transcriptional feedback loop that controls the circadian clock, with an enzymatic loop in which SIRT1 regulates the levels of its own cofactor.

The oscillation of NAD^+ levels begs the question of whether the activity of other NAD^+-dependent enzymes may be regulated in a circadian manner. In this respect, one class of enzymes appears to occupy a privileged position: the poly(ADP-ribose) polymerases (PARPs), which have been shown to functionally interact with SIRT1.[57] PARP-1, the most well characterized PARP, is activated by DNA damage and plays a role in DNA repair. Since increased activity of PARP depletes the intracellular pool of NAD^+, this may lead to reduced SIRT1 activity and cell death.[57] Aside from potential effects on SIRT1, the activity of PARP-1 was shown to be rhythmic over the circadian day/night cycle,[58] resulting in a number of molecular consequences. PARP-1 was shown to directly bind CLOCK and BMAL1 and, subsequently, poly(ADP-ribosyl)ate CLOCK, which also modulated the ability of the circadian transcription factors to bind target DNA consensus sites.[58] Furthermore, data also suggested that PARP-1 is a critical regulator of feeding entrainment on peripheral circadian clocks. PARP-1–deficient mice,

compared to wild-type control animals, exhibited a phase delay in circadian gene expression in response to altered feeding regimens, implying that a link exists between PARP-1 and metabolic cues that signal to peripheral circadian clocks.[58] Increasing evidence indicates that circadian regulatory enzymes link metabolism with clock-timing systems, yet the extent to which these circuits are regulated by the cellular metabolic state and how they potentially feedback to the central clock is an intriguing concept that requires further investigation.

Given the direct control of SIRT1 deacetylase activity, as well as the control of PARP-1 activity by NAD^+, circadian regulation of NAD^+ levels appears to be a critical regulatory mechanism controlling circadian rhythms, metabolism, and cell growth. Interestingly, altered NAMPT levels have been implicated in metabolic disorders and cancer, and FK866, a highly specific NAMPT inhibitor that abolishes NAD^+ circadian oscillations and thereby SIRT1 cyclic activity, is used to control cell death in human cancer tissues. These results suggest that a direct molecular coupling exists between the circadian clock, energy metabolism, and cell survival. Future studies will reveal the precise function of SIRT1-directed circadian control in regulation of metabolism.

The evidence that NAD^+ intracellular levels are under control of the circadian clock begs the question regarding what other metabolites also follow a circadian oscillation. Recent studies have reported that in human[59] and mouse blood plasma,[60] as well as in mouse liver,[61] a number of metabolites are under the control of the circadian clock. In mouse liver, varying metabolites peak at different times during the circadian cycle, including nucleotide, carbohydrate, and lipid metabolite peaks at zeitgeber time (ZT) 9, versus amino acid and xenobiotic metabolite peaks at ZT 15–21.[61] Amino acids and metabolites of the urea cycle were reported to follow a circadian oscillation in mouse plasma.[60] Also, approximately ~15% of identified metabolites in human blood plasma or saliva followed a rhythmic pattern, including fatty acids in plasma and amino acids in saliva.[59] These studies reveal the vast amount of data that are currently unexplored related to circadian metabolomics, and suggest a number of mechanistic pathways remain to the elucidated that connect the circadian clock with metabolic state. Also, considering a number of enzymes (including histone modifiers) that use varying cofactors and metabolites, a detailed analysis of such metabolites that are currently not known to oscillate is needed.

Concluding remarks

The circadian clock comprises a hierarchical network of transcriptional, translational, and post-translational events that govern a tightly controlled timekeeping system. The intricate interrelationship between the central and peripheral clocks is a highly regulated system that requires precise specificity to maintain proper biological rhythms. Yet emerging evidence suggests that the circadian clock machinery is extremely plastic and can respond to external cues, suggesting an intimate link between the circadian cycle and environmental state. An additional level of complexity exists in the number of metabolic processes that are emerging as direct targets of the circadian machinery. New evidence also suggests that the metabolic state of the cell can directly modulate the circadian epigenome and transcriptional control. We are only beginning to understand the complexity of the circadian clock and appreciate the scope of this network, how it is regulated, and the extent to which it governs systemic physiological state.

Acknowledgments

We wish to thank all members of our laboratory for discussions and support. Work in the laboratory is supported by the National Institutes of Health USA, by the Inserm (Institut National de la Sante et la Recherche Medicale), France, and Sirtris Pharmaceutical Inc., a GSK Company.

Conflicts of interest

The authors declare no conflicts of interest.

References

1. Schibler, U. & P. Sassone-Corsi. 2002. A web of circadian pacemakers. *Cell* **111:** 919–922.
2. Ueda, H.R. *et al.* 2005. System-level identification of transcriptional circuits underlying mammalian circadian clocks. *Nat. Genet.* **37:** 187–192.
3. Green, C.B., J.S. Takahashi & J. Bass. 2008. The meter of metabolism. *Cell* **134:** 728–742.
4. Sahar, S. & P. Sassone-Corsi. 2009. Metabolism and cancer: the circadian clock connection. *Nat. Rev. Cancer* **9:** 886–896.
5. Ralph, M.R. *et al.* 1990. Transplanted suprachiasmatic nucleus determines circadian period. *Science* **247:** 975–978.
6. Welsh, D.K. *et al.* 1995. Individual neurons dissociated from rat suprachiasmatic nucleus express independently phased circadian firing rhythms. *Neuron* **14:** 697–706.
7. Morse, D. & P. Sassone-Corsi. 2002. Time after time: inputs to and outputs from the mammalian circadian oscillators. *Trends Neurosci.* **25:** 632–637.
8. Yamazaki, S. *et al.* 2000. Resetting central and peripheral circadian oscillators in transgenic rats. *Science* **288:** 682–685.
9. Yoo, S.H. *et al.* 2004. PERIOD2: LUCIFERASE real-time reporting of circadian dynamics reveals persistent circadian oscillations in mouse peripheral tissues. *Proc. Natl. Acad. Sci. USA* **101:** 5339–5346.
10. Quintero, J.E., S.J. Kuhlman & D.G. McMahon. 2003. The biological clock nucleus: a multiphasic oscillator network regulated by light. *J. Neurosci.* **23:** 8070–8076.
11. Bellet, M.M. & P. Sassone-Corsi. 2010. Mammalian circadian clock and metabolism—the epigenetic link. *J. Cell Sci.* **123:** 3837–3848.
12. Freedman, M.S. *et al.* 1999. Regulation of mammalian circadian behavior by non-rod, non-cone, ocular photoreceptors. *Science* **284:** 502–504.
13. Gooley, J.J. *et al.* 2001. Melanopsin in cells of origin of the retinohypothalamic tract. *Nat. Neurosci.* **4:** 1165.
14. Crosio, C. *et al.* 2000. Light induces chromatin modification in cells of the mammalian circadian clock. *Nat. Neurosci.* **3:** 1241–1247.
15. Lehman, M.N. *et al.* 1987. Circadian rhythmicity restored by neural transplant. Immunocytochemical characterization of the graft and its integration with the host brain. *J. Neurosci.* **7:** 1626–1638.
16. Pando, M.P. *et al.* 2002. Phenotypic rescue of a peripheral clock genetic defect via SCN hierarchical dominance. *Cell* **110:** 107–117.
17. Antle, M.C. & R. Silver. 2005. Orchestrating time: arrangements of the brain circadian clock. *Trends Neurosci.* **28:** 145–151.
18. Cheng, M.Y. *et al.* 2002. Prokineticin 2 transmits the behavioural circadian rhythm of the suprachiasmatic nucleus. *Nature* **417:** 405–410.
19. Kramer, A. *et al.* 2001. Regulation of daily locomotor activity and sleep by hypothalamic EGF receptor signaling. *Science* **294:** 2511–2515.
20. Stokkan, K.A. *et al.* 2001. Entrainment of the circadian clock in the liver by feeding. *Science* **291:** 490–493.
21. Damiola, F. *et al.* 2000. Restricted feeding uncouples circadian oscillators in peripheral tissues from the central pacemaker in the suprachiasmatic nucleus. *Genes Dev.* **14:** 2950–2961.
22. Eckel-Mahan, K. & P. Sassone-Corsi. 2009. Metabolism control by the circadian clock and vice versa. *Nat. Struct. Mol. Biol.* **16:** 462–467.
23. Roenneberg, T. & M. Merrow. 2005. Circadian clocks—the fall and rise of physiology. *Nat. Rev. Mol. Cell Biol.* **6:** 965–971.
24. Brown, S.A. *et al.* 2002. Rhythms of mammalian body temperature can sustain peripheral circadian clocks. *Curr. Biol.* **12:** 1574–1583.
25. Preitner, N. *et al.* 2002. The orphan nuclear receptor REV-ERBalpha controls circadian transcription within the

positive limb of the mammalian circadian oscillator. *Cell* **110:** 251–260.

26. Sato, T.K. *et al.* 2004. A functional genomics strategy reveals Rora as a component of the mammalian circadian clock. *Neuron.* **43:** 527–537.

27. Cardone, L. *et al.* 2005. Circadian clock control by SUMOylation of BMAL1. *Science* **309:** 1390–1394.

28. Stephan, F.K. & I. Zucker. 1972. Circadian rhythms in drinking behavior and locomotor activity of rats are eliminated by hypothalamic lesions. *Proc. Natl. Acad. Sci. USA* **69:** 1583–1586.

29. Masri, S. & P. Sassone-Corsi. 2010. Plasticity and specificity of the circadian epigenome. *Nat. Neurosci.* **13:** 1324–1329.

30. Tousson, E. & H. Meissl. 2004. Suprachiasmatic nuclei grafts restore the circadian rhythm in the paraventricular nucleus of the hypothalamus. *J. Neurosci.* **24:** 2983–2988.

31. Moore, R.Y. & V.B. Eichler. 1972. Loss of a circadian adrenal corticosterone rhythm following suprachiasmatic lesions in the rat. *Brain Res.* **42:** 201–206.

32. Akhtar, R.A. *et al.* 2002. Circadian cycling of the mouse liver transcriptome, as revealed by cDNA microarray, is driven by the suprachiasmatic nucleus. *Curr. Biol.* **12:** 540–550.

33. Panda, S. *et al.* 2002. Coordinated transcription of key pathways in the mouse by the circadian clock. *Cell* **109:** 307–320.

34. Yan, J. *et al.* 2008. Analysis of gene regulatory networks in the mammalian circadian rhythm. *PLoS Comput. Biol.* **4:** e1000193.

35. Baggs, J.E. *et al.* 2009. Network features of the mammalian circadian clock. *PLoS Biol.* **7:** e52.

36. Travnickova-Bendova, Z. *et al.* 2002. Bimodal regulation of mPeriod promoters by CREB-dependent signaling and CLOCK/BMAL1 activity. *Proc. Natl. Acad. Sci. USA* **99:** 7728–7733.

37. Cheung, P., C.D. Allis & P. Sassone-Corsi. 2000. Signaling to chromatin through histone modifications. *Cell* **103:** 263–271.

38. Felsenfeld, G. & M. Groudine. 2003. Controlling the double helix. *Nature* **421:** 448–453.

39. Cheung, P. *et al.* 2000. Synergistic coupling of histone H3 phosphorylation and acetylation in response to epidermal growth factor stimulation. *Mol. Cell* **5:** 905–915.

40. Borrelli, E. *et al.* 2008. Decoding the epigenetic language of neuronal plasticity. *Neuron.* **60:** 961–974.

41. Grunstein, M. 1997. Histone acetylation in chromatin structure and transcription. *Nature* **389:** 349–352.

42. Kuo, M.H. & C.D. Allis. 1998. Roles of histone acetyltransferases and deacetylases in gene regulation. *Bioessays* **20:** 615–626.

43. Struhl, K. 1998. Histone acetylation and transcriptional regulatory mechanisms. *Genes. Dev.* **12:** 599–606.

44. Wade, P.A. & A.P. Wolffe. 1997. Histone acetyltransferases in control. *Curr. Biol.* **7:** R82–R84.

45. Workman, J.L. & R.E. Kingston. 1998. Alteration of nucleosome structure as a mechanism of transcriptional regulation. *Annu. Rev. Biochem.* **67:** 545–579.

46. Doi, M., J. Hirayama & P. Sassone-Corsi. 2006. Circadian regulator CLOCK is a histone acetyltransferase. *Cell* **125:** 497–508.

47. Hirayama, J. *et al.* 2007. CLOCK-mediated acetylation of BMAL1 controls circadian function. *Nature* **450:** 1086–1090.

48. Nakahata, Y. *et al.* 2008. The NAD+-dependent deacetylase SIRT1 modulates CLOCK-mediated chromatin remodeling and circadian control. *Cell* **134:** 329–340.

49. Katada, S. & P. Sassone-Corsi. 2010. The histone methyltransferase MLL1 permits the oscillation of circadian gene expression. *Nat. Struct. Mol. Biol.* **17:** 1414–1421.

50. DiTacchio, L. *et al.* 2011. Histone lysine demethylase JARID1a activates CLOCK-BMAL1 and influences the circadian clock. *Science* **333:** 1881–1885.

51. Haigis, M.C. & L.P. Guarente. 2006. Mammalian sirtuins–emerging roles in physiology, aging, and calorie restriction. *Genes Dev.* **20:** 2913–2921.

52. Asher, G. *et al.* 2008. SIRT1 regulates circadian clock gene expression through PER2 deacetylation. *Cell* **134:** 317–328.

53. Feng, D. *et al.* 2011. A circadian rhythm orchestrated by histone deacetylase 3 controls hepatic lipid metabolism. *Science* **331:** 1315–1319.

54. Betz, A. & B. Chance. 1965. Phase relationship of glycolytic intermediates in yeast cells with oscillatory metabolic control. *Arch. Biochem. Biophys.* **109:** 585–594.

55. Nakahata, Y. *et al.* 2009. Circadian control of the NAD+ salvage pathway by CLOCK-SIRT1. *Science* **324:** 654–657.

56. Ramsey, K.M. *et al.* 2009. Circadian clock feedback cycle through NAMPT-mediated NAD+ biosynthesis. *Science* **324:** 651–654.

57. Kolthur-Seetharam, U. *et al.* 2006. Control of AIF-mediated cell death by the functional interplay of SIRT1 and PARP-1 in response to DNA damage. *Cell Cycle* **5:** 873–877.

58. Asher, G. *et al.* 2010. Poly(ADP-ribose) polymerase 1 participates in the phase entrainment of circadian clocks to feeding. *Cell* **142:** 943–953.

59. Dallmann, R. *et al.* 2012. The human circadian metabolome. *Proc. Natl. Acad. Sci. USA* **109:** 2625–2629.

60. Minami, Y. *et al.* 2009. Measurement of internal body time by blood metabolomics. *Proc. Natl. Acad. Sci. USA* **106:** 9890–9895.

61. Eckel-Mahan, K.L. *et al.* 2012. Coordination of the transcriptome and metabolome by the circadian clock. *Proc. Natl. Acad. Sci. USA* **109:** 5541–5546.

62. Etchegaray, J.P. *et al.* 2006. The polycomb group protein EZH2 is required for mammalian circadian clock function. *J. Biol. Chem.* **281:** 21209–21215.

Ann. N.Y. Acad. Sci. ISSN 0077-8923

ANNALS OF THE NEW YORK ACADEMY OF SCIENCES
Issue: *The Brain and Obesity*

Interacting epidemics? Sleep curtailment, insulin resistance, and obesity

Eliane A. Lucassen,[1,2] Kristina I. Rother,[3] and Giovanni Cizza[4]

[1]Immunogenetics Section, Clinical Center, National Institutes of Health, Bethesda, Maryland. [2]Laboratory of Neurophysiology, Department of Molecular Cell Biology, Leiden University Medical Center, Leiden, the Netherlands. [3]Section on Pediatric Diabetes and Metabolism, [4]Section on Neuroendocrinology of Obesity, National Institute of Diabetes and Digestive and Kidney Diseases, National Institutes of Health, Bethesda, Maryland

Address for correspondence: Eliane A. Lucassen, 10 Center Drive, Bethesda, Maryland 20892. eliane.lucassen@nih.gov

In the last 50 years, the average self-reported sleep duration in the United States has decreased by 1.5–2 hours in parallel with an increasing prevalence of obesity and diabetes. Epidemiological studies and meta-analyses report a strong relationship between short or disturbed sleep, obesity, and abnormalities in glucose metabolism. This relationship is likely to be bidirectional and causal in nature, but many aspects remain to be elucidated. Sleep and the internal circadian clock influence a host of endocrine parameters. Sleep curtailment in humans alters multiple metabolic pathways, leading to more insulin resistance, possibly decreased energy expenditure, increased appetite, and immunological changes. On the other hand, psychological, endocrine, and anatomical abnormalities in individuals with obesity and/or diabetes can interfere with sleep duration and quality, thus creating a vicious cycle. In this review, we address mechanisms linking sleep with metabolism, highlight the need for studies conducted in real-life settings, and explore therapeutic interventions to improve sleep, with a potential beneficial effect on obesity and its comorbidities.

Keywords: sleep; obesity; insulin resistance; diabetes; appetite

Introduction

According to the National Health and Nutrition Examination Survey (NHANES) from 2009 to 2010, 33% of adult Americans were overweight (body mass index (BMI) of 25.0–29.9 kg/m^2), and an even larger proportion (36%) were obese (BMI $\geq$ 30.0 kg/m^2).[1] The report contained good and bad news: for more than a decade, the age-adjusted prevalence of obesity had not changed in women overall, but there were marked ethnic, gender, and socioeconomic differences. Mexican American and Black women and men continued to show a significant rise in the prevalence of obesity. These figures are reflected in important obesity-related complications, such as hypertension, hyperlipidemia, and type 2 diabetes, all leading causes of cardiovascular disease.[2] Type 2 diabetes affects 8.3% of the U.S. population (27% of U.S. residents aged 65 years and older), whereas 35% of U.S. adults have fasting glucose and glycosylated hemoglobin (HbA1c) levels

in the prediabetic range.[3] Thus, it is of great importance to characterize contributing etiological factors in order to effectively counteract the obesity and diabetes epidemics. The exact mechanisms explaining the relationship between short sleep, obesity, and diabetes remain unclear.

Accumulating evidence has pointed to an association between sleep curtailment and obesity,[4,5] type 2 diabetes,[6–9] and possibly mortality (see Ref. 10 for a review). In the last 50 years, self-reported sleep duration in the United States has decreased by 1.5–2 hours due to lifestyle changes in our society.[11] Worldwide, secular trends in adult sleep duration may be more variable. A recent review article summarizing 12 studies from 15 countries from 1960 to 2000 did not find a consistent decrease in self-reported adult sleep duration: self-reported sleep duration had increased in seven countries, decreased in six, and results were inconsistent for the United States and Sweden.[12] Imprecision in assessing sleep duration, as well as socioeconomic

doi: 10.1111/j.1749-6632.2012.06655.x

Ann. N.Y. Acad. Sci. 1264 (2012) 110–134 © 2012 New York Academy of Sciences. No claim to original U.S. Government works.

and demographic factors may play a role in this heterogeneity.

In this paper, we will first present the epidemiologic evidence for these relationships; we will then focus on sleep physiology and its role in metabolic functions. Then, we will critically evaluate the studies of sleep manipulation on glucose metabolism and energy expenditure (EE). We will highlight endocrine parameters that may influence these processes, including thyroid hormones, growth hormone (GH), cortisol, sympathovagal balance, and the immune system. We will also delineate the relationships between short sleep and some of the hormones that control appetite, including leptin, ghrelin, orexin, and neuropeptide Y (NPY). In addition, we will present evidence for the reverse effect: of obesity/diabetes on sleep. Finally, we will describe emerging evidence for a role of melatonin in the sleep–metabolism relationship and its therapeutic potential in metabolic disturbances.

Epidemiologic evidence for an association among sleep duration, obesity, and diabetes

A meta-analysis of 45 cross-sectional studies, including 604,509 adults and 30,002 children, confirmed the relationship between short sleep (generally fewer than five hours per night in adults and fewer than 10 hours per night in children) and obesity, and quantified the associated risk (OR 1.55; 1.43–1.68; OR 1.89; 1.46–2.43, for adults and children, respectively).[13] A similar OR of 1.58 (1.26–1.98) was found in children in another meta-analysis.[14] Most of the included studies used actigraphy or self-reported sleep duration. A polysomnography study, not included in these meta-analyses, of 2,700 men above the age of 65 years revealed an inverse relationship between slow-wave sleep (SWS) duration and BMI or waist circumference.[15] Another study reported that obese adults (BMI $41 \pm 1 \, kg/m^2$) without sleep apnea slept 88 min fewer than lean subjects.[16] The relationship between short sleep and BMI was also confirmed by actigraphy in 612 participants (35–50 years old).[17]

In principle, causality is best addressed in prospective randomized controlled studies. This ideal approach may not be practical in this case: studies of sleep extension or curtailment cannot be blinded, and participants may not behave according to the allocation group (i.e., control subjects changing sleep duration or, vice versa, intervention subjects not changing sleep duration). For these reasons, the existing prospective studies have all been observational in nature. Short sleepers experience greater weight increases over time according to most,[4,18–22] but not all studies.[17,23,24] However, average weight gain was modest and of relative clinical meaning. Short sleepers (fewer than five to six hours) gained 2 kg more in a 6-year study,[19] and 0.4 kg more in another 16-year study.[18] In addition, subjects had a 35% and 31% greater chance of gaining more than 5 kg in 6 years, and gaining more than 15 kg in 16 years, respectively.[18,19] Thus, there is a large variability among studies in weight gain associated with short sleep for unknown reasons: the magnitude of this effect may be gender dependent or depend on other yet-to-be identified genetic and environmental factors. For example, short sleep may affect metabolism differently in natural "short sleepers" versus chronically sleep-deprived individuals. A large study including 35,247 subjects found that lean men, but not lean women, who slept fewer than five hours had increased odds of becoming overweight within one year (OR 1.91; 1.36–2.67).[25] In contrast, in a cohort of 3576 elderly subjects, women, but not men, who slept fewer than five hours, had a higher risk of gaining 5 kg in two years (OR 3.41; 1.31–8.69).[20] Interestingly, in this study, BMI was also increased in women sleeping more than eight hours. A similar *U*-shaped association has been reported in other studies.[19,25] Collectively, these studies indicate that self-reported short or long sleep is associated with increased BMI. The amount of weight gain over time, however, varied greatly, ranging from less than 1 kg to several kg, and age may be one of the parameters modulating this association.

Many mechanisms may be responsible for the frequently observed insulin resistance in obese individuals, including increased release of free fatty acids, leptin, and TNF-α from adipose tissue.[26] We suggest that short sleep (fewer than five to six hours) may now qualify as an additional clinical factor in the development of insulin resistance and diabetes, as shown in several cross-sectional studies.[27–29] A recent study of 174,542 middle-aged subjects showed that short sleepers had an OR of 1.46 (1.31–1.63) for developing diabetes after 3–10 years of follow-up.[30] A meta-analysis of 10 prospective studies reported a pooled OR of 1.28 (1.03–1.60) for diabetes in short sleepers.[31] Interestingly, this correlation was only present in men (OR 2.07; 1.16–3.72). Both long sleep (OR 1.48; 1.13–1.96) and decreased sleep efficiency also resulted in greater chances of developing diabetes.[31] Difficulty in falling asleep resulted in a greater incidence of diabetes after 4.2–14.8 years of follow up.[32–34] In smaller studies, the association between sleep duration and diabetes is more variable.[7–9,35] In addition, this variability may be real and differ on the basis of ethnic group and socioeconomic status: a study found increased risk of developing diabetes only in Whites and Hispanics who slept fewer than seven hours.[36] Besides sleep-duration, poor sleep quality relates to insulin resistance. Frequency of nocturnal awakenings, as measured by actigraphy, correlated positively with HbA1c levels in patients with type 2 diabetes.[37] Furthermore, sleep disordered breathing was related to worse blood glucose control.[38] Taken together, there is solid epidemiologic evidence for an interconnection between sleep duration (too short or too long), insulin resistance, and obesity.

Physiological changes of metabolism during normal sleep

Sleep architecture, the structure and pattern of sleep, is best described by a combined approach that uses electroencephalographic, electromyographic, and other criteria[39] (Fig. 1). Many metabolic pa-

rameters display diurnal rhythms, which are the resultant of both homeostatic and circadian mechanisms. From a homeostatic point of view, sleep pressure increases with sustained wakefulness. Circadian rhythms are endogenously generated in the suprachiasmatic nucleus of the hypothalamus by feedback loops of transcripts of a group of "core clock" genes, including the CLOCK gene.[40,41] Timekeeping signals are passed on from the suprachiasmatic nucleus to oscillators in peripheral tissues also containing core clock genes. Endogenous rhythms, in turn, are synchronized to the outside world by various environmental clues, mostly by light via the retina.

During sleep, energy requirements are lower as metabolic needs for processes such as breathing, gut motility, heart rate and muscular activity, are decreased.[42] Indeed, resting EE, as determined by indirect calorimetry, was lower in healthy volunteers during sleep compared to wakefulness.[43] During sleep glucose levels remain stable or fall only minimally in spite of fasting, mostly because of decreased energy needs.[44] In contrast, if subjects are kept awake in bed while fasting for a comparable number of hours, glucose levels do fall, approximately 15 mg/dl.[45] During the first part of the night, when most slow wave sleep (SWS) occurs, glucose production is diminished[44] (Fig. 2). Brain

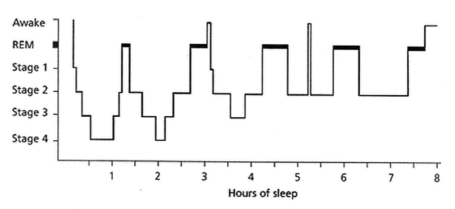

Figure 1. Representative sleep architecture during eight hours of uninterrupted sleep (adapted from Ref. 39). Sleep can be divided into rapid eye movement (REM) and nonrapid eye movement (NREM) sleep. On average, adults spend 75–80% in NREM sleep, which can be further subdivided into stages 1 to 4. Stages 3 and 4 are often referred to as slow-wave sleep (SWS) because of the appearance of well-defined waves of 0.5–4.0 Hz. In REM sleep, a phasic and a tonic phase can be distinguished. Skeletal muscles are atonic or hypotonic during REM sleep (except for the diaphragm, extraocular, and sphincter muscles), but bursts of muscle activity can occur during the phasic phase. Furthermore, dreams mostly take place during REM sleep and are more complex during this stage. Eye movement is only present during the phasic phase. Throughout a normal night of sleep, there are three to five sleep cycles, from NREM (stage 1 to 4) to REM, each cycle taking 90–120 minutes. Humans display less SWS and more REM sleep toward the end of the night.

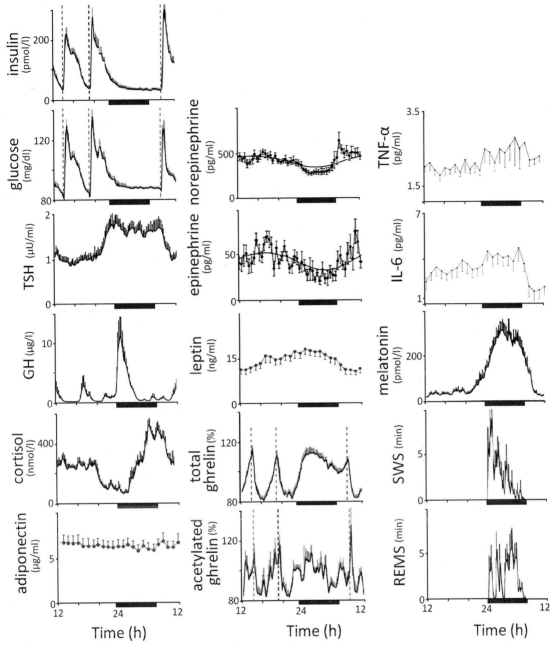

Figure 2. Mean levels of metabolic parameters and sleep stages in lean young men ($n = 8$ for TSH, cortisol, GH, melatonin, SWS, and REM sleep;[48] $n = 14$ for insulin, glucose, total and acetylated ghrelin;[76] $n = 8$ for catecholamines[73]); in 23 lean women for leptin and adiponectin;[69] and in 25 individuals (13 females, 12 males) for IL-6 and TNF-α.[83] Ghrelin levels are indicated as a percent of mean 24-h values (1027 pg/mL for total ghrelin; 80 pg/mL for acetylated ghrelin); leptin levels are shown as percent change from levels at 08:00. Dark bars below each plot indicate bedtimes. Sleep was monitored polysomnographically during all measurements. Striped bars indicate meal times. Modified, with permission Refs. 48, 69, 73, 76, and 83

glucose metabolism, as measured by PET, decreases by 30–40%.[46,47] During the second half of the night, glucose consumption increases due to increased REM sleep requiring more energy.[46,47]

EE and glucose metabolism are modulated by a number of hormones. Thyroid hormones are important determinants of EE; cortisol and GH are powerful modulators of circadian rhythms of glucose metabolism. The sympathovagal system and adiponectin influence both EE and glucose metabolism. Thyroid stimulating hormone (TSH) levels rise at night, mostly due to increased hypothalamic secretion of thyroid releasing hormone (TRH) (Fig. 2).[48]

GH secretion is under dual hypothalamic control: somatostatin is inhibitory and growth hormone releasing hormone (GHRH) is stimulatory. Somatostatin and GHRH are secreted in alternation, thus generating the typical pulsatile pattern of GH secretion. GH typically peaks at sleep onset during SWS (Fig. 2).[49] GH levels were lower in SWS-deprived individuals[50] and higher in individuals with pharmacologically induced SWS.[51] In addition, REM sleep deprivation did not affect plasma GH levels.[52] Intravenous GH administration decreases SWS, suggesting a modulatory role for GH on SWS.[53,54] The GH peak at sleep onset has insulin lowering effects.[55] GHRH injection increased glucose levels at awakening by almost 50%[56] and GH administration rapidly decreased muscular glucose uptake.[55,57]

Cortisol concentrations reach their zenith in the morning, experience a gradual fall during the day, which is briefly interrupted by meals, and have their nadir around 3 am (Fig. 2).[48] Cortisol impairs insulin sensitivity with a latency of four to six hours.[56,58,59] Plasma insulin levels also display a modest diurnal variation with a 10% excursion, a nadir between midnight and 6 am and a peak between noon and 6 pm.[60]

Adiponectin is a hormone secreted by the adipocytes in an inverse fashion compared to fat mass.[61] Its role in energy homeostasis is unclear: increases, decreases, or no changes in energy balance are derived from peripheral adiponectin administration in rodents.[62] Adiponectin knockout mice develop insulin resistance and increased TNF-α levels when fed a high fat diet. In humans, plasma adiponectin levels correlated negatively with insulin resistance, and higher levels correlated with future weight gain in 1,063 women.[63,64] Individuals with type 2 diabetes typically have low adiponectin levels, even when adjusted for body weight.[65] In contrast, patients with type 1 diabetes have higher than expected adiponectin levels: an observation that is poorly understood.[66] Sleep duration was inversely related to serum adiponectin in 109 lean Japanese males.[67] Adiponectin has anti-inflammatory properties: *in vitro* exposure to adiponectin decreases macrophage activation and TNF-α production. Finally, adiponectin displays some diurnal variability, with lower levels at night (Fig. 2).[68,69]

During sleep, vagal tone increases.[70] Sympathovagal balance is lowest during REM sleep, when the locus coeruleus, the main brain source of catecholamine biosynthesis, is silent.[71] Plasma norepinephrine (NE) and epinephrine (EPI) are produced by the sympathetic system: EPI is secreted exclusively from the adrenal medulla, whereas NE can also be released from postganglionic sympathetic nerves.[72] In a small study of eight lean males, plasma catecholamines were consistently lower during the night with no differences among sleep stages (Fig. 2).[73] Heart rate and mean blood pressure were lower during NREM sleep, compared to during awakenings and REM sleep.[70]

Several hormones involved in appetite regulation display circadian rhythms. Leptin is secreted by the white adipose tissue in a highly pulsatile fashion.[74] Leptin reflects fat stores: its levels increase after meals and during the night and are associated with decreased appetite (Fig. 2).[74] Ghrelin, produced in the stomach, stimulates appetite and circulates mainly in a nonacetylated form.[75] Acetylated ghrelin, the active form, is necessary for most of its endocrine actions, including GH release, appetite stimulation, and gastric emptying. Total and acetylated ghrelin levels rise with fasting and during sleep (Fig. 2).[76] Leptin and ghrelin exert their effects in the arcuate nucleus of the hypothalamus, through anorexigenic proopiomelanocortin/cocaine- and amphetamine-regulated transcript (POMC/CART) and orexigenic neuropeptide Y/agouti-related protein (NPY/AgRP) neurons. Leptin activates POMC/CART neurons, which induces hunger and inhibits NPY/AgRP neurons,[77,78] which induces satiety. Ghrelin has the opposite effect on the arcuate nucleus, stimulating appetite, and prolonging postprandial glucose responses, while stimulating GH release.[79] Insulin, while peripherally lowering blood glucose and stimulating appetite, inhibits appetite centrally in a leptin-like manner.[78]

Orexin A and B, two neuropeptides released by lateral hypothalamic neurons, stimulate appetite. Orexin neurons are active during waking and quiescent during sleep; consistently, orexin levels in the CSF are maximal at the end of the waking period.[80,81] The orexin system is activated by ghrelin and inhibited by leptin and glucose. Projections from orexin neurons in the lateral hypothalamus activate NPY neurons in the arcuate nucleus.[80] Melatonin is a sleep-promoting hormone that is produced by the pineal gland. Melatonin is exclusively secreted during the dark, as light inhibits its production (Fig. 2).

Interleukin 6 (IL-6), interleukin 1β (IL-1β), TNF-α, and C-reactive protein (CRP) are proinflammatory factors. IL-6 displays a circadian rhythm, with a first peak around 2 am and a second peak around 5 am (Fig. 2).[82,83] IL-1β decreases during the night and reaches its nadir at 8 am in the morning.[82] TNF-α also displays a circadian rhythm, peaking close to the awakening (6 am) and reaching a nadir around 3 pm (Fig. 2). IL-6 and TNF-α stimulate secretion of cortisol and, in turn, cortisol inhibits their secretion.[84] Reports on diurnal variability of CRP are few and inconsistent: some report no diurnal variance,[85] whereas others find increased morning levels.[86] In 10 healthy, lean men, white blood cell counts peaked around 11 pm, decreased throughout the night and reached a nadir at 8 am.[82]

The relationships here described between hormone secretion and sleep stages are necessarily descriptive. In theory, as new pharmacological tools aimed at manipulating hormonal status (i.e., selective hormone agonists/antagonists) and/or sleep become available for investigation, we may be able to better understand the numerous causal links interconnecting these phenomena. In summary, many metabolic parameters involved in EE regulation, glucose metabolism, and appetite control display diurnal rhythms, reflecting the different metabolic needs during the sleep and wake states. In the next section, we will report the effect of sleep deprivation on these metabolic parameters.

Effects of sleep restriction on metabolic parameters

The relationship between sleep and metabolism has been extensively studied in human subjects under varying conditions, such as total sleep deprivation,[87–100] partial sleep deprivation,[101–122] and sleep fragmentation[123–125] (Table 1). Partial sleep deprivation affects SWS less than other sleep stages. Most of the experiments on the effects of acute sleep deprivation were conducted in healthy, lean volunteers. Food intake and physical activity were often, but not always, strictly controlled (Table 1). It is important to note that experimentally induced sleep alterations may not be representative of changes occurring in clinical condition of chronic sleep deprivation. In addition, important differences exist between acute and chronic sleep deprivation. We will describe effects of sleep curtailment on important metabolic parameters that may contribute to insulin resistance and obesity, including effects on appetite and EE (Fig. 3).

Effects on glucose metabolism

In the course of the night, glucose supply to the brain must remain adequate in the setting of slightly declining plasma glucose levels. Infusion of [^{3}H3]-labeled glucose in healthy, lean subjects revealed decreased glucose usage during the night, whereas in subjects kept awake glucose usage was increased (Table 1).[44] Total and partial sleep deprivation reduced glucose tolerance in lean individuals. This has been documented by intravenous glucose tolerance tests (ivGTT),[101,109,114,123,124] oral glucose tolerance tests (OGTT),[87,89,109] insulin suppression tests with octreotide,[90] as well as hypoglycemic and euglycemic hypoinsulinemic clamps[97,114] (Table 1). Impaired glucose metabolism was not only shown in healthy study participants, but also in individuals with type 1 diabetes: after a four-hour night in adults with type 1 diabetes, glucose infusion rate decreased by 21%, as assessed by hyperinsulinemic euglycemic clamp[115] (Table 1). In addition to altered peripheral glucose metabolism during total sleep deprivation, cerebral glucose uptake decreased, especially in the prefrontal and posterior parietal cortex and in the thalamus of healthy volunteers[91] (Table 1). Interestingly, most studies found no compensatory rise in plasma insulin secondary to increased insulin resistance after short sleep,[109,114]

Table 1. Results of 37 human experimental studies assessing the influence of sleep curtailment on metabolism

Study design Energy intake Behavioral activity	Gender/sample size Age ± SEM (years) Mean weight ± SEM Prestudy conditions	Experimental sleep protocol	Techniques used for blood sampling (time, frequency)	Main findings (sleep deprived vs. longer sleep)

Benedict *et al.*[100]
Randomized
  cross-over study
Controlled,
  postintervention
  ad libitum
Bed rest
	Men (*n* = 14)
	22.6 ± 0.8
	23.9 ± 0.5 kg/m^2
	6 weeks regular bedtimes[c]
		1: 1 night of 8-h
		  bedtime[e]
		2: 1 night of TSD
		Conditions ≥ 4 weeks
		  apart
			Indirect calorimetry
			  pre- and
			  postbreakfast;
			  morning VAS
			  hunger ratings 24 h,
			  every 1.5–3 h
				↑ postprandial glucose (8%)
				↓ RMR (5%) and
				  postprandial (20%)
				  metabolic rate
				↑ cortisol (7%)[a]
				↑ NE (12%)[a]
				~ food intake; ↑ hunger
				  (60%)[a]
				↑ morning ghrelin (11%)
				~ leptin

Born *et al.*[82]
Randomized
  cross-over study
Standardized meals
Synchronous
  inpatient activities
	Men (*n* = 10)
	24.7 (21–29)
	NA
	NA
		1: 1 night of 8-h
		  bedtimes[e] then 1
		  night of *ad libitum*
		  sleep[e]
		2: 24 h TSD[e]
		Conditions ≥ 10 days
		  apart
			LPS-stimulated whole
			  blood and ELISA
			  (TNF-α; IL-1β;
			  IL-6)
			51 h, every 3 h
				~ IL-6
				~ IL-1β; ↑ nocturnal whole
				  blood
				~ TNF-α
				↑ white blood cells (8%)[1]

Brondel *et al.*[113]
Randomized
  cross-over study
Ad libitum
Free physical activity[d]
	Men (*n* = 12)
	22 ± 3 22.3 ± 1.83 kg/m^2
	2 days sleep/diet diaries[c]
		1: 1 night of 8-h
		  bedtime[c]
		  and 1 night *ad*
		  libitum sleep at
		  home[c,d]
		2: 1 night of 4-h
		  bedtime[e] and 1
		  night of free
		  bedtimes at home[c,d]
		Conditions ≥ 5 days
		  apart
			Actigraphy; 12
			  daytime VAS
			  hunger ratings
				~ EE; ↑ physical activity
				  (13%)
				↑ hunger; ↑ food intake
				  (22%)

Buxton *et al.*[114]
Consecutive phases
  study
Controlled isocaloric
  (finishing meal
  required)
Controlled sedentary
  living
	Men (*n* = 20)
	26.8 ± 5.2
	  23.3 ± 3.1 kg/m^2
	≥ 5 days of 10-h
	  bedtimes[c,d]
		1: 3 nights of 10-h
		  bedtimes[d]
		2: 7 nights of 5-h
		  bedtimes[d]
			ivGTT penultimate
			  day; clamp last day;
			  salivary sampling
			  15:00–21:00 last 2
			  days, every hour,
			  24-h urine
			  sampling last 2
			  days; indirect
			  calorimetry
				↑ insulin resistance (20% by
				  ivGTT; 11% by clamp)
				~ fasting RMR
				↑ salivary cortisol (51%)
				↑ urinary NE (18%) and
				  EPI (22%)[1]

Clore *et al.*[44]
Comparative study
Similar evening meals
Bed rest
	5 females, 12 males
	  (*n* = 17)
	25 ± 1
	24.1 ± 0.6 kg/m^2
	NA
		1. Group 1 (*n* = 11): 1
		  night of 8-h, 30 min
		  bedtime
		2. Group 2 (*n* = 6): 1
		  night of TSD
			Continuous [^{3}−3H]
			  glucose infusion
				↓ nocturnal fall in glucose
				  use
				↓ nocturnal fall in glucose
				  production

Donga *et al.*[115]
Randomized
  cross-over study
NA
	3 females, 4 males
	  (*n* = 9)
	44 ± 7
	23.5 ± 0.9 kg/m^2
		1: 3 nights of 8-h,
		  30-min bedtimes[d]
			Hyperinsulinemic
			  euglycemic clamp
				↑ insulin resistance (21%)

Continued

Table 1. *Continued*

Study design Energy intake Behavioral activity	Gender/sample size Age ± SEM (years) Mean weight ± SEM Prestudy conditions	Experimental sleep protocol	Techniques used for blood sampling (time, frequency)	Main findings (sleep deprived vs. longer sleep)
NA	Type 1 diabetes, HbA1c ≤ 8.5%, 1 week of regular bedtimes[c,d]	2: 3 nights of 4-h bedtimes[e] Conditions ≥ 3 weeks apart		
Dzaja *et al.*[96] Randomized cross-over study 1,800 kcal/24 h Bed rest	Men (*n* = 10) 28 ± 3.1 24.0 ± 2.9 kg/m² 1 week of regular bedtimes[c]	1: 1 night of 8-h bedtime[e] 2: 1 night of TSD Conditions ≥ 2 weeks apart	24 h, every h	↓ GH peak (71%)[b] ~ mean cortisol levels ↓ nocturnal total ghrelin increase (39%)[b]
Frey *et al.*[96] Consecutive phases study Hourly controlled intake Bed rest	9 females, 10 males (*n* = 19) 28.1 ± 8.6 18.5–24.5 kg/m² 3 weeks of regular bedtimes[c,d]	1: 3 nights of 8-h bedtimes[e] 2: 1 night of TSD	Saliva 36 h, every h; hsELISA (IL-6; IL-1β; CRP) 40 h, every 30 min	~ saliva cortisol; ↓ at 13:00 (53%)[b] and 20:00 (17%)[b] ↓ daytime IL-6 ↑ IL-1β ↓ morning hsCRP
Gonzáles-Ortiz *et al.*[90] Consecutive phases study Controlled isocaloric Free inpatient movement	7 females, 7 males (*n* = 14) 21.0 ± 2.1 22.1 ± 1.3 kg/m² 3 days diet diary	1: baseline "normal sleep" 2: 1 night of TSD	Insulin suppression test with octreotride Once, morning	↑ insulin resistance (18%)[a] ~ morning cortisol
Haack *et al.*[108] Comparative study Controlled isocaloric Free inpatient movement	6 females, 12 males (*n* = 18) 27.3 ± 5.8 23.1 ± 3.3 <30 kg/m² ≥ 10 days regular bedtimes[c]	(*n* = 8) 1: 10 nights of 8-h bedtimes[d] (*n* = 10) 2: 10 nights of 4-h bedtimes[d]	24-h urine 24-h blood, every 4 h (high sensitivity ELISA IL-6 and hsCRP)	~ food intake ↑ IL-6 (62%)[a] ~ TNF-receptor 1 ↑ hsCRP (102%; *P* = 0.11)[1]
Hursel *et al.*[125] Randomized cross-over study Controlled isocaloric Sedentary activity	Men (*n* = 15) 23.7 ± 3.5 24.1 ± 1.9 kg/m² 2 days controlled diet	1. 2 nights of 8-h bedtimes[e] 2: 2 nights of sleep fragmentation hourly 2-min–long alarms[e] Conditions ≥ 2 weeks apart	Indirect calorimetry	~ total EE ↑ physical activity (8%)[a] ↑ RQ (3%)[a]
Irwin *et al.*[107] Consecutive phases study NA No exercise	13 females, 17 males (*n* = 30) 37.6 ±9.8 <30 kg/m² 2 weeks regular bedtimes[c]	1: 3 nights of 8-h bedtimes[e] 2: 1 night of 4-h bedtime[e]	Flow cytometry; real time PCR (*n* = 10); high-density oligonucleotide array (*n* = 5) Daytime (08:00–23:00), every 3 to 4 h	↑ IL-6 monocyte expression (344%)[a]; ↑ IL-6 mRNA (250%)[a] ↑ TNF-α monocyte expression (87%)[a]; ↑ TNF-α mRNA (100%)[a]

Continued

Table 1. *Continued*

Study design Energy intake Behavioral activity	Gender/sample size Age ± SEM (years) Mean weight ± SEM Prestudy conditions	Experimental sleep protocol	Techniques used for blood sampling (time, frequency)	Main findings (sleep deprived vs. longer sleep)
Jung *et al.*[43] Consecutive phases study Controlled isocaloric Sedentary	2 females, 5 males ($n = 7$) 22.4 ± 4.8 22.9 ± 2.4 kg/m^2 1 week 8-h sleep[c,d]; 3 days diet	1: 2 nights of 8-h bedtimes[e] 2: 40 h of TSD	Indirect calorimetry	↑ total EE (7%); ↑ nocturnal EE (32%); ∼ daytime EE ∼ RQ
Kuhn *et al.*[87] Consecutive phases study Balanced diet NA	Human subjects ($n = 28$) 20–30 NA	1: 4–5 days "control period" 2: 72–126 h of TSD 3: 3 days "control period"	OGTT Once daily	↓ glucose tolerance after 3–4 days ↑ 17-OH corticosteroids on Day 2 and 3, then return to baseline ↑ urinary catecholamines
Mullington *et al.*[94] Consecutive phases study Controlled isocaloric (finishing not required) Sedentary	Men ($n = 10$) 27.2 26.1 ± 1.9 kg/m^2 1 week of regular bedtimes[c,d]	1. 3 nights 8-h bedtimes[e] 2. 3 nights of TSD	120 h, every 90 min	∼ leptin mesor, ↓ leptin circadian amplitude
Nedeltcheva *et al.*[109] Kessler *et al.*[116] Randomized cross-over study *Ad libitum* Sedentary	5 females, 6 males ($n = 11$) 39 ± 5 26.5 ± 1.5 kg/m^2 NA	1: 14 nights of 8.5-h bedtimes[e] 2: 14 nights of 5.5-h bedtimes[e] Conditions ≥ 3 months apart	OGTT; ivGTT 24 h, every 15–30 min	↓ insulin sensitivity (18%)[a] ↓ TSH (7%; especially in females)[a] ↓ free T4 (8%)[a] ∼ GH; ↓ during first 4-h of bedtime period in males (31%)[a] ∼ cortisol; ↓ and 44 min later acrophase (10%)[a]; 71 min later nadir ↑ EPI (21%)[a] ∼ mean NE; ↑ nocturnal (24%)[a]
Nedeltcheva *et al.*[118] Randomized cross-over study Controlled restricted (90% RMR) Sedentary	3 females, 7 males ($n = 10$) 41 ± 5 27.4 ± 2 kg/m^2 NA	1: 14 nights of 8.5-h bedtimes[e] 2: 14 nights of 5.5-h bedtimes[e] Conditions ≥ 3 months apart	Indirect calorimetry; VAS hunger rating scale before meal/at 22:30, 24 h, every hour	∼ total EE; ↓ RMR (8%)[a]; ↑ fasting and postprandial RQ (4%)[a] ∼ thyroid hormones ∼ GH ∼ cortisol ↓ EPI (12%)[a]; ∼NE ∼ food intake; ↑ hunger ∼ total weight loss (3 kg), ↓ fat weight loss (55%), ↑ fat-free weight loss (60%) ↑ mean acylated ghrelin (7%)[a] ∼ leptin

Continued

Table 1. *Continued*

Study design Energy intake Behavioral activity	Gender/sample size Age ± SEM (years) Mean weight ± SEM Prestudy conditions	Experimental sleep protocol	Techniques used for blood sampling (time, frequency)	Main findings (sleep deprived vs. longer sleep)

Omisade *et al.*[117]
Consecutive phases
  study
Controlled
Free inpatient
  movement
	Women ($n = 15$)
	21.6 ± 2.2
	24.5 ± 8 kg/m^2
	1 week regular bedtimes
	and diet[c,d]
		1: 1 night of 10-h
		  bedtime[d]
		2: 1 night of 3-h
		  bedtime[d]
			VAS hunger ratings;
			salivary cortisol
			daytime, every 1–2
			h; salivary leptin
			morning and
			afternoon
				↓ median morning cortisol
				  (19%)[a]; slower decline; ↑
				  afternoon/evening
				  (44%)[a]
				~ hunger
				↑ morning leptin (8%)[a]

Parker *et al.*[88]
Consecutive phases
  study
Controlled
Sedentary
	Men ($n = 4$)
	21–30
	NA
	NA
		1: 1 night of 8-h
		  bedtime[e]
		2: 2 nights of TSD
			72 h, every 30 min	↑ TSH; ↑ amplitude, longer
				  peak, later
				  nadir/acrophase

Pejovic *et al.*[99]
Consecutive phases
  study
Ad libitum
Ambulatory
	11 females, 10 males
	  ($n = 21$)
	24.1 ± 3.1
	24.1 ± 2.6 kg/m^2
	2 weeks regular
	  bedtimes[d]
		1: 4 nights of 8-h
		  bedtimes[e]
		2: 1 night of TSD, 50%
		  had 2-h afternoon
		  nap
			VAS hunger ratings
			24 h, every 30 min
				~ cortisol
				~ adiponectin
				~ hunger
				↑ leptin (14%)[a]

Schmid *et al.*[97]
Randomized
  cross-over study
None
Bed rest
	Men ($n = 10$)
	25.3 ± 1.4
	23.8 ± 0.5 kg/m^2
	Light dinner before
	  arrival in the lab
	  (21:00)
		1: 1 night of 8-h
		  bedtime[e]
		2: 1 night of TSD
		Conditions ≥ 2 weeks
		  apart
			Hypoglycemic clamp,
			Two morning samples
				↓ glucagon levels (16%)[a]; ↑
				  response to hypoglycemia
				↓ cortisol levels (16%)[a];
				  ~ ACTH
				~ catecholamines
				↑ hunger

Schmid *et al.*[98]
Randomized
  cross-over study
None
Bed rest
	Men ($n = 9$)
	24.2 ± 1.0
	23.8 ± 0.6 kg/m^2
	NA
		1: 1 night of 7-h,
		  30-min bedtime[e]
		2: 1 night of 5-h
		  bedtime[e]
		3: 1 night of TSD
		Conditions ≥ 2 weeks
		  apart
			Hunger graded 0–9,
			Two morning samples
				↑ hunger (129%)[a]
				~ leptin
				↑ total ghrelin (22%)

Schmid *et al.*[110]
Schmid *et al.*[119]
Randomized
  cross-over study
Ad libitum
Free outpatient
  movement
	Men ($n = 15$)
	27.1 ± 1.3
	22.9 ± 0.3 kg/m^2
	4 weeks regular
	  bedtimes[c]
		1: 2 nights of 8-h
		  bedtimes[e]
		2: 2 nights of 4-h
		  bedtimes[e]
		Conditions ≥ 6 weeks
		  apart
			Actigraphy, buffet on
			day 2, ELISA (IL-6)
			08:00–23:00 (every h)
				↑ insulin (40%) and glucose
				  (11%) peak response to
				  breakfast
				↓ physical activity (13%)[a]
				~ cortisol; ~ ACTH
				~ total energy intake; ↑ fat
				  intake (23%)[a]
				~ hunger; ~ intake
				~ leptin; ~ ghrelin
				~ IL-6

Schmid *et al.*[111]
Randomized
  cross-over study
None
Bed rest
	Men ($n = 10$)
	25.3 ± 1.4
	23.8 ± 0.5 kg/m^2
	2 weeks regular
	  bedtimes[c]
		1: 1 night of 7-h
		  bedtime[e]
		2: 1 night of 4.5-h
		  bedtime[e]
		Conditions ≥ 2 weeks
		  apart
			Hypoglycemic clamp
			2 morning samples
				~ insulin resistance
				↓ glucagon (8%)[b]
				~ GH
				↓ ACTH (44%)[a]; ↓ cortisol
				  (44%)[a]
				~ EPI; ~ NE

Continued

Table 1. *Continued*

Study design / Energy intake / Behavioral activity	Gender/sample size / Age ± SEM (years) / Mean weight ± SEM / Prestudy conditions	Experimental sleep protocol	Techniques used for blood sampling (time, frequency)	Main findings (sleep deprived vs. longer sleep)
Shearer *et al.*[93] Comparative study Controlled isocaloric No exercise	Men ($n = 42$) 28.7 (21–47) NA 2 weeks regular bedtimes[d]	($n = 21$) 1: 3 nights of 8-h bedtimes, then 4 nights of 2-h bedtimes ($n = 21$) 2: 3 nights of 8-h bedtimes, then 4 nights of TSD	5 days, every 6 h (ELISA)	No changes in condition 1; in condition 2: ↑ IL-6 ~ TNF-α ↑ sTNF-α receptor I ~ sTNF-α receptor II
Simpson *et al.*[120] Consecutive phases study *Ad libitum* Free inpatient movement	67 females, 69 males ($n = 136$) 30.4 (22–45) 24.7, 17.7–32.6 kg/m^2 1 week regular bedtimes[c,d]	1: 2 nights of 10-h bedtimes 2: 5 nights of 4-h bedtimes	Once, morning (10:30–12:00)	↑ leptin (33%)[a]
Simpson *et al.*[121] Consecutive phases study *Ad libitum* Free inpatient movement	33 females, 41 males ($n = 74$) 29.9, 22–45 24.6, 17.7–33.1 kg/m^2 1 week regular bedtimes[c,d]	1: 2 nights of 10-h bedtimes[e] 2: 5 nights of 4-h bedtimes[e]	Once, morning (10:30–12:00)	~ adiponectin in men; ↑ in African American women; ↓ in Caucasian women
Spiegel *et al.*[101] Spiegel *et al.*[102] Spiegel *et al.*[103] Spiegel *et al.*[104] Consecutive phases study Controlled Free outpatient movement	Men ($n = 11$) 22 ± 1 23.4 ± 0.5 kg/m^2 ≥ 1 week of 8-h bedtimes, diet[d,e]	1: 6 nights of 4-h bedtimes[e] 2: 6 nights of 12-h bedtimes[e]	ivGTT; 24-h heart rate variability; salivary sampling 15:00-bedtime, every 30 min; blood sampling 24 h, every 30 min	↓ glucose clearance (30%) ↓ TSH (26%), ↑fT4 (7%)[a] ~ GH, different profile ↑ afternoon saliva/plasma cortisol (25%)[a] ↑ sympathovagal balance (17%)[a] ↓ leptin (19%); ↓ and 2.5 h earlier acrophase (26%); ↓ amplitude (20%) ↓ melatonin; later onset; ↓ acrophase
Spiegel *et al.*[105] Randomized cross-over study Controlled Free outpatient movement	Men ($n = 12$) 22 ± 2 23.6 ± 2.0 ≥ 1 week of 8-h bedtimes	1: 2 nights of 10 h bedtimes[e] 2: 2 nights of 4-h bedtimes[e] Conditions ≥ 6 weeks apart	VAS hunger ratings 08:00–21:00, every 20 min	↑ hunger (23%; 33–45% for sweets/salty foods) ↓ leptin (18%) ↑ ghrelin (28%)
Stamatakis *et al.*[123] Consecutive phases study *Ad libitum* NA	2 females, 9 males ($n = 11$) 23.2 (18–29) 24.3 ± 0.9 kg/m^2 3 days diet and >7-h bedtimes[a,b]	1: 1 night of unfragmented sleep[e] 2: 2 nights of fragmented sleep (~ 31.4 auditory/mechanical stimuli per hour)[e]	ivGTT; heart rate variability; enzyme-linked immunosorbent assay techniques (IL-6; hsCRP) 2 morning samples (08:00 and 16:00)	↑ insulin resistance (25%) ↑ morning cortisol (13%) ~ adiponectin ↑ sympathetic tone (17%) ~ leptin ~ Il-6; ~hsCRP

Continued

Table 1. *Continued*

Study design Energy intake Behavioral activity	Gender/sample size Age $\pm$ SEM (years) Mean weight $\pm$ SEM Prestudy conditions	Experimental sleep protocol	Techniques used for blood sampling (time, frequency)	Main findings (sleep deprived vs. longer sleep)
St-Onge *et al.*[122] Randomized cross-over study Controlled isocaloric, self-selected on day 5 Free inpatient movement	15 females; 15 males ($n = 30$) 33.9 ± 4.3; 36.6 ± 5.6 23.0 ± 1.1; 24.1 ± 1.1 kg/m^2 2 weeks regular bedtimes[c,d]	1: 5 nights of 9-h bedtimes[e] 2: 5 nights of 4-h bedtimes[e] Conditions $\geq$ 4 weeks apart	Double-labeled water (EE); indirect calorimetry (RMR); actigraphy; VAS hunger ratings	$\sim$ RMR; $\sim$ EE; $\sim$ physical activity $\sim$ hunger; $\uparrow$ food intake (12%)[a]; especially $\uparrow$ saturated fat intake in females (62%)
Tasali *et al.*[123] Randomized cross-over study Controlled Sedentary activities	4 females, 5 males ($n = 9$) 20–31 64.2 kg; 19–24 kg/m^2 NA 1 week regular bedtimes[d]	1: 2 nights of undisturbed sleep[e] 2: 3 nights of disturbed sleep by acoustic stimuli during SWS ($\sim$33 micro-arousals/h)[e] Conditions $\geq$ 4 weeks apart	ivGTT; heart rate variability 24 h, every 20 min	$\uparrow$ insulin resistance (25%) $\sim$ cortisol; $\sim$ daytime; $\sim$ nocturnal $\uparrow$ sympathovagal balance (14%)
Thomas *et al.*[91] Consecutive phases study NA NA	Men ($n = 17$) 24.7 ± 2.8 NA $\geq$ 7 days of 7-h 45 min bedtimes[d]	1: 3 nights of 7-h 45-min bedtimes[e] 2: 85 h of TSD (only results from 24 h shown)	[18]FDG and PET, every 24 h	$\downarrow$ global cerebral glucose metabolic rate (8%), especially the thalamus, prefrontal and posterior parietal cortices
Van Cauter *et al.*[45] Consecutive phases study Only breakfast on study day Bed rest	Men ($n = 8$) 22 – 27 22.7 $\pm$ 0.7 kg/m^2 1 week 23:00–07:00 bedtimes[c]	1: 2 nights of 8-h bedtimes[e] 2: 28 h of TSD	Continuous glucose infusion during phase 2 24 h, every 30 min	$\downarrow$ nocturnal GH pulses (70%)[a] $\sim$ cortisol, $\downarrow$ acrophase (4%);[1] 1 hour earlier nadir
VanHelder *et al.*[89] Consecutive phases study Isocaloric controlled Sedentary and exercise	Men ($n = 10$) 22 ± 3 74.5 $\pm$ 11.8 kg 1 night of 7-h bedtimes[e]	60 h of TSD followed by 7-h of recovery sleep[e] 1: sedentary activities 2: daily exercise Conditions $\geq$ 10 days apart	OGTT at 10 h, 60 h, and after recovery sleep	$\uparrow$ insulin response (21% in group 1)
Van Leeuwen *et al.*[112] Consecutive phases study Isocaloric controlled Free outpatient movement	Men ($n = 13$) 23.1 ± 2.5 NA 2 weeks regular bedtimes[c]	1: 2 nights of 8-h bedtimes[e] 2: 5 nights of 4-h bedtimes[e]	Flow cytometry, real time PCR (1 morning sample), salivary cortisol (10 times a day)	$\sim$ cortisol $\uparrow$ IL-6 (63%) $\uparrow$ IL-1β (37%) $\sim$ TNF-α $\uparrow$ hsCRP (45%)

Continued

Table 1. *Continued*

Study design Energy intake Behavioral activity	Gender/sample size Age ± SEM (years) Mean weight ± SEM Prestudy conditions	Experimental sleep protocol	Techniques used for blood sampling (time, frequency)	Main findings (sleep deprived vs. longer sleep)

Vgontas *et al.*[106]
Consecutive phases
  study
Usual diet
Free outpatient
  movement
  (sedentary during
  sampling)
	13 females, 12 males
	  (n = 25)
	25.2 ± 3.8
	23.8 ± 2.3 kg/m²
	2 weeks regular
	  bedtimes[d]
		1: 4 nights of 8-h
		  bedtimes[e]
		2: 8 nights of 6-h
		  bedtimes[e]
			24 h, every 30 min
			  (ELISA, TNF-α,
			  and IL-6)
				∼ cortisol; ↓ and 2 h earlier
				  peak
				↑ IL-6
				↑ TNF-α in males; ∼ in
				  females

NOTE: Sleep curtailment was verified by [c]questionnaires/diaries, [d]by actigraphy, or [e]by polysomnography. The percentage of change in metabolic parameters was given in the article or calculated from [a]absolute levels in the text/tables or from [b]graph values of the original article. Metabolic measurements are in plasma/serum, unless otherwise specified. Bed rest was always required during the sleep-deprived state. ∼ similar; TSD: total sleep deprivation; VAS: visual analogue scale; RMR: resting metabolic rate; IL-6: interleukin 6; IL-1β: interleukin 1β; TNF-α: tumor necrosis factor-α; EE: energy expenditure; ivGTT: intravenous glucose tolerance test; RQ: respiratory quotient; OGTT: oral glucose tolerance test; REM: rapid-eye movement; GH: growth hormone; ACTH: adrenocorticotropic hormone; FDG: fludeoxyglucose; PET: position emission tomography.

fragmented sleep,[123] or total sleep deprivation[90] (Table 1). However, increased insulin secretion in response to ivGTTs was found after 60 hours of sleep deprivation[89] and two nights of sleep fragmentation.[124] Thus, almost all studies find decreased glucose tolerance and a dampened compensatory insulin response after sleep curtailment.

Effects on EE

Two groups found that one night of total sleep deprivation increased EE during the night, whereas the following day, resting and postprandial EE were reduced by 5% and 20%, respectively, overall leading to a positive energy balance[43,100] (Fig. 3 and Table 1). Similarly, Schmid *et al.* reported that volunteers had decreased spontaneous physical activity after a night of short sleep compared to a night with restful sleep[110] (Table 1). Fragmenting sleep for two consecutive nights lead to an increase in the respiratory quotient (RQ), indicating a shift from fat toward carbohydrate oxidation, without affecting total EE[125] (Table 1). High RQ predicts weight gain and insulin resistance.[126] Two weeks of partial sleep deprivation (5.5 hours) decreased resting metabolic rate and increased RQ in a randomized, cross-over study[109] (Table 1). In contrast, subjects with chronic insomnia had increased metabolic rate.[127] We have recently shown that poor sleep quality was associated with higher REE, a higher RQ, and an activation of the stress system in obese subjects with short sleep duration.[128] In summary, EE may be differentially affected by chronic versus acute sleep curtailment and may also depend on differences in reasons of sleep curtailment (experimentally induced short sleep vs. chronic insomnia).

Additional endocrine mechanisms linking sleep, glucose metabolism, and energy expenditure

As mentioned above, thyroid hormones, GH, cortisol, adipokines (e.g., adiponectin), and sympathovagal balance modulate glucose metabolism and energy homeostasis (Fig. 3). Acute total sleep deprivation increased TSH and free T_4 levels,[88] whereas partial sleep deprivation of 5.5-h nights for two weeks or six 4-h nights decreased TSH and T_4 levels, suggesting a dose-related effect[104,116] (Table 1). Decreased peripheral thyroid hormone levels are likely secondary to suppression of the nocturnal TSH peak, possibly secondary to decreased drive from TRH neurons in the PVN.[116] Sleep-deprived rodents show reduced levels of hypothalamic TRH mRNA in the PVN of the hypothalamus.[129] Of note, a different stressor such as acute immobilization also decreases TRH mRNA hypothalamic levels and plasma TSH levels in rodents.[130]

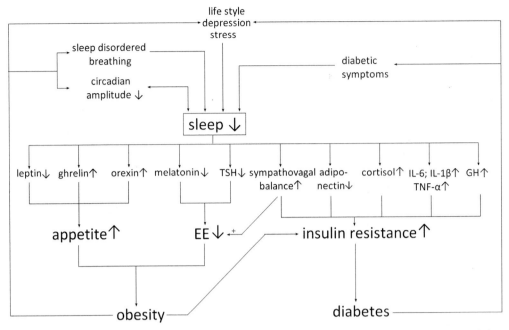

Figure 3. A simplified schematic representation of putative pathways of sleep curtailment leading to obesity and diabetes, via endocrine mechanisms that stimulate appetite, decrease energy expenditure, and increase insulin resistance. The direction of change (an increase or decrease) of each mechanism due to sleep loss is displayed; increased sympathovagal balance could stimulate EE, depicted by +, while a decreased EE would contribute to obesity. For readability, the relationships between decreases in leptin levels associated with IL-6 and TNF-α production are not displayed. Cortisol and proinflammatory cytokines display a positive bidirectional relationship. Increased insulin levels stimulate orexin secretion and orexic arcuate neurons, while decreasing the activity of anorexic arcuate neurons, leading to a further increase in appetite. Adiponectin influences energy homeostasis, possibly via modulating appetite through arcuate neurons and modulating EE.

Insulin resistance can result from increases in insulin counter-regulatory hormones, including GH and cortisol (Fig. 3). One night of total sleep deprivation markedly decreased the GH peak that usually occurs around sleep onset[45,95] (Table 1). Partial sleep deprivation did not affect 24-h levels of GH in lean or obese individuals,[101,109,118] but six 4-h nights changed GH nocturnal secretory patterns: sleep deprivation was associated with a biphasic pattern compared to the usual single large GH peak at sleep onset.[102] This extended exposure to GH may decrease glucose uptake in muscles, thus contributing to insulin resistance. However, the threshold of GH levels contributing to insulin resistance is unknown. In contrast to the previous study, decreased GH levels during the first four hours of sleep were reported after a two-week partial sleep-deprivation period, in association with more SWS.[109,118] GH is usually secreted during SWS, thus this unexpected finding may be due to a dissociation of GH secretion from SWS. In summary, the effects of sleep deprivation on GH plasma levels are variable.

Partial and total sleep deprivation resulted in elevated salivary and plasma cortisol the following afternoon and evening[100,101,109,114,117] (Table 1), while morning cortisol levels may be unchanged or even decreased.[90,97,117] As awakening is usually associated with a cortisol surge, this finding probably reflects the lack of the awakening response. In addition, a delay and/or a reduction in the acrophase of cortisol have been reported after partial and total sleep deprivation.[95,102,109] In a sleep-deprived state, cortisol levels decreased more slowly after reaching their acrophase, possibly indicating decreased sensitivity to the negative feedback effects on the hypothalamic–pituitary–adrenal axis. Mean 24-h levels of plasma cortisol often remain unchanged by sleep deprivation,[43,95,96,99,106,109,123] but the circadian cortisol-pattern changes, as described above.

Adiponectin levels were similar after partial and total sleep deprivation and sleep fragmentation[99,121,124] (Table 1). However, sub-group analysis revealed that effects of sleep restriction on adiponectin levels may be gender and race dependent: adiponectin decreased in Caucasian women and increased in African American women, while adiponectin levels in males did not change after short sleep[121] (Table 1). Besides influencing energy homeostasis, decreased adiponectin levels worsen insulin resistance.

Sympathovagal balance, defined as the ratio between the activity of the sympathetic and the parasympathic nervous system, can be indirectly assessed with heart rate variability analysis. Since the activity of this system is very diverse, on the basis of the different anatomic branches, it is unclear to which extent perturbations in heart-rate variability may be predictive of changes in basal metabolic rate and insulin sensitivity. Furthermore, heart rate, EPI and NE levels are markers of sympathetic activity. Acute sleep deprivation is associated with increased sympathetic activity, decreased parasympathetic tone and therefore, with increased sympathovagal balance[101,109,123,124] (Table 1). Total EE is usually decreased in acute sleep deprivation; thus, other mechanisms, such as decreased thyroid hormones or adiponectin levels, must prevail over increased sympathovagal tone (Fig. 3). Increased sympathovagal balance also adds to insulin resistance[131] (Fig. 3). Furthermore, activation of the sympathetic nervous system and inhibition of the parasympathetic nerves decreases insulin secretion; thus, changes in the autonomic nervous system could explain the absence of a compensatory insulin response after sleep curtailment.[132]

In summary, sleep loss impairs glucose metabolism and may decrease EE. This is mediated by multiple factors, such as decreased thyroid hormones and, possibly, adiponectin levels; increased sympathovagal balance; and altered patterns of cortisol secretion, while the role of GH is still unclear (Fig. 3).

Effects on appetite

It is not clear to what extent obesity in subjects with short sleep is caused by increased energy intake versus decreased EE[133] (Fig. 3 and Table 1). If food is provided *ad libitum*, partial or total sleep deprivation may not change hunger ratings, while likely increasing food intake.[99,117,120] On the other hand, when food intake is controlled, sleep deprivation usually increases appetite. Self-reported sleep quality was inversely related to appetite in 53 first-degree, normal weight relatives of subjects with diabetes.[134] After two 4-h nights, there was a 24% increase in daytime appetite ratings;[105] in addition, sleep deprivation specifically increased appetite for sweet and salty foods. The effects of sleep deprivation on appetite may not be immediate: one 4-h night of sleep deprivation was associated with increases in food intake 36 h later, while this parameter was unaffected immediately following sleep deprivation.[113] In addition, the amount of sleep deprivation needed to increase hunger ratings may be variable. One night of total, but not partial, sleep deprivation, significantly increased hunger ratings in normal weight males.[98,113] No differences in total food intake were observed after two 4-h, 15 min-versus 8-h, 15 min- nights, but subjects ate relatively more fat when sleeping less.[119] In a randomized, cross-over, five-day study of 4 h versus 9 h of sleep, with a controlled diet on days 1–4 versus an *ad libitum* diet on day five, women, but not men, exhibited increased total caloric intake and a preference for fatty foods, especially if saturated, suggesting that women may be more susceptible to the orexic effects of sleep deprivation than men.[122]

Chronotype, the individual attribute determining morning or evening preference, modulates metabolism. Obese subjects carrying the CLOCK 3111TC/CC polymorphism, a variant of one of the core clock genes associated with being an "owl" or an evening person, sleep 20 minutes less.[135] In addition, these subjects had a higher intake of trans-fatty acids and proteins, and consumed relatively later in the day, while having the same total energy intake of noncarriers. Subjects carrying the CLOCK 3111TC/CC genotype are also more resistant to weight loss: they lost about 2 kg less while on a hypocaloric diet for 30 weeks.[135]

Leptin and ghrelin inhibit and stimulate appetite, respectively (Fig. 3). Following partial sleep deprivation, mean daytime values of leptin were approximately 18% lower compared to the rested state.[104,105] In addition, the circadian amplitude of leptin was dampened and the acrophase was decreased and phase advanced by two hours. A decreased circadian amplitude was also observed after three nights of total sleep deprivation.[94] In a large cohort study of 1,024 subjects, shorter habitual sleep duration as assessed by sleep diaries related to lower morning values of leptin.[5] Some studies found no association between short sleep and leptin levels.[94,98,100,119,136] On the contrary, morning leptin levels were found to be increased after five nights of four hours,[120] after one night of three hours[117] and after one night of total sleep deprivation.[99] Short sleep duration, as assessed by one night of polysomnography in 561 subjects, also related to higher morning leptin levels.[137] It is difficult to explain the inconsistent results of these different studies. Of note, studies that found decreased leptin after short sleep controlled food intake, whereas when sleep-deprived subjects were allowed to eat *ad libitum*, leptin levels were unchanged or even increased.

In obese subjects, leptin levels are elevated but several of its actions are impaired, a concept referred to as *leptin resistance*. In a cross-over study design, no association among habitual sleep duration, measured actigraphically, and morning leptin levels was observed in obese subjects.[138] Mean 24-h leptin levels in obese subjects on a controlled diet were also unchanged after 14 nights of 5.5 h versus 8.5 hours.[118]

The physiological increase in total ghrelin levels in the early night was blunted when individuals were kept awake.[95] A similar phenomenon is reported in chronic insomniacs.[135] However, morning total ghrelin levels were increased after one night of sleep deprivation[95,98] and in individuals with shorter habitual sleep durations.[5] Mean daytime total ghrelin levels were higher after two nights of 4-h sleep.[105] CLOCK 3111TC/CC carriers, who display an evening preference and sleep 20 min shorter, also had increased morning ghrelin levels.[134] Mean 24-h levels of the metabolically active acetylated ghrelin were higher in obese subjects after two weeks of 5.5-h nights.[118] Overall, sleep curtailment is consistently associated with higher daytime ghrelin.

The effect of sleep deprivation on the orexin system in normal subjects is not well studied. However, it is known that patients with narcolepsy have abnormally low orexin-A CSF levels, which correlate both with body weight and sleep abnormality.[139] Total sleep deprivation increased orexin levels in the CSF of squirrel monkeys, and increased cFos expression in orexin neurons in rats.[81,140] It is challenging to assess the activity of the orexin system in humans. Central determination requires collection of CSF,

an invasive technique. Determination of orexin in plasma is hampered by the lack of adequate assays, in terms of sensitivity and specificity. Hyperactivity of the orexin system could contribute to increased food intake during sleep deprivation. In summary, appetite and food intake are often increased following sleep curtailment, possibly via decreased leptin, increased ghrelin and a hyperactive orexin system (Fig. 3).

Effects on the immune system

Enhanced levels of circulating leucocytes, proinflammatory cytokines (e.g., IL-6, TNF-α, IL-1β) and CRP have been associated with an increased risk of cardiovascular disease and type 2 diabetes in long-term human and experimental studies.[141–144] Increased plasma levels of IL-6 were reported after partial and total sleep deprivation in most cases,[93,106,107,108,112] although two studies only reported a loss of circadian rhythms[82,96] (Table 1). IL-1β was elevated after partial,[106,112] and total sleep deprivation,[96] but one study did not find any changes after total sleep deprivation[82] (Table 1). TNF-α levels are most often,[82,93,112] but not always,[107] unchanged after partial sleep restriction. One study only found increases in men, but not in women, after partial sleep deprivation[106] (Table 1). Another study reported higher levels of soluble TNF-α receptor I, through which TNF-α exerts its sleep promoting function.[93] Effects of sleep curtailment on CRP levels are inconsistent: they have been reported lower,[96] unchanged,[124] or higher.[108,112] Prolonged total sleep deprivation has been associated with leucocytosis[82] (Table 1). Thus, differences in study outcomes may be influenced by individual or gender differences. In addition, sensitivity and specificity of commercial cytokine assays are highly variable. Highly sensitive analytical methods, such as mass spectrometry, may be needed for accurate cytokine measurements.[145] We conclude that short sleep promotes a proinflammatory state, which, in turn, exerts its negative consequences on insulin resistance.

The other side of the coin: metabolic dysfunction influencing sleep

Most of the studies of sleep deprivation have been performed in healthy, lean subjects and therefore could not address the effect of sleep deprivation on subjects with obesity and diabetes. Obesity and diabetes are often associated with increased stress or differences in life style that could negatively influence sleep[146] (Fig. 3). Depression and obesity may occur together as well: obesity may negatively affect mood and both pharmacologic antidepressant therapy and major depression itself are associated with weight gain.[147] We recently reported that obese subjects often suffer from sleep-disordered breathing, which induces frequent microawakenings and loss of SWS.[148] Similar effects of BMI on sleep have also been observed in youth. Decreased total sleep time and less sleep efficiency were observed in severely obese adolescents, of which 74% displayed sleep apnea (16.5 years; BMI 60.3 ± 2.1 kg/m^2).[149] Changes in sleep architecture in obesity have been reported: severely obese adolescents (12–15 years; BMI 50.9; 39.8–63.0 kg/m^2, apnea index 14 episodes/h) have a decrease in REM sleep.[150]

Symptoms associated with diabetes, such as thirst, nocturia, extreme glucose excursions, and mood alterations, may also interfere with sleep, independent of obesity. To this point, adults with type 1 diabetes, who are typically not obese, report a poorer sleep quality than their matched controls without diabetes.[151] These awakenings were associated with a rapid decline of glucose levels, but not with the absolute values of hypoglycemia.[152] Furthermore, studies in animals suggest that diet composition may directly influence sleep: mice fed a high-fat diet display an attenuated rhythm in the circadian expression of core clock genes.[40] Of note, augmented amplitude of the peripheral and central circadian rhythms favors wakefulness during the day and sleep at night, whereas an attenuated circadian amplitude disrupts the sleep–wake cycle.[153] Likewise, studies in rodents demonstrate that physical activity at the congruent time (i.e., nocturnal for mice) increases the central circadian amplitude, suggesting that active behavior during the day and sleep at night in humans may be beneficial for high-amplitude circadian rhythms.[154] Thus, sleep and circadian rhythm are bidirectionally connected (Fig. 3). Leptin influences sleep architecture in normal fed rodents, increasing SWS by 13% and decreasing REM by 30%.[155] Furthermore, *ob/ob* mice that lack leptin and *db/db* mice that lack the leptin receptor, both display sleep disturbances that are reverted by leptin replacement. In summary, metabolic dysfunction, altered hormone levels, and sleep abnormalities probably all influence each other.[156,157]

In summary, sleep deprivation and obesity may potentiate each other in a vicious reverberating circuit; short sleep may induce weight gain, via the mechanisms outlined in this paper, whereas obesity, in turn, via sleep apnea and other symptoms, may disrupt sleep. Observational studies, especially if cross-sectional, may not be well suited to disentangle these aggregate, causative factors. This underlies the need for behavioral, life-style intervention studies.

Possible therapeutic role of melatonin

In nonequatorial areas, the duration, intensity, and spectral quality of natural light vary across seasons.[158] For example, during a summer in England, there was relatively more blue light versus green and red light exposure, in addition to increased light intensity and duration.[158] These are important signals for seasonal rhythms, transmitted to the body via the duration of nocturnal melatonin secretion. As a consequence of artificial lighting and other environmental modifications, seasonality is greatly attenuated in modern humans. Furthermore, because sleep is usually the only time during which humans do not experience light exposure, shorter sleep duration will result in longer light exposure. Light suppresses the pineal production of melatonin in a dose-related fashion. White artificial light greater than 200 lux completely suppressed nocturnal plasma melatonin production, dim room light (106 lux) reduced melatonin by 88% and light intensities lower than 80 lux did not affect melatonin levels.[159] As the night progresses, the intensity of continuous artificial light needed to suppress circulating melatonin increases.[160] Of note, light of shorter wavelength, such as the light that is typically emitted by electronic devices such as televisions and computers, induces relatively more melatonin suppression and likely leads to greater sleep disruption.[160]

Urinary 24-h melatonin concentrations decreased by more than 50% when subjects slept five hours versus eight hours, while being continuously exposed to white light (700 lux) during wake hours.[161] Diminished levels of melatonin were likely due to three additional hours of light exposure. Likewise, mean 24-h plasma melatonin concentrations were decreased after 6 nights of 4-h versus 12-h bedtimes.[103] The onset of melatonin secretion was delayed, due to four hours longer light exposure before bedtime (300 lux) and the acrophase was reduced in sleep-deprived subjects.

Recent studies suggest that melatonin may have a physiological role in modulating metabolism. For example, removal of the pineal gland decreased the responsiveness of several insulin-dependent hepatic kinases during the dark phase in rats.[162] This effect was reversed by nocturnal supplementation of melatonin in the drinking water. Chronic melatonin supplementation in drinking water also reduced body weight in rodents.[163,164] Patients with metabolic syndrome have disturbances in melatonin production, characterized by an alteration of the night melatonin-insulin ratio of unclear clinical significance.[165] In a large genome-wide association studies conducted in individuals of European origin, certain genetic variants of the melatonin receptor 1B were associated with an increased risk of type 2 diabetes.[166–168] However, the most common receptor variant did not influence the relationship between sleep abnormalities and type 2 diabetes.[166] Interestingly, the melatonin receptor 1B is present on pancreatic β-cells, and melatonin has an inhibitory effect on insulin secretion. Thus, it was feared that treatment with melatonin in subjects with diabetes would increase blood glucose levels. On the other hand, since type 2 diabetes is associated with reduced melatonin levels toward the end of the night;[169] thus, there may be a physiological rationale for melatonin supplementation at bedtime. Few studies have addressed these questions. Sleep efficiency improved after five months of treatment with 2 mg prolonged-release melatonin administered two hours before bedtime, while total sleep time was unaffected in 26 subjects with type 2 diabetes.[170] HbA1c levels also improved (8.47% vs. 9.13% at baseline). Lack of a control group prevented understanding whether improved HbA1c levels were in part due to other factors than melatonin administration, such as better pharmacological and diet compliance due to study participation. In lean subjects, melatonin administration at night for 2 months was associated with decreased LDL levels and improved oxidative status.[171] Another study found a positive correlation between plasma melatonin levels at 2 am and HDL levels in 36 women treated with 1 mg melatonin for one month.[172] As melatonin promotes sleep, the decrease in obesity and improvement in insulin resistance could be due to longer and better sleep per se, but it could also relate to other melatonin-dependent pathways yet to be identified (Fig. 3).

Discussion and future directions

Epidemiological and laboratory data prove the existence of a relationship between short or disturbed sleep and adverse metabolic outcomes. In general, these effects are small in size, but certain specific populations may be more susceptible than others. For example, age plays a clear role as the relationship between short sleep and obesity is stronger in children. Sleep duration and sleep architecture change with age in healthy individuals, with SWS becoming less common with age.[173] It is also well known that sex and reproductive hormones affects sleep regulatory mechanisms[174] and that sleep disorder affects women differently from men.[175] It is unfortunate that experimental studies, in both humans and in rodents, are mostly conducted in the male gender, thus unduly discriminating the female gender.

Similarly, sleep deprivation is more common in minorities but the effects of socioeconomic status per se versus ethnicity remain to be determined. Individual susceptibility within these categories varies greatly: identification of contributing factors will greatly help prevention and treatment and be one of the tenets of individualized medicine.[176] Few negative effects are expected to be observed in individuals with voluntary sleep curtailment (natural short sleepers), in contrast to individuals with short sleep due to social or work-related pressure or chronic insomnia. Current epidemiologic studies often lack an objective measurement of sleep duration, such as actigraphy or repeated sleep diaries. Thus, prospective studies with objective sleep duration measurements are warranted in different subgroups to further explore this relationship.

In addition, the causality of the relationship between sleep and metabolic disorders remains unclear, even in prospective studies. Obese/diabetic individuals may be sleeping less than lean subjects because of differences in life style or stress; their sleep could be interrupted more often by symptoms of disturbed glucose levels or sleep apnea; evidence arises that possibly circulating hormones affect the brain inducing worsened sleep (Fig. 3). As already mentioned, causality can be addressed best in long-term studies where "prescribed" sleep duration is randomized, but this may present with practical difficulties and significant challenges, including the fact that the biological need for optimal sleep duration varies and cannot be estimated at an individual level. Sleep is regulated by circadian and homeostatic mechanisms. As far as homeostatic mechanisms are concerned, it has been the matter of a long scientific debate whether during wakefulness, single or rather multiple metabolites accumulate, progressively increasing the pressure to sleep during the day. Classic experiments of parabiosis seem to support the notion of endogenous "sleep-promoting" substances. Recently, the importance of adenosinergic neurotransmission in the regulation of sleep, especially non-REM sleep, has been underlined.[177] Nevertheless, it is certain that there is a large individual variability in sleep needs. Sensitive and specific biomarkers of sleep deprivation are needed for various reasons, including clinical and medico-legal reasons.[178]

The effect of sleep extension in subsets of short sleepers on metabolism can be addressed in randomized controlled trials. Currently, we are conducting such a study in short-sleeping obese individuals.[179] Investigating the effects of sleep extension in other subgroups of varying age, BMI, and ethnicity will also be of interest. The effect of metabolic disturbances on sleep, such as hyperinsulinemia or increased circulating levels of proinflammatory cytokines, may be empirically addressed in polysomnographical studies. As it is difficult to artificially mimic the internal milieu of obese/diabetic individuals, effects of metabolism may be modeled in animal studies using parabiosis, connecting circulations of lean and obese animals and comparing sleep characteristics.

As indicated by studies of sleep deprivation, the mechanisms connecting short sleep and obesity/insulin resistance are probably mediated by three pathways: increases in appetite, decreases in EE, and influences on glucose metabolism (Fig. 3). Endocrine mechanisms influencing these pathways are complex and interconnected. The existing studies of sleep deprivation have generated variable results; small sample size, differences in the control of food-intake and physical activity levels, and inherent individual differences may have all contributed to this variability. In addition, these studies only addressed the effects of acute sleep deprivation and were not performed in real-life situations. We suggest that future studies should focus on the effects of sleep deprivation in specific patient popula-

tions, including obese/insulin-resistant individuals and chronic insomniacs versus natural short sleepers, and should mimic real-life conditions.

Most studies find worsened glucose tolerance after sleep curtailment, which may be beneficial from a teleological perspective, at least in the short term. When our ancestors remained awake during the night, they were probably experiencing a threat (e.g., absence of food, presence of predators), so inducing a state in which glucose availability was advantageous. Likewise, increases in appetite result in more food intake. Together with reduced EE, this will further increase blood glucose availability. In modern societies, however, prolonged exposure to such a state may result in obesity and type 2 diabetes.

At this point, there is a general consensus that short sleep has negative effects on health, but the amount of sleep curtailment causing metabolic disturbances is unclear, and specific groups at greater risk are yet to be identified. Whether melatonin would benefit obese/diabetic subjects, or a subset of these individuals, remains to be proven in rigorous randomized clinical trials. Supplementation of melatonin has been shown to improve sleep efficiency, and HbA1c levels in patients with type 2 diabetes.[170] Lean subjects may also profit from melatonin administration as well, as it improved lipid profiles.[171,172] Thus, melatonin appears to be a promising supplement in battling the obesity/diabetes epidemics, but controlled studies need to confirm its positive effect on metabolism. Furthermore, the long-term effects and side effects of chronic melatonin supplementation need to be examined in different subgroups.

Overall, improving sleep duration and quality is a potential tool to counteract the epidemics of obesity and diabetes. However, before sleep extension advice is translated to the clinic, more research is needed.

Acknowledgments

This study was supported by the National Institutes of Health (NIH), Intramural Research Program: National Institute of Diabetes and Digestive and Kidney Diseases (NIDDK). Eliane A. Lucassen was the recipient of a Fulbright Scholarship and received scholarships from VSBfonds and the Leiden University Fund.

Conflicts of interest

The authors declare no conflicts of interest.

References

1. Flegal, K.M., M.D. Carroll, B.K. Kit & C.L. Ogden. 2012. Prevalence and trends in obesity in the distribution of body mass index among US adults, 1999–2010. *JAMA* **307:** 491–497.
2. Korner, J., S.C. Woods & K.A. Woodworth. 2009. Regulation of energy homeostasis and health consequences in obesity. *Am. J. Med.* **122:** S12–S18.
3. Centers for Disease Control and Prevention. 2011. National diabetes fact sheet: national estimates and general information on diabetes and prediabetes in the United States, 2011. Available from: http://www.cdc.gov/diabetes/pubs/pdf/ndfs_2011.pdf.
4. Hasler, G., D.J. Buysse, R. Klaghofer, *et al.* 2004. The association between short sleep duration and obesity in young adults: a 13-Year prospective study. *Sleep* **27:** 661–666.
5. Taheri, S., L. Lin, D. Austin, T. Young & E. Mignot. 2004. Short sleep duration is associated with reduced leptin, elevated ghrelin, and increased Body Mass Index (BMI). *Sleep* **27:** A146–A147.
6. Ayas, N.T., D.P. White, W.K. Al-Delaimy, *et al.* 2003. A prospective study of sleep duration and coronary heart disease in women. *Arch. Intern. Med.* **163:** 205–209.
7. Björkelund, C., D. Bondyr-Carlsson, L. Lapidus, *et al.* 2005. Sleep disturbances in midlife unrelated to 32-year diabetes incidence: the prospective population study of women in Gothenburg. *Diabetes Care* **28:** 2739–2744.
8. Yaggi, H.K., A.B. Araujo & J.B. McKinlay. 2006. Sleep duration as a risk factor for the development of type 2 diabetes. *Diabetes Care* **29:** 657–661.
9. Gangwisch, J.E., S.B. Heymsfield, B. Boden-Albala, *et al.* 2007. Sleep duration as a risk factor for diabetes incidence in a large US sample. *Sleep* **30:** 1667–1673.
10. Gallicchio, L. & B. Kalesan. 2009. Sleep duration and mortality: a systematic review and meta-analysis. *J. Sleep Res.* **18:** 148–158.
11. National Sleep Foundation. 2005 *Sleep in America Poll*. National Sleep Foundation. Washington DC.
12. Bin, Y.S., N.S. Marshall & N. Glozier. 2012. Secular trends in adult sleep duration: A systematic review. *Sleep Med. Rev.* **16:** 223–230.
13. Cappuccio, F.P., F.M. Taggart, N.B. Kandala, *et al.* 2008. Meta-analysis of short sleep duration and obesity in children and adults. *Sleep* **31:** 619–626.
14. Chen, X., M.A. Beydoun & Y. Wang. 2008. Is sleep duration associated with childhood obesity? A systematic review and meta-analysis. *Obesity* **16:** 265–274.
15. Rao, M.N., T. Blackwell, S. Redline, *et al.* 2009. Association between sleep architecture and measures of body composition. *Sleep* **32:** 483–490.
16. Rasmussen, M.H., G. Wildschiodtz, A. Juul & J. Hilsted. 2008. Polysomnographic sleep, growth hormone, insulin-like growth factor-I axis, leptin, and weight loss. *Obesity* **16:** 1516–1521.
17. Lauderdale, D.S, K.L. Knutson, P.J. Rathouz, *et al.* 2009. Cross-sectional and longitudinal associations between objectively measured sleep duration and body mass index: the CARDIA Sleep Study. *Am. J. Epidemiol.* **170:** 805–813.
18. Patel, S.R., A. Malhotra, D.P. White, D.J. Gottlieb & F.B. Hu. 2006. Association between reduced sleep and weight gain in women. *Am. J. Epidemiol.* **164:** 947–954.
19. Chaput, J.P., J.P. Després, C. Bouchard & A. Tremblay. 2008. The association between sleep duration and weight gain in adults: a 6-year prospective study from the Quebec Family Study. *Sleep* **31:** 517–523.
20. López-García, E., R. Faubel, L. León-Muñoz, *et al.* 2008. Sleep duration, general and abdominal obesity, and weight change among the older adult population of Spain. *Am. J. Clin. Nutr.* **87:** 310–316.
21. Gunderson, E.P., S.L. Rifas-Shiman, E. Oken, *et al.* 2008. Association of fewer hours of sleep at 6 months postpartum with substantial weight retention at 1 year postpartum. *Am. J. Epidemiol.* **167:** 178–187.
22. Hairston, K.G., M. Bryer-Ash, J.M. Norris, *et al.* 2010. Sleep duration and five-year abdominal fat accumulation in a minority cohort: the IRAS family study. *Sleep* **33:** 289–295.
23. Gangwisch, J.E., D. Malaspina, B. Boden-Albala & S.B. Heymsfield. 2005. Inadequate sleep as a risk factor for obesity: analyses of the NHANES I. *Sleep* **28:** 1289–1296.
24. Stranges, S., F.P. Cappuccio, N.B. Kandala, *et al.* 2008. Cross-sectional versus prospective associations of sleep duration with changes in relative weight and body fat distribution: the Whitehall II Study. *Am. J. Epidemiol.* **167:** 321–329.
25. Watanabe, M., H. Kikuchi, K. Tanaka & M. Takahashi. 2010. Association of short sleep duration with weight gain and obesity at 1-year follow-up: a large-scale prospective study. *Sleep* **33:**161–167.
26. Boden, G. & G.I. Shulman. 2002. Free fatty acids in obesity and type 2 diabetes: defining their role in the development of insulin resistance and beta-cell dysfunction. *Eur. J. Clin. Invest.* **32:** 14–23.
27. Knutson, K.L., A.M. Ryden, B.A. Mander & E. van Cauter. 2006. Role of sleep duration and quality in the risk and severity of type 2 diabetes mellitus. *Arch. Intern. Med.* **166:** 1768–1774.
28. Choi, K.M., J.S. Lee, H.S. Park, *et al.* 2008. Relationship between sleep duration and the metabolic syndrome: Korean National Health and Nutrition Survey 2001. *Int. J. Obes.* **32:** 1091–1097.
29. Tuomilehto, H., M. Peltonen, M. Partinen, *et al.* 2008. Sleep duration is associated with an increased risk for the prevalence of type 2 diabetes in middle-aged women – The FIN-D2D survey. *Sleep Med.* **9:** 221–227.
30. Xu, Q., Y. Song, A. Hollenbeck, *et al.* 2010. Day napping and short night sleeping are associated with higher risk of diabetes in older adults. *Diabetes Care* **33:** 78–83.
31. Cappuccio, F.P., L. D'Elia, P. Strazzullo & M.A. Miller. 2010. Quantity and quality of sleep and incidence of type 2 diabetes: a systematic review and meta-analysis. *Diabetes Care* **33:** 414–420.
32. Kawakami, N., N. Takatsuka & H. Shimizu. 2004. Sleep disturbance and onset of type 2 diabetes. *Diabetes Care* **27:** 282–283.
33. Nilsson, P.M., M. Rööst, G. Engström, B. Hedblad & G. Berglund. 2004. Incidence of diabetes in middle-aged men is related to sleep disturbances. *Diabetes Care* **27:** 2464–2469.

34. Chaput, J.P., J.P. Després, C. Bouchard, A. Astrup & A. Tremblay. 2009. Sleep duration as a risk factor for the development of type 2 diabetes or impaired glucose tolerance: analyses of the Quebec Family Study. *Sleep Med.* **10:** 919–924.

35. Meisinger, C., M. Heier & H. Loewel. 2005. Sleep disturbance as a predictor of type 2 diabetes mellitus in men and women from the general population. *Diabetologia* **48:** 235–241.

36. Beihl, D.A., A.D. Liese & S.M. Haffner. 2009. Sleep duration as a risk factor for incident type 2 diabetes in a multiethnic cohort. *Ann. Epidemiol.* **19:** 351–357.

37. Trento, M., F. Broglio, F. Riganti, *et al.* 2008. Sleep abnormalities in type 2 diabetes may be associated with glycemic control. *Acta Diabetol.* **45:** 225–229.

38. Aronsohn, R.S., H. Whitmore, E. van Cauter & E. Tasali. 2010. Impact of untreated obstructive sleep apnea on glucose control in type 2 diabetes. *Am. J. Res. Crit. Care Med.* **181:** 507–513.

39. Neubauer, D.N. 1999. Sleep problems in the Elderly. *Am. Fam. Physician* **59:** 2551–2558.

40. Kohsaka, A., A.D. Laposky, K.M. Ramsey, *et al.* 2007. High-fat diet disrupts behavioral and molecular circadian rhythms in mice. *Cell Metab.* **6:** 414–421

41. Asher, G. & U. Schibler. Crosstalk between components of circadian and metabolic cycles in mammals. *Cell Metab.* **13:** 125–137.

42. Cizza, G., M. Requena, G. Galli & L. de Jonge. 2011. Chronic sleep deprivation and seasonality: implications for the obesity epidemic. *J. Endocrinol. Invest.* **34:** 793–800.

43. Jung, C.M., E.L. Melanson, E.J. Frydelhall, *et al.* 2011. Energy expenditure during sleep, sleep deprivation and sleep following sleep deprivation in adult humans. *J. Physiol.* **589:** 235–244.

44. Clore, J.N., J.E. Nestler & W.G. Blackard. 1989. Sleep-associated fall in glucose disposal and hepatic glucose output in normal humans. Putative signaling mechanism linking peripheral and hepatic events. *Diabetes* **38:** 285–290.

45. Van Cauter, E., J.D. Blackman, D. Roland, *et al.* 1991. Modulation of glucose regulation and insulin secretion by circadian rhythmicity and sleep. *J. Clin. Invest.* **88:** 934–942.

46. Boyle, P.J., R.J. Nagy, A.M. O'Connor, *et al.* 1994. Adaptation in brain glucose uptake following recurrent hypoglycemia. *Proc. Natl. Acad. Sci. USA* **91:** 9352–9356.

47. Maquet, P. 1997. Positron emission tomography studies of sleep and sleep disorders. *J. Neurol.* **244:** S23–S28.

48. Van Coevorden, A., J. Mockel, E. Laurent, *et al.* 1991. Neuroendocrine rhythms and sleep in aging men. *Am. J. Physiol.* **260:** E651–E661.

49. Holli, R.W., M.L. Hartman, J.D. Veldhuis, W.M. Taylor & M.O. Thorner. 1991. Thirty second sampling of plasma growth hormone in man: correlation with sleep stages. *J. Clin. Endocrinol. Metab.* **72:** 854–861.

50. Karacan, I., A.L. Rosenbloom, R.L. Williams, W.W. Finley & C.J. Hursch. 1971. Slow wave sleep deprivation in relation to plasma growth hormone concentration. *Behav. Neuropsychiatry* **2:** 11–14.

51. Van Cauter, E., L. Plat, M.B. Scharf, *et al.* 1997. Simultaneous stimulation of slow-wave sleep and growth hormone secretion by gamma-hydroxybutyrate in normal young men. *J. Clin. Invest.* **100:** 745–753.

52. Othmer, E., V. Daughaday & S. Guze. 1969. Effects of 24 hour REM deprivation on serum growth hormone levels in humans. *Electroencephalogr. Clin. Neurophysiol.* **27:** 685.

53. Mendelson, W.B., S. Slater, P. Gold & J.C. Gillin. 1980. The effect of growth hormone administration on human sleep: a dose-response study. *Biol. Psychiatry* **15:** 613–618.

54. Kern, W., R. Halder, S. al-Reda, *et al.* 1993. Systemic growth hormone does not affect human sleep. *J. Clin. Endocrinol. Metab.* **76:** 1428–1432.

55. Møller, N., J.O.L. Jorgensen, O. Schmitz, *et al.* 1990. Effects of a growth hormone pulse on total and forearm substrate fluxes in humans. *Am. J. Physiol. Endocrinol. Metab.* **258:** E86–E91.

56. Byrne, M., J. Sturis, J.D. Blackman, K.S. Polonsky & E. van Cauter. 1992. Decreased glucose tolerance during sleep may be partially mediated by GH. Program of the 74th Annual Meeting of The Endocrine Society, San Antonio, TX (Abstract 40).

57. Møller, N., P.C. Butler, M.A. Antsiferov & K.G.M.M. Alberti. 1989. Effects of growth hormone on insulin sensitivity and forearm metabolism in normal man. *Diabetologia* **32:** 105–110.

58. McMahon, M., J. Gerich & R. Rizza. 1988. Effects of glucocorticoids on carbohydrate metabolism. *Diabetes Metab. Rev.* **4:** 17–30.

59. Dinneen, S., A. Alzaid, J. Miles & R. Rizza. 1993. Metabolic effects of the nocturnal rise in cortisol on carbohydrate metabolism in normal humans. *J. Clin. Invest.* **92:** 2283–2290.

60. Boden, G., J. Ruiz, J. Ubrain & X. Chen. 1996. Evidence for a circadian rhythm of insulin secretion. *Am. J. Physiol.* **271:** E246–E252.

61. Stanley, K., K. Wynne, B. McGowan & S. Bloom. 2005. Hormonal regulation of food intake. *Physiol. Rev.* **85:** 1131–1158.

62. Dridi, S. & M. Taouis. 2009. Adiponectin and energy homeostasis: consensus and controversy. *J. Nutr. Biochem.* **20:** 831–839.

63. Hivert, M.F., L.M. Sullivan, C.S. Fox, *et al.* 2008. Associations of adiponectin, resistin, and tumor necrosis factor-alpha with insulin resistance. *J. Clin. Endocrinol. Metab.* **93:** 3165–3172.

64. Hivert, M.F., Q. Sun, P. Shrader, *et al.* 2011. Higher adiponectin levels predict greater weight gain in healthy women in the Nurses' Health Study. *Obesity* **19:** 409–415.

65. Lilja, M., O. Rolandsson, M. Norberg & S. Soderberg. 2012. The impact of leptin and adiponectin on incident type 2 diabetes is modified by sex and insulin resistance. *Metab. Syndr. Relat. Disord.* **10:** 143–151.

66. Maahs, D.M., L.G. Ogden, J.K. Snell-Bergeon, *et al.* 2007. Determinants of serum adiponectin in persons with and without type 1 diabetes. *Am. J. Epidemiol.* **166:** 731–740.

67. Kontani, K., N. Sakane, K. Saiga, *et al.* 2007. Serum adiponectin levels and lifestyle factors in Japanese men. *Heart Vessels* **22:** 291–296.

68. Scheer, F.A., J.L. Chan, J. Fargnoli, *et al.* 2010. Day/night variations of high-molecular-weight adiponectin and lipocalin-2 in healthy men studied under fed and fasted conditions. *Diabetologia* **53**: 2401–2405.

69. Cizza, G., V.T. Nguyen, F. Eskandari, *et al.* 2010. Low 24-hour adiponectin and high nocturnal leptin concentrations in a case-control study of community-dwelling pre-menopausal women with major depressive disorder: the Premenopausal, Ostopenia/Osteoporosis, Women, Alendronate, Depression (POWER) study. *J. Clin. Psychiatry* **71**: 1079–1087.

70. Somers, V.K., M.E. Dyken, A.L. Mark & F.M. Abbound. 1993. Sympathetic-nerve activity during sleep in normal subjects. *N. Engl. J. Med.* **328**: 303–307.

71. Aston-Jones, G., S. Chen, Y. Zhu & M.L. Oshinsky. 2001. A neural circuit for circadian regulation of arousal. *Nat. Neurosci.* **4**: 732–738.

72. Hjemdahl, P. 1993. Plasma catecholamines—analytical challenges and physiological limitations. *Baillieres Clin. Endocrinol. Metab.* **7**: 307–353.

73. Linsell, C.R., S.L. Lightman, P.E. Mullen, M.J. Brown & R.C. Causon. 1985. Circadian rythms of epinephrine and norepinephrine in man. *J. Clin. Endocrinol. Metab.* **60**: 1210–1205.

74. Licinio, J., C. Mantzoros, A.B. Negrao, *et al.* 1997. Human leptin levels are pulsatile and inversely related to pituitary-adrenal function. *Nat. Med.* **3**: 575–579.

75. Badman, M.K. & J.S. Flier. 2005. The gut and energy balance: visceral allies in the obesity wars. *Science* **307**: 1909–1914.

76. Spiegel, K., E. Tasali, R. Leproult, N. Scherberg & E. van Cauter. 2011. Twenty-four-hour profiles of acylated and total ghrelin: relationship with glucose levels and impact of time of day and sleep. *J. Clin. Endocrinol. Metab.* **96**: 486–493.

77. Elias, C.F, C. Aschkenasi, C. Lee, *et al.* 1999. Leptin differentially regulates NPY and POMC neurons projecting to the lateral hypothalamic area. *Neuron* **23**: 775–786.

78. Morton, G.J. & M.W. Schwartz. 2011. Leptin and the central nervous system control of glucose metabolism. *Physiol. Rev.* **91**: 389–411.

79. Hataya, Y., T. Akamizu, K. Takaya, *et al.* 2001. A low dose of ghrelin stimulates growth hormone (GH) release synergistically with GH-releasing hormone in humans. *J. Clin. Endocrinol. Metab.* **86**: 4552.

80. Van den Top, M., K. Lee, A.D. Whyment, A.M. Blanks & D. Spanswick. 2004. Orexigensensitive NPY/AgRP pacemaker neurons in the hypothalamic arcuate nucleus. *Nat. Neurosci.* **7**: 493–494.

81. Zeitzer, J.M., C.L. Buckmaster, D.M. Lyons & E. Mignot. 2007. Increasing length of wakefulness and modulation of hypocretin-1 in the wake-consolidated squirrel monkey. *Am. J. Physiol. Regul. Integr. Comp. Physiol.* **293**: R1736–R1742.

82. Born, J., T. Lange, K. Hansen, *et al.* 1997. Effects of sleep and circadian rhythm on human circulating immune cells. *J. Immul.* **158**: 4454–4464.

83. Vgontzas, A.N., M. Zoumakis & D.A. Papanicolaou. 2002. Chronic insomnia is associated with a shift of interleukin-6 and tumor necrosis factor secretion from nighttime to daytime. *Metabolism* **51**: 887–892.

84. Tsigos, C., D.A. Papanicolaou, R. Defensor, *et al.* 1997. Dose effects of recombinant human interleukin-6 on pituitary hormone secretion and energy expenditure. *Neuroendocrinology* **66**: 54–62.

85. Meier-Ewert, H.K., P.M. Ridker, N. Rifai, *et al.* 2001. Absence of diurnal variation of C-reactive protein concentration in healthy human subjects. *Clin. Chem.* **47**: 426–430.

86. Koc, M., O. Karaarslan, G. Abali & M.K. Batur. 2010. Variations in high-sensitivity C-reactive protein levels over 24 hours in patients with stable coronary artery disease. *Tex. Heart. Inst. J.* **37**: 42–48.

87. Kuhn, E., V. Brodan, M. Brodanova & K. Rysanek. 1969. Metabolic reflection of sleep deprivation. *Act. Nerv. Super* **11**: 165–174.

88. Parker, D.C., L.G. Rossman, A.E. Pekary & J.M. Hershman. 1987. Effect of 64-hour sleep deprivation on the circadian waveform of thyrotropin (TSH): further evidence of sleep-related inhibition of TSH release. *J. Clin. Endocrinol. Metab.* **64**: 157–161.

89. VanHelder, T., J.D. Symons & M.W. Radomski. 1993. Effects of sleep deprivation and exercise on glucose tolerance. *Aviat. Space Environ. Med.* **64**: 487–492.

90. Gonzáles-Ortiz, M., E. Martínez-Abundis, B.R. Balcázar-Muñoz & S. Pascoe-González. 2000. Effect of sleep deprivation on insulin sensitivity and cortisol concentration in healthy subjects. *Diabetes Nutr. Metab.* **13**: 80–83.

91. Thomas, M., H. Sing, G. Belenky, *et al.* 2000. Neural basis of alertness and cognitive performance impairments during sleepiness: I. Effects of 24 h of sleep deprivation on waking human regional brain activity. *J. Sleep Res.* **9**: 335–352.

92. Garfinkel, D., M. Zorin, J. Wainstein, *et al.* 2011. Efficacy and safety of prolonged-release melatonin in insomnia patients with diabetes: a randomized, double-blind, crossover study. *Diabetes Metab. Syndr. Obes.* **4**: 307–313.

93. Shearer, W.T., J.M. Reuben, J.M. Mullington, *et al.* 2001. Soluble TNA-alpha receptor 1 and IL-6 plasma levels in humans subjected to the sleep deprivation model of spaceflight. *J. Allergy Clin. Immunol.* **107**: 165–170.

94. Mullington, J.M., J.L. Chan, H.P. van Dongen, *et al.* 2003. Sleep loss reduces diurnal rhythm amplitude of leptin in healthy men. *J. Neuroendocrinol.* **15**: 851–854.

95. Dzaja, A., M.A. Dalal, H. Himmerich, *et al.* 2004. Sleep enhances nocturnal plasma ghrelin levels in healthy subjects. *Am. J. Physiol. Endocrinol. Metab.* **286**: E963–E967.

96. Frey, D.J., M. Fleshner & K.P. Wright. 2007. The effects of 40 hours of total sleep deprivation on inflammatory markers in healthy young adults. *Brain Behav. Immun.* **21**: 1050–1057.

97. Schmid, S.M., M. Hallschmid, K. Jauch-Chara, *et al.* 2007. Sleep loss alters basal metabolic hormone secretion and modulates the dynamic counterregulatory response to hypoglycemia. *J. Clin. Endocrinol. Metab.* **92**: 3044–3051.

98. Schmid, S.M., M. Hallschmid, K. Jauchchara, J. Born & B. Schultes. 2008. A single night of sleep deprivation increases ghrelin levels and feelings of hunger in normal-weight healthy men. *J. Sleep Res.* **17**: 331–334.

99. Pejovic, S., A.N. Vgontzas, M. Basta, *et al.* 2010. Leptin and hunger levels in young healthy adults after one night of sleep loss. *J. Sleep. Res.* **19**: 552–558.

100. Benedict, C., M. Hallschmid, A. Lassen, *et al.* 2011. Acute sleep deprivation reduces energy expenditure in healthy men. *Am. J. Clin. Nutr.* **93:** 1229–1236.

101. Spiegel, K., R. Leproult & E. Van Cauter. 1999. Impact of sleep debt on metabolic and endocrine function. *Lancet* **354:** 1435–1439.

102. Spiegel, K., R. Leproult, E.F. Colecchia, *et al.* 2000. Adaptation of the 24 h growth hormone profile to a state of sleep debt. *Am. J. Psychiol.* **279:** R874–R883.

103. Spiegel, K., R. Leproult & E. van Cauter. 2003. Impact of sleep debt on physiological rhythms. *Rev. Neurol.* **159:** 6S11–6S20.

104. Spiegel, K., E. Tasali, P. Penev & E. Van Cauter. 2004a. Brief communication: sleep curtailment in healthy young men is associated with decreased leptin levels, elevated ghrelin levels, and increased hunger and appetite. *Ann. Intern Med.* **141:** 846–850.

105. Spiegel, K., R. Leproult, M. L'Hermite-Baleriaux, *et al.* 2004b. Leptin levels are dependent on sleep duration: relationships with sympathovagal balance, carbohydrate regulation, cortisol, and thyrotropin. *J. Clin. Endocrinol. Metab.* **89:** 5762–5771.

106. Vgontzas, A.N., E. Zoumakis, E.O. Bixler, *et al.* 2004. Adverse effects of modest sleep restriction on sleepiness, performance, and inflammatory cytokines. *J. Clin. Endocrinol. Metab.* **89:** 2119–2126.

107. Irwin, M.R., M. Wang, C.O. Campomayor, *et al.* 2006. Sleep deprivation and activation of morning levels of cellular and genomic markers of inflammation. *Arch. Intern. Med.* **166:** 1756–1762.

108. Haack, M., E. Sanchez & J.M. Mullington. 2007. Elevated inflammatory markers in response to prolonged sleep restriction are associated with increased pain experience in healthy volunteers. *Sleep* **30:** 1145–1152.

109. Nedeltcheva, A.V., L. Kessler, J. Imperial & P.D. Penev. 2009. Exposure to recurrent sleep restriction in the setting of high caloric intake and physical inactivity results in increased insulin resistance and reduced glucose tolerance. *J. Clin. Endocrinol. Metab.* **94:** 3242–3250.

110. Schmid, S.M., M. Hallschmid, K. Jauch-Chara, *et al.* 2009a. Short-term sleep loss decreases physical activity under free-living conditions but does not increase food intake undertime-deprived laboratory conditions in healthy men. *Am. J. Clin. Nutr.* **90:** 1476–1482.

111. Schmid, S.M., K. Jauch-Chara, M. Hallschmid & B. Schultes. 2009b. Mild sleep restriction acutely reduces plasma glucagon levels in healthy men. *J. Clin. Endocrinol. Metab* **94:** 5169–5173.

112. van Leeuwen, W.M., M. Lehto, P. Karisola, *et al.* 2009. Sleep restriction increases the risk of developing cardiovascular diseases by augmenting proinflammatory responses through IL-17 and CRP. *PLoS One* **4:** e4589.

113. Brondel, L., M.A. Romer, P.M. Nougues, P. Touyarou & D. Davenne. 2010. Acute partial sleep deprivation increases food intake in healthy men. *Am. J. Clin. Nutr.* **91:** 1550–1559.

114. Buxton, O.M. & E. Marcelli. 2010. Short and long sleep are positively associated with obesity, diabetes, hypertension, and cardiovascular disease among adults in the United States. *Soc. Sci. Med.* **71:** 1027–1036.

115. Donga, E., M. van Dijk, J.G. van Dijk, *et al.* 2010. Partial sleep restriction decreases insulin sensitivity in type 1 diabetes. *Diabetes Care* **33:** 1573–1577.

116. Kessler, L., A. Nedeltcheva, J. Imperial & P.D. Penev. 2010. Changes in serum TSH and free T4 during human sleep restriction. *Sleep* **33:** 1115–1118.

117. Omisade, A., O.M. Buxton & B. Rusak. 2010. Impact of acute sleep restriction on cortisol and leptin levels in young women. *Physiol. Behav.* **99:** 651–656.

118. Nedeltcheva, A.V., J.M. Kilkus, J. Imperial, D.A. Schoeller & P.D. Penev. 2010. Insufficient sleep undermies dietary efforts to reduce adiposity. *Ann. Intern. Med.* **153:** 435–441.

119. Schmid, S.M., M. Hallschmid, K. Jauch-Chara, *et al.* 2011. Disturbed glucoregulatory response to food intake after moderate sleep restriction. *Sleep* **34:** 371–377.

120. Simpson, N.S., S. Banks & D.F. Dinges. 2010. Sleep restriction is associated with increased morning plasma leptin concentrations, especially in women. *Biol. Res. Nurs.* **12:** 47–53.

121. Simpson, N.S., S. Banks, S. Arroyo & D.F. Dinges. 2010. Effects of sleep restriction on adiponectin levels in healthy men and women. *Physiol. Behav.* **101:** 693–698.

122. St-Onge, M., A.L. Roberts, J. Chen, *et al.* 2011. Short sleep duration increases energy intakes but does not change energy expenditure in normal-weight individuals. *Am. J. Clin. Nutr.* **94:** 410–416.

123. Tasali, E., R. Leproult, D.A. Ehrmann & E. Van Cauter. 2008. Slow-wave sleep and the risk of type 2 diabetes in humans. *Proc. Natl. Acad. Sci. USA* **105:** 1044–1049.

124. Stamatakis, K.A. & N.M. Punjabi. 2010. Effects of sleep fragmentation on glucose metabolism in normal subjects. *Chest* **137:** 95–101.

125. Hursel, R., F. Rutters, H.K. Gonnissen, E.A. Martens & M.S. Westerterp-Plantenga. 2011. Effects of sleep fragmentation in healthy men on energy expenditure, substrate oxidation, psysical activity, and exhaustion measured over 48 h in a respiratory chamber. *Am. J. Clin. Nut.r* **94:** 804–808.

126. Ravussin, E. 1995. Metabolic differences and the development of obesity. *Metabolism* **44:** 12–14.

127. Bonnet, M.H. & D.L. Arand. 1995. 24-hour metabolic rate in insomniacs and matched normal sleepers. *Sleep* **18:** 581–588.

128. De Jonge, L., X. Zhao, M.S. Mattingly, *et al.* 2012. Poor sleep quality and sleep apnea are associated with higher energy expenditure in obese individuals with shot sleep duration. *J. Clin. Endocrinol. Metab.,* in press Jun 11. [Epub ahead of print].

129. Everson, C.A. & T.S. Nowak, Jr. 2002. Hypothalamic thyrotropin-releasing hormone mRNA responses to hypothyroxinemia induced by sleep deprivation. *Am. J. Physiol. Endocrinol. Metab.* **283:** E85–E93.

130. Cizza, G., L.S. Brazy, M.E. Esclapes, *et al.* 1996. Age and gender influence basal and stress-modulated hypothalamic-pituitary-thyroidal function in Fischer 344/N rats. *Neuroendocrinology* **64:** 440–448.

131. Deibert, D.C. & R.A. DeFronzo. 1980. Epinephrine-induced insulin resistance in man. *J. Clin. Invest.* **65:** 717–721.

132. Avogaro, A., G. Toffolo, A. Valerio & C. Cobelli. 1996. Epinephrine exerts opposite effects on peripheral glucose disposal and glucose stimulated insulin secretion. A stable label intravenous glucose tolerance test minimal model study. *Diabetes* **45:** 1373–1378.

133. Cizza, G. & K.I. Rother. 2012. Beyond fast food and slow motion: weighty contributors to the obesity epidemic. *J. Endocrinol. Invest.* **35:** 236–242.

134. Kilkus, J.M., J.N. Booth, L.E. Bromley, *et al.* 2012. Sleep and eating behavior in adults at risk for type 2 diabetes. *Obesity* **20:** 112–117.

135. Garaulet, M., C. Moreno, C.E. Smith, *et al.* 2011. Ghrelin, sleep reduction and evening preference: relationships to CLOCK3111 T/C SNP and weight loss. *PLoS One* **6:** e17435.

136. Motivala, S.J., J. Tomiyama, M. Ziegler, S. Khandrika & M.R. Irwin. 2009. Nocturnal levels of ghrelin and leptin and sleep in chronic insomnia. *Psychoneuroendocrinology* **34:** 540–545.

137. Hayes, A.L., F. Xy, D. Babineau & S.R. Patel. 2011. Sleep duration and circulatin adipokine levels. *Sleep* **34:** 147–152.

138. Knutson, K.L., G. Galli, X. Zhao, M. Mattingly & G. Cizza. 2011. No association between leptin levels and sleep duration or quality in obese adults. *Obesity* **19:** 2433–2435.

139. Nakamura, T. Kanbayashi, T. Sugiura & Y. Inoue. 2011. Relationship between clinical characteristics of narcolepsy and CSF orexin-A levels. *J. Sleep Res.* **20:** 45–49.

140. Estabrooke, I.V., M.T. McCarthy, E. Ko, *et al.* 2001. Fos expression in orexin neurons varies with behavioral state. *J. Neurosci.* **21:** 1656–1662.

141. Hotamisligil, G.S., A. Budavari, D. Murray & B.M. Spiegelman. 1994. Reduced tyrosine kinase activity of the insulin receptor in obesity-diabetes. Central role of tumor necrosis factor-alpha. *J. Clin. Invest.* **94:** 1543–1549.

142. Kristiansen, O.P & T. Mandrup-Poulsen. 2005. Interleukin-6 and diabetes: the good, the bad, or the indifferent? *Diabetes* **54:** S114–S124.

143. Rother, K.I. 2007. Diabetes treatment – bridging the divide. *New Eng. J. Med.* **356:** 1499–1501.

144. Nieto-Vazquez, I., S. Fernández-Veledo, D.K. Krämer, *et al.* 2008. Insulin resistance associated to obesity: the link TNF-alpha. *Arch. Physiol. Biochem.* **114:** 183–194.

145. Cizza, G., A.H. Marques, F. Eskandari, *et al.* 2008. Elevated neuroimmune biomarkers in sweat patches and plasma of premenopausal women with major depressive disorder in remission: the POWER study. *Biol. Psychiatry* **64:** 907–911.

146. Siervo, M., J.C. Wells & G. Cizza. 2009. The contribution of phychosocial stress to the obesity epidemic: an evolutionary approach. *Horm. Metab. Res.* **41:** 261–270.

147. Patten, S.B., J.V. Williams, D.H. Lavorato, S. Khaled & A.G. Bulloch. 2011. Weight gain in relation to major depression and antidepressant medication use. *J. Affect. Disord.* **134:** 288–293.

148. Knutson, K.L., X. Zhao, M. Mattingly, G. Galli & G. Cizza. 2012. Predictors of sleep-disordered breathing in obese adults who are chronic short sleepers. *Sleep Med.* **13:** 484–489.

149. Kalra, M., M. Manna, K. Fitz, *et al.* 1998. Effect of surgical weight loss on sleep architecture in adolescents with severe obesity. *Pediatrics* **101:** 61–67.

150. Willi, S.M., M.J. Oexmann, N.M. Wright, N.A. Collop & L. Lindon Key. 2008. The effects of a high-protein, low-fat, ketogenic diet on adolescents with morbid obesity: body composition, blood chemistries, and sleep abnormalities. *Obes. Surg.* **18:** 675–679.

151. M. van Dijk, E. Donga, J.G. van Dijk, *et al.* Disturbed subjective sleep characteristics in adult patients with long-standing type 1 diabetes mellitus. *Diabetologia* **54:** 1967–1976.

152. Pillar, G., G. Schuscheim, R. Weiss, *et al.* 2003. Interactions between hypoglycemia and sleep architecture in children with type 1 diabetes mellitus. *J. Pediatrics* **142:** 163–168.

153. Pandi-Perumal, S.R., A. Moscovitch, V. Srinivasan, *et al.* 2009. Bidirectional communication between sleep and circadian rhythms and its implications for depression: lessons from agomelatine. *Prog. Neurobiol.* **88:** 264–271.

154. van Oosterhout, F., E.A. Lucassen, T. Houben, *et al.* 2012. Amplitude of the SCN clock enhanced by the behavioral activity rhythm. *PLoS One.* In press.

155. Sinton, C.M., T.E. Fitch & H.K. Gershenfeld. 1999. The effects of leptin on REM sleep and slow wave delta in rats are reversed by food deprivation. *J. Sleep Res.* **8:** 197–203.

156. Laposky, A.D., J. Shelton, J. Bass, *et al.* 2006. Altered sleep regulation in leptin-deficient mice. *Am. J. Physiol. Regul. Integr. Comp. Physiol.* **290:** R894–R903.

157. Laposky, A.D., M.A. Bradley, D.L. Williamd, J. Bass & F.W. Turek. 2008. Sleep-wake regulation is altered in leptin-resistance (db/db) genetically obese and diabetic mice. *Am. J. Physiol. Regul. Integr. Comp. Physiol.* **295:** R2059–R2066.

158. Thorne, H.C., K.H. Jones, S.P. Peters, S.N. Archer & D.J. Dijk. 2009. Daily and seasonal variation in the spectral composition of light exposure in humans. *Chronobiol. Int.* **26:** 854–866.

159. Zeitzer, J.M., D.J. Dijk, R. Kronauer, E. Brown & C. Czeisler. 2000. Sensitivity of the human circadian pacemaker to nocturnal light: melatonin phase resetting suppression. *J. Physiol.* **526:** 695–702.

160. Gooley, J.J., S.M. Rajaratnam, G.C. Brainard, *et al.* 2010. Spectral responses of the human circadian system depend on the irradiance and duration of exposure to light. *Sci. Trans. Med.* **2:** 31ra33.

161. Minami, M., H. Takahashi, T. Sasaki, *et al.* 2012. The effect of sleep restriction and psychological load on the diurnal metabolic changes in tryptamine-related compounds in human urine. *Environ. Health Prev. Med.* **17:** 87–97.

162. Nogueira, T.C., C. Lellis-Santos, D.S. Jesus, *et al.* 2011. Absence of melatonin induces night-time hepatic insulin resistance and increased gluconeogenesis due to stimulation of nocturnal unfolded protein response. *Endocrinology* **152:** 1253–1263.

163. Wolden-Hanson, T., D.R. Mitton, R.L. McCants, *et al.* 2000. Daily melatonin administration to middle-aged rats suppresses body weight, intraabdominal adiposity, and plasma leptin and insulin dependent of food intake and total body fat. *Endocrinology* **141:** 487–497.

164. Cardinali, D.P., E.S. Pagano, P.A. Scacchi Bernasconi, R. Reynoso & P. Scacchi. 2011. Disrupted chronobiology of sleep and cytoprotection in obesity: possible therapeutic value of melatonin. *Neuro. Endocrinol. Lett.* **32:** 588–606.

165. Robeva, R., G. Kirilov, A. Tomova & P. Kumanov. 2008. Melatonin-insulin interactions in patients with metabolic syndrome. *J. Pineal Res.* **44:** 52–56.

166. Olsson, L., E. Pettersen, A. Ahlbom, *et al.* 2011. No effect by the common gene variant rs10830963 of the melatonin receptor 1B on the association between sleep disturbances and type 2 diabetes: results from the Nord-Trøndelag Health Study. *Diabetologia* **54:** 1375–1378.

167. Prokopenko, I., C. Langenberg, J.C. Florez, *et al.* 2009. Variant in MTNR1B influence fasting glucose levels. *Nat. Genet.* **41:** 77–81.

168. Bonnefond, A., N. Clément, K. Fawcett, *et al.* 2012. Rare MTNR1B variants impairing melatonin receptor 1B function contribute to type 2 diabetes. *Nat. Genet.* **44:** 297–301.

169. Peschke, E., T. Frese, E. Chankiewitz, *et al.* 2006. Diabetic Goto Kakizaki rats as well as type 2 diabetic patients show a decreased diurnal serum melatonin level and an increased pancreatic melatonin-receptor status. *J. Pineal Res.* **40:** 135–143.

170. Garfinkel, D., M. Zorin, J. Wainstein, *et al.* 2011. Efficacy and safety of prolonged-release melatonin in insomnia patients with diabetes: a randomized, double-blind, crossover study. *Diabetes Metab. Syndr. Obes.* **4:** 307–313.

171. Kozirog, M., A.R. Poliwczak, P. Duchnowicz, *et al.* 2011. Melatonin treatment improves blood pressure, lipid profile, and parameters of oxidative stress in patients with metabolic syndrome. *J. Pineal. Res.* **50:** 261–266.

172. Tamura, H., Y. Nakamura, A. Narimatsu, *et al.* 2008. Melatonin treatment in peri- and postmenopausal women elevates serum high density lipoprotein cholesterol levels without influencing total cholesterol. *J. Pineal. Res.* **45:** 101–105.

173. Ohayon, M.M., M.A. Carskadon, C. Guilleminault & M.V. Vitiello. 2004. Meta-analysis of quantitative sleep parameters from childhood to old age in healthy individuals: developing normative sleep values across the human lifespan. *Sleep* **27:** 1255–1273.

174. Paul, K.N., F.W. Turek & M.H. Kryger. 2008. Influence of sex on sleep regulatory mechanisms. *J. Womens Health* **17:** 1201–1208.

175. Philips, B.A., N.A. Collop, C. Drake, *et al.* Sleep disorders and medical conditions in women. 2008. Proceedings of the Women & Sleep Workshop, National Sleep Foundation, Washington, DC, March 5–6, 2007. *J. Womens Health* **17:** 1191–1199.

176. Ershow, A.G. 2009. Environmental influences on development of type 2 diabets and obesity: challenges in personalizing prevention and management. *J. Diabetes Sci. Technol.* **2:** 727–734.

177. Landolt, H.P. 2008. Sleep homeostasis: a role for adenosine in humans? *Biochem. Pharmacol.* **75:** 2070–2079.

178. Czeisler, C.A. 2011. Impact of sleepiness and sleep deficiency on public health-utility of biomarkers. *J. Clin. Sleep Med.* **7**(5 Suppl): S6–S8.

179. Cizza, G., P. Marincola, M. Mattingly, *et al.* 2010. Treatment of obesity with extension of sleep duration: a randomized, prospective, controlled trial. *Clinical Trials* **7:** 274–285.

Ann. N.Y. Acad. Sci. ISSN 0077-8923

ANNALS OF THE NEW YORK ACADEMY OF SCIENCES

Issue: *The Brain and Obesity*

Childhood obesity and sleep: relatives, partners, or both?—a critical perspective on the evidence

David Gozal and Leila Kheirandish-Gozal

Department of Pediatrics and Comer Children's Hospital, Pritzker School of Medicine, The University of Chicago, Chicago, Illinois

Address for correspondence: David Gozal, M.D., The Herbert T. Abelson Professor and Chair, Department of Pediatrics, Physician-in-Chief, Comer Children's Hospital, The University of Chicago, 5721 S. Maryland Avenue, MC 8000, Suite K-160 Chicago, IL 60637. dgozal@uchicago.edu

In modern life, children are unlikely to obtain sufficient or regular sleep and waking schedules. Inadequate sleep affects the regulation of homeostatic and hormonal systems underlying somatic growth, maturation, and bioenergetics. Therefore, assessments of the obesogenic lifestyle, including as dietary and physical activity, need to be coupled with accurate evaluation of sleep quality and quantity, and coexistence of sleep apnea. Inclusion of sleep as an integral component of research studies on childhood obesity should be done as part of the study planning process. Although parents and health professionals have quantified normal patterns of activities in children, sleep has been almost completely overlooked. As sleep duration in children appears to have declined, reciprocal obesity rates have increased. Also, increases in pediatric obesity rates have markedly increased the risk of obstructive sleep apnea syndrome (OSAS) in children. Obesity and OSAS share common pathways underlying end-organ morbidity, potentially leading to reciprocal amplificatory effects. The relative paucity of data on the topics covered in the perspective below should serve as a major incentive toward future research on these critically important concepts.

Keywords: sleep; obesity; children; sleep apnea

Sleep trends in children and potential associations with obesity

Although sleep assessments of children have largely relied on parental reports,[1–7] it has become apparent that children are highly unlikely to obtain sufficient sleep on a stable and regular schedule. Polls by the National Sleep Foundation[1,2] show that parents routinely overestimate their children's sleep duration, and that in fact children sleep much less than what is deemed appropriate for their age. At the society level, and likely across all pediatric ages, children sleep less than what they did one century ago.[8] Despite the compelling evidence supporting a vital role for healthy sleep in brain maturation, somatic growth, information processing, memory consolidation, learning, and other important neurobehavioral functions, parents and professionals instead focus their attention on children's accomplishments (e.g., first steps, first words, school grades, extracurricular activity

performances), and often treat sleep as a tradable commodity.

Clearly, disturbed sleep patterns can lead to multidimensional adverse effects. For example, 43% of school-aged children and 57% of adolescents have a TV in their bedroom;[1,2,9] and as many as 42% have mobile phones in their bedroom and many other electronic devices, such as computers, video games, and other electronic gadgetry are also frequently present in bedrooms.[9,10] These devices account for multiple intrusion patterns that can curtail or disrupt sleep in children and may result in the development of decreased opportunities for sleep in children over time—put differently, it could be said that modern life has "polluted" the opportunity to sleep because of a variety of intrusions, which may further lead to reduced sleep duration or sleep disruption.

In the last 20 years, a large number of studies have been published on sleep duration in children, and most have aimed to evaluate the potentially adverse

doi: 10.1111/j.1749-6632.2012.06723.x
Ann. N.Y. Acad. Sci. 1264 (2012) 135–141 © 2012 New York Academy of Sciences.

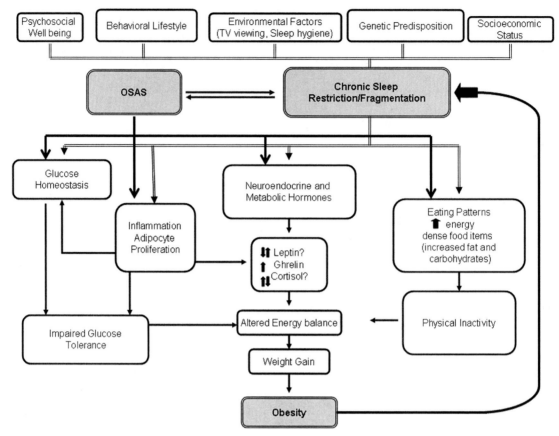

Figure 1. Putative interactions between reduced or fragmented sleep, sleep apnea, and obesity in children.

impact of poor sleep on health outcomes. Although these studies are heterogeneous in their methodology and scope, they all point to the critical need for urgent large-scale representative and longitudinal studies on objective sleep–wake patterns in children, particularly to explore the impact of sleep on health and body weight in a valid ecological and contextual setting. For example, the family domain, important with respect to sleep–wake patterns,[11,12] influences food and health-related behaviors in the developing child. Not surprisingly, the presence of parental beliefs about being overweight and nutrition can have significant impact on the risk of childhood obesity.[13,14] Because daily activities and sleep patterns evolve in a circadian continuum, parenting feeding styles[15] may interact with sleep. For instance, when U.S. preschoolers were exposed to three household routines, namely, evening family meal for more than 5 nights per week, sleeping ≥ 10.5 h/night on weekdays, and ≤ 2 h/day television, video, or other screen-

viewing behavior, obesity prevalence decreased by 40% compared with when no such simple routines were present.[16]

The estimated dose–response relationship between sleep and obesity is highly variable,[17] with pooled odd ratios ranging from 1.15 to 11.0, which likely represents heterogeneities in methodologies and the diversity of contributing factors (Fig. 1).[18] In a recent study in which sleep and weight were carefully monitored using objective assessments, we found that irregular and shorter sleep is a significant risk factor for the occurrence of weight problems in children, and that a nonlinear trend between sleep and weight was present.[19] We further found that obese children are less likely to catch-up in their sleep "debt" during weekends, and that the combination of shorter sleep duration and more variable sleep patterns was associated not only with increased weight risk but also with adverse metabolic outcomes (i.e., insulin resistance, elevated serum lipids, and increased high-sensitivity C reactive protein levels).

The biological correlates of the association between sleep patterns and body mass index (BMI) have been partially elucidated and would support the assumption that either inadequate amounts of sleep or disrupted sleep architecture lead to alterations in some of the neuropeptides that regulate appetite, such as increased levels of ghrelin, reduced levels of leptin, and reduced central biological activity of orexin, with the anticipated cumulative effect of promoting increased food intake (Fig. 1).[20,21]

The association between short sleep duration and increased risk for obesity is, as mentioned, somewhat conflictive, whereby discrepancies also emerge in relation to age. For example, although a significant contribution of sleep duration to obesity risk is present in adults, the associations are, as indicated above, highly variable and could reflect a multitude of possible confounders that may be operational in early life.[22–28] Furthermore, in an elegant and important cross-sectional and longitudinal study, Chaput and colleagues reported that only those adults manifesting short sleep duration, highly disinhibited eating behaviors, and/or low dietary calcium intake had significantly higher BMI compared with corresponding controls. Indeed, over the six-year follow-up period, such high-risk adults were significantly more likely to gain weight and develop obesity.[18]

Notwithstanding these considerations, a recent meta-analysis of the literature appears to support the conclusion that the strength of the association between sleep duration and obesity is actually stronger in children and adolescents, and that it declines over time.[29] Furthermore, some degree of predisposition for the existence of such association has been proposed in light of the findings that sleep-associated changes in BMI appear to be primarily occurring in children whose BMI was already elevated.[30]

We should emphasize that the majority of these studies have, as mentioned above, relied on subjective, parentally reported estimates of sleep duration,[31–33] and that the effects of variability of sleep schedules on BMI has not been explored. Also, parental reports generally overestimate sleep duration of children.[32–36] In addition, we are unaware of experimental studies that aim to characterize the effect of sleep manipulations on metabolic homeosta-sis in children. Indeed, the definition of what constitutes "short sleep" is arbitrary and highly variable across different people and across various studies, further adding to the complexity of the critical assessment of the association between sleep and BMI. A given "sleep duration" over a relatively large range may in fact be considered long and sufficient sleep in one child, although short and insufficient sleep in another.

Regardless of the aforementioned limitations, the association of sleep duration with BMI generally exhibits about a 1.5- to twofold increase in the likelihood of being a short sleeper when obesity is present.[37–39] And although interventional studies aiming to modify sleep patterns in children are clearly needed, they may be fraught with substantial failure rates, particularly considering that both sleep regularity and sleep duration are maintained across long periods of time during childhood; and thus any intervention will likely need to be initiated very early in life if the effect is to be measurable.[39] Accordingly, identification of particular young children at high risk and prospective interventions aiming to prolong and regularize sleep in these children will provide more definitive evidence regarding the role of sleep in BMI and metabolic regulation. Even if such studies are conducted, it will take a long time before their findings can be incorporated into clinical practice. Accordingly, based on the current, albeit deficient level of knowledge, we feel that it might be worth advocating for implementing educational campaigns aimed at families and health professionals, campaigns that target the promotion of longer and regular sleep habits among toddlers and beyond.

Although not exactly within the scope of this paper, it is worthwhile mentioning that strong evidence supporting the biological plausibility of a strong link between sleep, appetite regulation, and adiposity has been made over the last few years. For example, circadian clocks are integral regulators of cellular metabolism that also modulate both appetite and food intake in both animals and humans.[40–42] Perturbations of the endogenous global or organ-specific circadian cycle, or alterations in the integrity of sleep homeostatic mechanisms, therefore increases the risk of altered energy intake and disposition, which may ultimately lead to increased propensity for developing obesity and metabolic disturbances.

Obesity and obstructive sleep apnea syndrome

Since its initial description in 1976, obesity and obstructive sleep apnea syndrome (OSAS) has become widely recognized as a highly prevalent condition in children.[43] OSAS has now been recognized as leading to a spectrum of potentially serious morbidities affecting the central nervous, cardiovascular, and metabolic systems, and studies exploring the potential mechanisms leading to end-organ dysfunction in the context of OSAS indicate that both oxidative stress and inflammatory processes are operational. The interplay of these processes with disease severity, environmentally related modifiers, and individual genetic susceptibility is clearly emerging as the optimal model accounting for phenotypic variance.[44–48] In parallel, the last two decades has witnessed a shift from the classic presentation of children with OSAS (i.e., adenotonsillar hypertrophy and failure to thrive) to a majority of children being overweight or obese, even though adenotonsillar hypertrophy continues to play a role in the latter group.[49]

OSAS in children is characterized by recurrent events of partial or complete upper airway obstruction during sleep, resulting in disruption of normal gas exchange (intermittent hypoxia and hypercapnia) and of sleep through multiple arousals, leading to sleep fragmentation. Although enlarged tonsils and adenoids in the upper airway clearly play a role,[50] it has now become evident that the interplay between alterations in structural and anatomical characteristics, upper airway mucosal properties and inflammation, and protective reflexes and neuromotor abnormalities of the upper airway are the major determinants of whether upper airway obstruction will develop during sleep, as well as its severity and frequency. The clinical spectrum of obstructive sleep-disordered breathing includes frank OSAS of varying severity, upper airway resistance syndrome (traditionally associated with low-frequency obstructive apneic events and globally preserved normal oxygenation patterns but increased respiratory-related sleep fragmentation), and, at the low end of the severity spectrum, a condition that has been termed either primary or habitual snoring (i.e., habitual snoring in the absence of apneas, gas exchange abnormalities, and/or disruption of sleep architecture). The prevalence of OSAS

in children is currently estimated to be ∼3% among 2- to 8-year-old children.[51] However, habitual snoring during sleep—the hallmark of increased upper airway resistance—is much more prevalent.[52,53]

More recently, obesity has been shown to markedly increase the risk of OSAS,[54–60] whereby upper airway narrowing may be the consequence of fatty infiltration of upper airway structures and the tongue, whereas subcutaneous fat deposits in the anterior neck region and other cervical structures also exert force vectors promoting increased pharyngeal collapsibility.[61,62] Increased adipose tissue mass in the abdominal wall and cavity, as well as in the thorax, increases the global respiratory load and reduces intrathoracic diaphragm excursion, particularly during the supine position, leading to decreased lung volumes and oxygen reserve, and increased work of breathing during sleep. Furthermore, obesity can be accompanied by poor quality sleep, which, in turn, can perturb arousal mechanisms and, therefore, delay upper airway opening, thus exacerbating the duration of apnea.[63]

As mentioned, the presence of OSAS may promote leptin resistance and enhance ghrelin levels, both of which can perpetuate the tendency for obesogenic behaviors.[64,65] The unidirectional drivers of obesity—increasing the probability of OSAS—appear to be functional in the reverse direction as well. In other words, OSAS may either promote or exaggerate the tendency for obesity, or its consequences (Fig. 1). For example, OSAS is associated with daytime sleepiness, and sleepiness is exacerbated when obesity is concurrently present.[66,67] In addition, sleepiness will reduce the likelihood of engaging in physical activity and enhance obesogenic eating behaviors that favor calorie-dense foods, particularly in those children at risk for obesity.[65] Finally, OSAS is a chronic low-grade inflammatory disease that interacts with and potentiates obesity-induced inflammatory processes.[68–71]

We should again emphasize that despite the rather compelling evidence supporting a bidirectional interaction between OSAS and obesity, there are no well-controlled interventional trials that have assessed whether effective treatment of OSAS will improve obesity risk and obesity-related outcomes, and conversely, whether improvements in obesity are associated with improvements in OSAS severity.[59,72,73]

Conclusions

In summary, sleep and body weight appear to share a constellation of contributing factors that originate in the spectrum encompassed by child, family, and society, and in which food intake patterns, physical activity, and sleep habits emerge as integral contributors to obesity risk. Unfortunately, the relative contribution of poor sleep remains virtually unexplored—a fact that should prompt renewed efforts to elucidate the mechanisms underlying metabolic dysregulation in the context of short sleep, irregular sleep, or disrupted sleep. The nature and scope of these interactions and the potential mechanisms underlying such putative associations will need to be explored much more thoroughly and carefully in the near future, to enable formulation of realistic and cogent public intervention programs aimed at reducing the unacceptably high rates of obesity in children. We have also discussed the putative presence of reciprocal contributions of OSAS and obesity that perpetuate and enhance each other.

There is little doubt that more precise understanding of the effects of sleep disruption and tissue hypoxia on the phenotypic expression of these diseases is warranted. The assessment and identification of unique genomic, proteomic, and metabolomic pathways underlying the antecedents and consequences of these highly prevalent pediatric diseases appears necessary and before implementing sound and valid treatment strategies.

Acknowledgments

D.G. is supported by National Institutes of Health Grants HL-065270 and HL-086662.

Conflicts of interest

The authors declare no conflicts of interest.

References

1. National Sleep Foundation 2004. Sleep in America poll–children and sleep. Available at http://www.sleepfoundation.org/article/sleep-america-polls/2004-children-and-sleep (last accessed May 3, 2012).
2. National Sleep Foundation 2006. Sleep in America poll—teens and sleep. Available at http://www.sleepfoundation.org/article/sleep-america-polls/2006-teens-and-sleep (last accessed May 1, 2012).
3. Knutson, K.L. & D.S. Lauderdale. 2009. Sociodemographic and behavioral predictors of bed time and wake time among US adolescents aged 15 to 17 years. *J. Pediatr.* 154: 426–430.
4. Dollman, J., K. Ridley, T. Olds & E. Lowe. 2007. Trends in the duration of school-day sleep among 10- to 15-year-old South Australians between 1985 and 2004. *Acta Paediatr.* 96: 1011–1014.
5. Sadeh, A., A. Raviv & R. Gruber. 2000. Sleep patterns and sleep disruptions in school-age children. *Dev. Psychol.* 36: 291–301.
6. Yang, C.K., J.K. Kim, S.R. Patel, & J.H. Lee. 2005. Age-related changes in sleep/wake patterns among Korean teenagers. *Pediatrics* 115: 250–206.
7. Warner, S., G. Murray & D. Meyer. 2008. Holiday and school-term sleep patterns of Australian adolescents. *J. Adolesc.* 31: 595–608.
8. Matricciani, L.A., T.S. Olds, S. Blunden, *et al.* 2012. Never enough sleep: a brief history of sleep recommendations for children. *Pediatrics* 129: 548–556.
9. Spruyt, K. & Slaapproblemen. bij. kinderen. 2007. *Basisgids Voor Ouders en Hulpverleners (Sleep problems in children. A clinical guide for parents and health care providers).* Lannoo. Tielt, Belgium.
10. Li, S., X. Jin, S. Wu, *et al.* 2007. The impact of media use on sleep patterns and sleep disorders among school-aged children. *Sleep* 30: 361–367.
11. El-Sheikh, M., J.A. Buckhalt, P.S. Keller, *et al.* 2007. Child emotional insecurity and academic achievement: the role of sleep disruptions. *J. Fam. Psychol.* 21: 29–38.
12. El-Sheikh, M., J.A. Buckhalt, E. Mark Cummings & P. Keller. 2007. Sleep disruptions and emotional insecurity are pathways of risk for children. *J. Child. Psychol. Psychiatry* 48: 88–96.
13. Cole, T.J. 2006. The international growth standard for preadolescent and adolescent children: statistical considerations. *Food. Nutr. Bull.* 27: S237–S243.
14. Baughcum, A.E., L.A. Chamberlin, C.M. Deeks, *et al.* 2000. Maternal perceptions of overweight preschool children. *Pediatrics* 106: 1380–1386.
15. Chassin, L., C.C. Presson, J. Rose, *et al.* 2005. Parenting style and smoking-specific parenting practices as predictors of adolescent smoking onset. *J. Pediatr. Psychol.* 30: 333–344.
16. Anderson, S.E. & R.C. Whitaker. 2010. Household routines and obesity in US preschool aged children. *Pediatrics* 125: 420–428.
17. Chen, X.B.M. & Y. Wang. 2008. Is sleep duration associated with childhood obesity? A systematic review and meta-analysis. *Obesity (Silver Spring)* 16: 265–274.
18. Cappuccio, F.P., F.M. Taggart, N.B. Kandala, *et al.* 2008. Meta-analysis of short sleep duration and obesity in children and adults. *Sleep* 31: 619–626.
19. Spruyt, K., D.L. Molfese & D. Gozal. 2011. Sleep duration, sleep regularity, body weight, and metabolic homeostasis in school-aged children. *Pediatrics* 127: e345–e352.
20. Zheng, H. & H.-R. Berthoud. 2008. Neural systems controlling the drive to eat: mind versus metabolism. *Physiology* 23(2): 75–83.
21. Mavanji, V., J.A. Teske, C.J. Billington & C.M. Kotz. 2010. Elevated sleep quality and orexin receptor mRNA in obesity-resistant rats. *Int. J. Obes. (Lond.)* 34: 1576–1588.
22. Hairston, K.G., M. Bryer-Ash, J.M. Norris, *et al.* 2010. Sleep duration and five-year abdominal fat accumulation in a minority cohort: the IRAS family study. *Sleep* 33: 289–295.

23. Grandner, M.A., N.P. Patel, P.R. Gehrman, *et al.* 2010. Problems associated with short sleep: bridging the gap between laboratory and epidemiological studies. *Sleep Med. Rev.* **14:** 239–247.

24. Monasta, L., G.D. Batty, A. Cattaneo, *et al.* 2010. Early-life determinants of overweight and obesity: a review of systematic reviews. *Obes. Rev.* **11:** 695–708.

25. Nishiura, C., J. Noguchi & H. Hashimoto. 2010. Dietary patterns only partially explain the effect of short sleep duration on the incidence of obesity. *Sleep* **33:** 753–757.

26. Taveras, E.M., S.L. Rifas-Shiman, J.W. Rich-Edwards, *et al.* 2011. Association of maternal short sleep duration with adiposity and cardiometabolic status at 3 years postpartum. *Obesity (Silver Spring)* **19:** 171–178.

27. Buxton, O.M. & E. Marcelli. 2010. Short and long sleep are positively associated with obesity, diabetes, hypertension, and cardiovascular disease among adults in the United States. *Soc. Sci. Med.* **71:** 1027–1036.

28. Taveras, E.M., S.L. Rifas-Shiman, J.W. Rich-Edwards & C.S. Mantzoros. 2011. Association of maternal short sleep duration with adiposity and cardiometabolic status at 3 years postpartum. *Metabolism* **60:** 982–986.

29. Nielsen, L.S., K.V. Danielsen & T.I. Sørensen. 2011. Short sleep duration as a possible cause of obesity: critical analysis of the epidemiological evidence. *Obes. Rev.* **12:** 78–92.

30. Bayer, O., A.S. Rosario, M. Wabitsch & R. von Kries. 2009. Sleep duration and obesity in children: is the association dependent on age and choice of the outcome parameter? *Sleep* **32:** 1183–1189.

31. Nixon, G.M., J.M.D. Thompson, D.Y. Han, *et al.* 2008. Short sleep duration in middle childhood: risk factors and consequences. *Sleep.* **31:** 71–78.

32. Marshall, N.S., N. Glozier & R.R. Grunstein. 2008. Is sleep duration related to obesity? A critical review of the epidemiological evidence. *Sleep Med. Rev.* **12**(4): 289–298.

33. Marshall, N.S., N. Glozier & R.R. Grunstein. 2008. Reply to Taheri and Thomas: is sleep duration associated with obesity-U cannot be serious. *Sleep Med. Rev.* **12:** 303–305.

34. Patel, S.R. & F. B. Hu. 2008. Short sleep duration and weight gain: a systematic review. *Obesity* **16:** 643–653.

35. Gupta, N.K., W.H. Mueller, W. Chan & J.C. Meininger. 2002. Is obesity associated with poor sleep quality in adolescents? *Am. J. Hum. Biol.* **14:** 762–768.

36. Dayyat, E.A., K. Spruyt, D.L. Molfese & D. Gozal. 2011. Sleep estimates in children: parental versus actigraphic assessments. *Nat. Sci. Sleep* **3:** 115–123.

37. Padez, C., I. Mourao, P. Moreira & V. Rosado. 2009. Long sleep duration and childhood overweight/obesity and body fat. *Am. J. Hum. Biol.* **21:** 371–376.

38. Taveras, E.M., S.L. Rifas-Shiman, E. Oken, *et al.* 2008. Short sleep duration in infancy and risk of childhood overweight. *Arch. Pediatr. Adolesc. Med.* **162:** 305–311.

39. Touchette, E., D. Petit, R.E. Tremblay, *et al.* 2008. Associations between sleep duration patterns and overweight/obesity at age 6. *Sleep* **31:** 1507–1514.

40. Huang, W., K.M. Ramsey, B. Marcheva & J. Bass. 2011. Circadian rhythms, sleep, and metabolism. *J. Clin. Invest.* **121:** 2133–2141.

41. Leproult, R. & E. Van Cauter. 2010. Role of sleep and sleep loss in hormonal release and metabolism. *Endocr. Dev.* **17:** 11–21.

42. Hanlon, E.C. & E. Van Cauter. 2011. Quantification of sleep behavior and of its impact on the cross-talk between the brain and peripheral metabolism. *Proc. Natl. Acad. Sci. USA* **108:** 15609–15616.

43. Guilleminault, C., F.L. Eldridge, F.B. Simmons & W.C. Dement. 1976. Sleep apnea in eight children. *Pediatrics* **58:** 23–30.

44. Capdevila, O.S., L. Kheirandish-Gozal, E. Dayyat & D. Gozal. 2008. Pediatric obstructive sleep apnea: complications, management, and long-term outcomes. *Proc. Am. Thorac. Soc.* **5:** 274–282.

45. Gozal, D. & L. Kheirandish-Gozal. 2008. The multiple challenges of obstructive sleep apnea in children: morbidity and treatment. *Curr. Opin. Pediatr.* **20:** 654–658.

46. Kheirandish-Gozal, L., R. Bhattacharjee & D. Gozal. 2010. Autonomic alterations and endothelial dysfunction in pediatric obstructive sleep apnea. *Sleep Med.* **11:** 714–720.

47. Kim, J., F. Hakim, L. Kheirandish-Gozal & D. Gozal. 2011. Inflammatory pathways in children with insufficient or disordered sleep. *Respir. Physiol. Neurobiol* **178:** 465–474.

48. Bhattacharjee, R., J. Kim, L. Kheirandish-Gozal & D. Gozal. 2011. Obesity and obstructive sleep apnea syndrome in children: a tale of inflammatory cascades. *Pediatr. Pulmonol.* **46**(4): 313–323.

49. Dayyat, E., L. Kheirandish-Gozal & D. Gozal. 2007. Childhood obstructive sleep apnea: one or two distinct disease entities? *Sleep Med. Clin.* **2:** 433–444.

50. Katz, E.S., & C.M. D'Ambrosio. 2008. Pathophysiology of pediatric obstructive sleep apnea. *Proc. Am. Thorac. Soc.* **5:** 253–262.

51. Lumeng, J.C. & R.D. Chervin. 2008. Epidemiology of pediatric obstructive sleep apnea. *Proc. Am. Thorac. Soc.* **5:** 242–252.

52. Ferreira, A.M., V. Clemente, D. Gozal, *et al.* 2000. Snoring in Portuguese primary school children. *Pediatrics* **106:** e64.

53. O'Brien, L.M., C.R. Holbrook, C.B. Mervis, *et al.* 2003. Sleep and neurobehavioral characteristics of 5- to 7-year-old children with parentally reported symptoms of attention-deficit/hyperactivity disorder. *Pediatrics* **111:** 554–563.

54. Arens, R. & H. Muzumdar. 2010. Childhood obesity and obstructive sleep apnea syndrome. *J. Appl. Physiol.* **108:** 436–444.

55. Redline, S., P.V. Tishler, M. Schluchter, *et al.* 1999. Risk factors for sleep-disordered breathing in children. Associations with obesity, race, and respiratory problems. *Am. J. Respir. Crit. Care. Med.* **159:** 1527–1532.

56. Wing, Y.K., S.H. Hui, W.M. Pak, *et al.* 2003. A controlled study of sleep related disordered breathing in obese children. *Arch. Dis. Child.* **88:** 1043–1047.

57. Kalra, M., T. Inge, V. Garcia, *et al.* 2005. Obstructive sleep apnea in extremely overweight adolescents undergoing bariatric surgery. *Obes. Res.* **13:** 1175–1179.

58. Bixler, E.O., A.N. Vgontzas, H.M. Lin, *et al.* 2009. Sleep disordered breathing in children in a general population sample: prevalence and risk factors. *Sleep* **32:** 731–736.

59. Verhulst, S.L., H. Franckx, L. Van Gaal, *et al.* 2009. The effect of weight loss on sleep-disordered breathing in obese teenagers. *Obesity (Silver Spring)* **17:** 1178–1183.

60. Dayyat, E., L. Kheirandish-Gozal, O. Sans Capdevila, *et al.* 2009. Obstructive sleep apnea in children: relative contributions of body mass index and adenotonsillar hypertrophy. *Chest* **136:** 137–144.

61. Horner, R.L., R.H. Mohiaddin, D.G. Lowell, *et al.* 1989. Sites and sizes of fat deposits around the pharynx in obese patients with obstructive sleep apnoea and weight matched controls. *Eur. Respir. J.* **2:** 613–622.

62. White, D.P., R.M. Lombard, R.J. Cadieux & C.W. Zwillich. 1985. Pharyngeal resistance in normal humans: influence of gender, age, and obesity. *J. Appl. Physiol.* **58:** 365–371.

63. Beebe, D.W., D. Lewin, M. Zeller, *et al.* 2007. Sleep in overweight adolescents: shorter sleep, poorer sleep quality, sleepiness, and sleep-disordered breathing. *J. Pediatr. Psychol.* **32:** 69–79.

64. Tauman, R., L.D. Serpero, O.S. Capdevila, *et al.* 2007. Adipokines in children with sleep disordered breathing. *Sleep* **30**(4): 443–449.

65. Spruyt, K., O. Sans Capdevila, L.D. Serpero, *et al.* 2010. Dietary and physical activity patterns in children with obstructive sleep apnea. *J. Pediatr.* **156:** 724–730, 730.e1–730.e3.

66. Tauman, R., L.M. O'Brien, C.R. Holbrook & D. Gozal. 2004. Sleep pressure score: a new index of sleep disruption in snoring children. *Sleep* **27:** 274–278.

67. Gozal, D. & L. Kheirandish-Gozal. 2009. Obesity and excessive daytime sleepiness in prepubertal children with obstructive sleep apnea. *Pediatrics* **123**(1): 13–18.

68. Gozal, D., L.D. Serpero, O. Sans Capdevila & L. Kheirandish-Gozal. 2008. Systemic inflammation in non-obese children with obstructive sleep apnea. *Sleep Med.* **9:** 254–259.

69. Kim, J., R. Bhattacharjee, A.B. Snow, *et al.* 2010. Myeloid-related protein 8/14 levels in children with obstructive sleep apnoea. *Eur. Respir. J.* **35:** 843–850.

70. Spruyt, K. & D. Gozal. 2012. A mediation model linking body weight, cognition, and sleep-disordered breathing. *Am. J. Respir. Crit. Care. Med.* **185:** 199–205.

71. Bhattacharjee, R., J. Kim, W.H. Alotaibi, *et al.* 2012. Endothelial dysfunction in children without hypertension: potential contributions of obesity and obstructive sleep apnea. *Chest* **141:** 682–691.

72. Gozal, D., O.S. Capdevila & L. Kheirandish-Gozal. 2008. Metabolic alterations and systemic inflammation in obstructive sleep apnea among nonobese and obese prepubertal children. *Am. J. Respir. Crit. Care. Med.* **177:** 1142–1149.

73. Bhattacharjee, R., L. Kheirandish-Gozal, K. Spruyt, *et al.* 2010. Adenotonsillectomy outcomes in treatment of obstructive sleep apnea in children: a multicenter retrospective study. *Am. J. Respir. Crit. Care. Med.* **182:** 676–683.